The CRISPR/Cas Tool Kit for Genome Editing

Aftab Ahmad • Sultan Habibullah Khan •
Zulqurnain Khan
Editors

The CRISPR/Cas Tool Kit for Genome Editing

Springer

Editors
Aftab Ahmad
Department of Biochemistry/US-Pakistan Center for Advanced Studies in Agriculture and Food Security
University of Agriculture
Faisalabad, Pakistan

Sultan Habibullah Khan
Center of Agricultural Biochemistry and Biotechnology (CABB)/US-Pakistan Center for Advanced Studies in Agriculture and Food Security
University of Agriculture
Faisalabad, Pakistan

Zulqurnain Khan
Institute of Plant Breeding and Biotechnology
Muhammad Nawaz Sharif University of Agriculture
Multan, Pakistan

ISBN 978-981-16-6304-8 ISBN 978-981-16-6305-5 (eBook)
https://doi.org/10.1007/978-981-16-6305-5

This Springer imprint is published by the registered company Springer Nature Singapore Pte Ltd.
The registered company address is: 152 Beach Road, #21-01/04 Gateway East, Singapore 189721, Singapore

Foreword

On the invitation of Dr. Aftab Ahmad, Dr. Sultan Habibullah Khan, and Dr. Zulqurnain, I am pleased to write introductory statements about *The CRISPR/Cas Tool Kit for Genome Editing*, a very informative book about developments in CRISPR/Cas and its applications.

The emergence of CRISPR/Cas and its re-engineering into a potent genome editing tool has revolutionized the life sciences with a broad range of applications in basic and applied research. It has brought much excitement and hope in medicine, agriculture, and food security with unprecedented control over the redesign of genomes, which has been recognized through the Nobel Prize for CRISPR in 2020. Derived from bacterial adaptive immune system against foreign DNA, CRISPR is a simple, efficient, and robust platform for genome editing in bacteria, plants, and animals. CRISPR has become a most favored method to create animal models for studying molecular mechanisms, fighting viral, bacterial, and fungal diseases in plants, eradicating vector-borne diseases such as Zika, dengue, and malaria, repairing disease-causing mutations, and many others. The CRISPR tool kit is continuously growing with the latest addition of Cas13 for transcriptional control, Cas14 for ssDNA, and EvolvR for rewriting genetic code. No doubt, the CRISPR-Cas system has transformed biological research with new tools, latest methods, and breakthrough discoveries of science. Although CRISPR has empowered scientists with unprecedented tools of precise genomic editing but CRISPR is at its developing stage and we can expect that best in CRISPR is yet to come.

This comprehensive book aims to provide the reader with basic and up-to-date knowledge about CRISPR/Cas-mediated genome editing and different techniques based on this cutting-edge technology. It contains ten chapters about various aspects of CRISPR/Cas from introduction to challenges and prospects of this revolutionary toolbox. It also contains a chapter discussing directed evolution focusing on rewriting of genetic code using CRISPR/Cas to evolve functions of proteins beyond natural evolution. In addition, one chapter highlights the importance of CRISPR/Cas13 with its detailed application in RNA world such as RNA editing, RNA imaging, and diagnostics. Moreover, a chapter provides insights into the applications of CRISPR/Cas beyond genome editing, which are exceeding genome editing aspects. The editors are also authors of most of the chapters, who have an established expertise on genome editing with CRISPR/Cas in plants.

I am happy to recognize this valuable work of Dr. A. Ahmad, Dr. S.H. Khan, and Dr. Z. Khan for academic and scientific readers with leading science publisher Springer Nature. I believe this volume provides a clear, informative, and an easy-to-understand text on CRISPR-mediated genome editing. I am happy to say that this book will be useful particularly for students, researchers, and scientists to gain insight about CRISPR/Cas.

State Key Laboratory of Plant Cell and Chromosome Engineering, Institute of Genetics and Developmental Biology, CAS
Beijing, China

Caixia Gao

Preface

CRISPR/Cas-based genome editing has become a new face of molecular biology as it has been adopted in almost every filed of biotechnology from agriculture to medicine. CRISPR/Cas is continuously evolving with new techniques and new Cas proteins with diverse functions. In the past few years, there have been rapid advancements in online tools, methods, protocols, and applications of CRISPR/Cas technology. Fine-tuning and continuous developments in CRISPR technology have allowed scientists to change single base pair with more precision and accuracy. In addition, CRISPR-based directed evolution has enabled us to rewrite genetic code and discover functions of proteins beyond natural evolution. This contributed volume, ***The CRISPR/Cas Tool Kit for Genome Editing***, provides a comprehensive overview of CRISPR-based tools from their discovery, development, and diversity to their applications in various organisms. Starting with an introductory chapter about CRISPR/Cas, the book aims to bring details about CRISPR applications in genome editing, which makes this volume equally helpful for readers of various levels. The chapters of this book not only summarize historical background and classification of CRISPR/Cas but also discuss a variety of CRISPR/Cas systems such as Cas9, Cas12, Cas13, CasX and their applications including multiplex genome editing, base editing, primer editing, RNA editing, and directed evolution. In addition, contents of this book also discuss design tools for determining on-targeting and off-targeting, and different delivery methods for efficient genome editing.

This book has been designed for students, researchers, and professionals interested in CRISPR/Cas and want to learn more about it. The book brings together different tools available in CRISPR/Cas technology for genome editing applications in bacteria, animals, plants, and model organisms. Re-engineering CRISPR/Cas system and its use for re-writing genetic codes provide an additional aspect to gain more attention of the advanced level researchers in the field of genome editing. Trends in the CRISPR technology, its applications and regulation are chalked out along with addressing several issues such as off-targeting, regulatory concerns, and biosafety. The work presented here may be a valuable resource for developing an academic course on CRISPR-based genome editing. Indeed, this nicely written and well-presented work is a valuable addition to the field of genome editing, which will help the scientific community to learn, understand, and apply the marvelous technology for the welfare of humankind. Having equal importance for undergraduate and postgraduate students to researchers and scientists, the book will also be found useful for policy makers and industrial stakeholders.

Faisalabad, Pakistan Aftab Ahmad
Faisalabad, Pakistan Sultan Habibullah Khan
Multan, Pakistan Zulqurnain Khan

Contents

About the Editors

Aftab Ahmad is currently working as Assistant Professor at the Department of Biochemistry/Center for Advanced Studies in Agriculture and Food Security (CAS-AFS), University of Agriculture, Faisalabad (UAF), Pakistan. Dr. Aftab's research interests include developing virus-resistant and salt-tolerant plants using CRISPR/Cas technology. He has published more than 20 research articles in international peer-reviewed journals. Dr. Aftab pursued his PhD and postdoc in Plant Molecular Biology from the University of Shizuoka, Japan, under Center of Excellence (COE, Twent-first Century) program. Dr. Aftab also served as visiting scientist in the Department of Plant Sciences at the University of California, Davis, USA. Dr. Aftab has been teaching biochemistry and molecular biology at UAF, and his research focuses on the area of CRISPR/Cas-based genome editing in plants.

Sultan Habibullah Khan was graduated from the University of Manchester-England in the field of Plant Sciences in 2004. During his predoctoral experience, he worked at the National Center of Excellence in Molecular Biology (NCEMB), Lahore, and National Institute for Biotechnology and Genetic Engineering (NIBGE), Faisalabad. After completing his PhD in the UK, Dr. Khan started his teaching career at the University of Agriculture-Faisalabad (UAF) in 2005. His postdoctoral experience also includes teaching and research at the University of California-Berkeley (as a visiting scholar) and Sultan Qaboos University-Oman (as a teaching consultant). Currently, he is working as Associate Professor of Plant Molecular Sciences at the Centre of Agricultural Biochemistry and Biotechnology (CABB), UAF. Over the past 20 years, Dr. Khan has been engaged in the fields of plant biotechnology, breeding, genetics, and molecular biology. His career goals center on tailoring crop varieties with better tolerance to the changing environment through breeding and genome engineering. Dr. Khan is the recipient of national/international scholarships and the Research Productivity Award (2014) and has over 60 peer-reviewed publications.

Zulqurnain Khan is currently working as Assistant Professor in the Institute of Plant Breeding and Biotechnology (IPBB), MNS University of Agriculture, Multan (MNSUAM), Pakistan. Dr. Khan earned his PhD in Biotechnology from the University of Agriculture, Faisalabad (UAF), Pakistan, during 2017. He is the first PhD candidate from Pakistan with a thesis on genome editing. He has been working in the field of genome editing since 2012. His research focuses on using genome editing tools (TALEs, TALENs, Cas9, dCas9, and multiplexed CRISPR/Cas9) for resistance against begomoviruses using model plants such as *Arabidopsis* and *Nicotiana benthamiana*. He is also using CRISPR/Cas9 technology for genetic improvement in cereal crops for abiotic and biotic stress resistance. He has published several book chapters, review articles, and research articles in international peer-reviewed journals in the field of genome editing. Dr. Khan received IRSIP fellowship from Higher Education Commission (HEC), Pakistan, during his PhD and worked as visiting researcher with Professor Caixia Gao at the Institute of Genetics and Developmental Biology (IGDB), Chinese Academy of Sciences (CAS), Beijing, China. Dr. Khan has been engaged in teaching various courses in the field of biotechnology, breeding, molecular biology, genetics, and genome editing since 2018. Dr. Khan has recently edited a book entitled *CRISPR Crops: The Future of Food Security*, published by Springer Nature.

An Introduction to Genome Editing Techniques

1

Nayla Munawar and Aftab Ahmad

Contents

Abstract

Understanding the genome and its function is crucial to understanding its role in health and disease. Genome editing—from inhibiting a gene to altering gene expression—provides an insight into a better understanding of genomic function. Genome editing techniques paved a way for manipulation of DNA by altering the function of specific genes. A specific, targeted change in DNA allowed scientists to precisely manipulate DNA both in vitro using cell line and in vivo in animal models. In the last two decades, advances in editing techniques have led to significant breakthroughs in genome editing science. Engineered nucleases

N. Munawar (✉)
Department of Chemistry, College of Science, United Arab Emirates University, Al-Ain, UAE
e-mail: nmunawar@uaeu.ac.ae

A. Ahmad
Department of Biochemistry/US-Pakistan Center for Advanced Studies in Agriculture and Food Security, University of Agriculture, Faisalabad, Pakistan
e-mail: aftab.ahmad@uaf.edu.pk

A. Ahmad et al. (eds.), *The CRISPR/Cas Tool Kit for Genome Editing*,
https://doi.org/10.1007/978-981-16-6305-5_1

enabled targeted DNA alterations in a wide variety of cells, leading to the treatment of genetic disorders which were previously impossible or difficult to treat. In this chapter, we discuss different modifiable endonuclease-based genome editing techniques, their mechanism of action, structure, and comparison of function, how these techniques have revolutionized genome editing, and their potential applications in different fields of biology.

Keywords

Genome editing · ZFNs · TALENs · CRISPR/Cas · Meganucleases · Cre-lox system · RNAi

Abbreviations

A	Adenine
Asn (N)	Asparagine
Asp (D)	Aspartate
bp	Base pair
C	Cytosine
crRNAs	CRISPR RNAs
CRISPR	Clustered regularly interspaced short palindromic repeats
DNA	Deoxyribonucleic acid
DSBs	Double-stranded breaks
G	Guanine
Gly (G)	Glycine
gRNA	Guide RNA
HDR	Homology-directed repair
HEs	Homing nucleases
His (H)	Histidine
Ile (I)	Isoleucine
IVC	In vitro compartmentalization
kDa	Kilodalton
KI	Knock-in
KO	Knockout
NHEJ	Non-homologous end joining
nt	Nucleotide
PAM	Protospacer adjacent motif
PCR	Polymerase chain reaction
RNA	Ribonucleic acid
RVDs	Repeat variable di-residues
sgRNA	Single-guide RNA
T	Thymine
TALE	Transcription activator-like effector
tracrRNA	Trans-activating CRISPR RNA
UV	Ultraviolet

ZnFs	Zinc fingers
ZFAs	Zinc finger activators
ZFNs	Zinc finger nucleases
ZFPs	Zinc finger proteins
ZFRs	Zinc finger repressors

1.1 What Is Genome Editing?

The term "genome" refers to the genetic material of an organism. The genome of an organism includes the complete nuclear DNA, including coding and non-coding regions, and the mitochondrial and chloroplast DNA present in a cell. The study of genetic material to understand the role of the DNA molecule in the cell is called "genomics." In 1910, Nobel laureate Albrecht Kossel discovered the five nucleotide bases (or "nucleobases") which form the basis of DNA and RNA, the genetic material for all living cells. This was followed in 1950 by Erwin Chargaff's discovery of the pairing pattern of four bases in DNA, leading to the determination of the double helix structure of DNA by Nobel laureates James Watson and Francis Crick in 1953. DNA sequencing techniques, first developed in 1977 by Frederick Sanger (also a Nobel laureate), and the discovery of the polymerase chain reaction (PCR) method by Kary Mullis (another Nobel laureate) in 1983, paved the way for gene sequencing and research to locate the genes involved in genetic diseases such as Huntington's disease, cystic fibrosis, and hemophilia in humans. The project to completely sequence the human genome was launched in 1990, while the first complete genome sequence of a free-living organism, the bacterium *Haemophilus influenza*, was achieved in 1995. Since the 1990s, with the aid of improved techniques, facilities, and knowledge about DNA, the complete genome sequences of several organisms, from prokaryotes to eukaryotes, have been published, including the human genome sequence, which was completed in 2003. The discovery of next-generation sequencing platforms in 2008 has dramatically reduced the cost of sequencing, while a combination of genome editing techniques has revolutionized genomic and genetic research.

The genome sequence formed as a pattern of the four bases—adenine, thymine, guanine, and cytosine (abbreviated as "ATGC")—in the DNA of an organism is specific and alterable by intrinsic or environmental factors. The DNA sequence of an organism may be altered by DNA replication during cell division or by environmental factors such as UV radiation in sunlight and exposure to chemicals able to create mutations in the DNA. Some inherited mutations may also be present in each cell. Not all DNA mutations are related to disease but instead form the genetic basis of diverse phenotypic characteristics such as blood type and skin, hair, and eye color. These DNA mutations, which do not cause abnormal proteins but are only responsible for phenotypic diversity, are known as examples of "polymorphism." However, some DNA mutations produce abnormal proteins and cause severe or lethal health issues. The identification of disease-causing DNA errors led scientists to think about ways to reverse these mistakes. To correct genome-related disorders, scientists began to look for genome editing strategies as early as the 1960s. Genome editing is defined as the "deliberate, precise change at some specific targeted region of DNA sequence in the cell" (Segal and Meckler 2013). Knowledge about the DNA molecule and availability of DNA sequencing techniques enabled ways to directly edit DNA at precise, predetermined locations among its bases. Genome editing was a difficult task before the discovery of programmable DNA nucleases in the 1990s. Precise alteration in the nucleotide (nt) sequence has since become a realistic objective through the use of commercially available reprogrammable DNA nuclease-based genome editing techniques, as described in this chapter. Because of rapid and novel advances in genome editing technology, we are now entering in an era in which genome editing is revealing a new horizon in medical and agricultural research.

1.2 History of Genome Editing

Genome editing is among the most valuable techniques available in biological research. It is important to understand the history of this technique to realize fully its current value and status. Several landmarks in genome editing history are shown

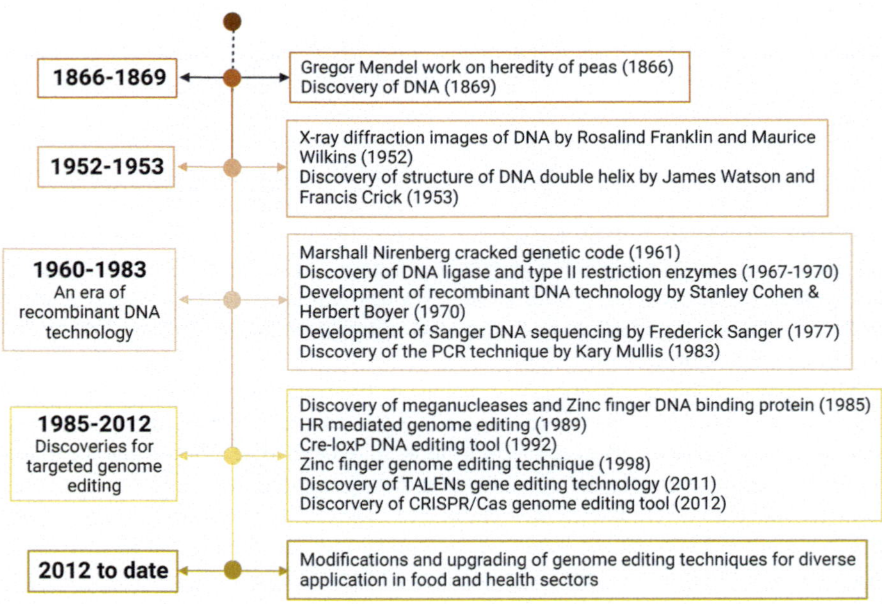

Fig. 1.1 Landmarks in genome editing history. The key events in the history of genomics from the discovery of DNA till targeted genome editing has been mentioned with important milestone that enabled conversion of DNA-binding proteins into targeted gene editing tools

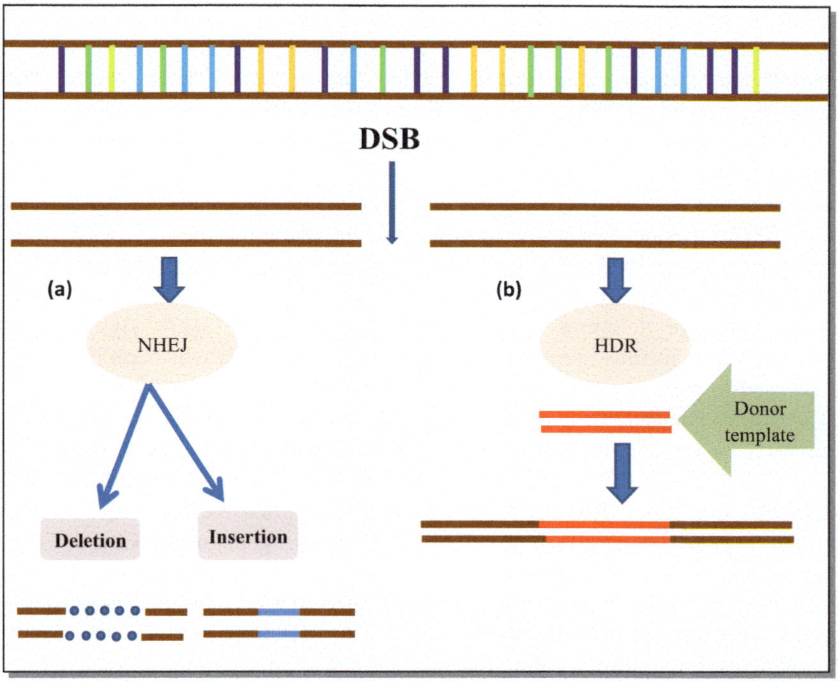

Fig. 1.2 Natural repair mechanisms of double-stranded DNA breaks in eukaryotic cells: (**a**) non-homologous end joining (NHEJ) repair; (**b**) homology-directed repair (HDR)

in Fig. 1.1. The discovery and subsequent understanding of the structure and function of the DNA molecule prior to recombinant DNA technology paved the way for many discoveries in genetics and genome editing. The discovery of DNA ligase and the discovery of the restriction enzymes in 1967 and 1968, respectively, were the two most significant events in recombinant DNA technology and genetic engineering; together they revolutionized the world of biology and created an opportunity to manipulate DNA in vitro (Smith and Wilcox 1970; Kelly Jr and Smith 1970; Danna and Nathans 1971). Subsequently, the establishment of recombinant DNA technology marked the beginning of genome manipulation and genome editing research (Rothstein 1983).

A foundation of gene editing methodologies was the hypothesis that targeted double-stranded breaks (DSBs) in DNA that stimulate endogenous cellular DNA repair pathways may be exploited to create targeted mutations or precise editing in the genome (Rouet et al. 1994; Kosicki et al. 2018). Normally, DNA DSBs are lethal for the cell. DSBs can repair naturally by one of the two major pathways: homology-directed repair (HDR) or non-homologous end joining (NHEJ) in each cell (Shalem et al. 2015). As its name indicates, HDR relies on recombination of undamaged homologous sequence in a chromatid and subsequent repair of the break in a template-dependent manner. This pathway can be exploited to incorporate a homologous donor template DNA provided by a researcher (Fig. 1.2b). Thus, precise and deliberate alterations may be introduced into the genome that is specific to the exogenous template DNA (Kim and Kim 2014; Verma and Greenberg 2016). NHEJ repairs DSBs by direct religation of the cleaved ends in a non-template-directed manner. This pathway is prone to errors and often results in variable lengths of insertion and deletion mutations (indels) at the site of the break (Fig. 1.2a). A site-specific DSB introduced in DNA can shift the open reading frames by indels if these breaks are repaired by NHEJ in the cell. Therefore, an erroneous repair created through NHEJ can be utilized for site-specific gene disruption by introducing small base insertions or deletions (Puchta 2005; Chang et al. 2017). By taking the advantage of the intrinsic repair machinery of the cell, tools which produce DSBs can be used to precisely alter the genome.

The issue of how DSBs can be introduced at a target site was solved with the discovery of natural rare-cutting endonucleases known as meganucleases in 1985 (Jacquier and Dujon 1985). The concept of genome editing was established in the late 1980s. In 1989, Capecchi (1989) used homologous recombination (HR) in mouse embryonic stem cells to target specific genes to generate knock-in (KI) and knockout (KO) cells. Although the process was highly inefficient because of the very low frequency of HR in mammalian cells, HR offered a revolutionary approach for targeting specific genes in eukaryotic cells. Since HR alone is not sufficient for efficient gene integration in mammalian cells, requiring extensive selection and screening to identify the one-in-a-million cell that expresses the modified gene, the introduction of DSBs facilitates recombination significantly. The Cre-lox technology, based on the site-specific DNA recombinase Cre, was used to develop a transgenic mouse model in the early 1990s. This technology enabled the knockout of the target gene and provided an opportunity to control gene expression, spatially and temporally, more effectively than HR alone (Orban et al. 1992). However, because of the genetic distance between loxP sites, Cre-lox was found to be less effective for controlling some target genes, and hence the technique failed to gain wide acceptance. However, via this technique, for the first time ever, scientists gained the ability to manipulate DNA and the possibility of gene editing using endonucleases. Henceforth, naturally occurring endonucleases were employed to create DSBs in DNA.

Rare-cutting meganucleases that recognize 14–40 base pairs (bp) in DNA increased genome editing efficiency, but naturally occurring meganucleases for each target site were not available. Therefore, scientists began to reengineer the existing meganucleases for target DNA, but the target efficiency of mutant nucleases was found to be very low (Sussman et al. 2004; Seligman et al. 2002; Rosen et al. 2006). Later, the discovery of zinc finger nucleases (ZFNs) in 1998 sidelined meganucleases as a gene editing technique. A ZFN is an artificial DNA cutter comprising zinc finger DNA-binding protein combined with the Fok1 endonuclease. This technique gained widespread popularity; however, another DNA-binding protein known as transcription activator-like effector (TALE) was found to be more easily manipulated than ZFNs. Hence, for the construction of targeted DNA nuclease, TALE was joining with Fok1 nuclease, with the resultant system known as transcription activator-like effector nucleases (TALENs). While both are effective, ZFNs and TALENs require expertise in protein engineering to modify their DNA-binding domain for each single target to create site-specific DSBs for genome editing. The discovery of clustered regularly interspaced short palindromic repeats (CRISPR)/Cas as a genetic engineering tool in 2012 transformed genome editing (Jinek et al. 2012). Unlike ZFNs and TALENs, CRISPR/Cas recognizes target sites with guide RNA instead of protein. The CRISPR/Cas known today consists of Cas nuclease and guide RNA, which is complementary of the target DNA. Because of its ease of design and flexibility, this technique has been broadly adopted by the scientific community and is utilized heavily in many fields of biological science.

1.3 Different Techniques of Genome Editing

Advanced genome editing techniques use designer nuclease and natural DNA repair systems present in cells to make precise changes in the genome. The mechanisms of action of some of the most powerful genome editing techniques are discussed here.

1.3.1 Cre-LoxP System

The Cre-lox technology was introduced as an efficient gene editing tool in the early 1990s. This technique is based on site-specific recombination which viruses normally adapt to integrate their DNA into host genome. The system consists of two functional components adapted from the P1 bacteriophage: (1) Cre, a recombinase, and (2) loxP, a recognition site of Cre. Cre is a 38-kilodalton (kDa) recombinase which recognizes specific 34-bp DNA fragment sequences known as loxP (**l**ocus **o**f **x**-over, **P**1) sites. LoxP sites contain two sets of 13-bp inverted and palindromic repeats separated by an 8-bp spacer region (Fig. 1.3). The bp sequence in the spacer region is variable (except for two bps in the middle), which gives the loxP sequence a certain direction or orientation.

The principles of the Cre-loxP system assembly and its mechanism of action are illustrated in Fig. 1.4. Two Cre recombinases bind to each 13-bp region of the loxP sequence and form a dimer (Fig. 1.4a). Two dimers on two loxP sites bind together

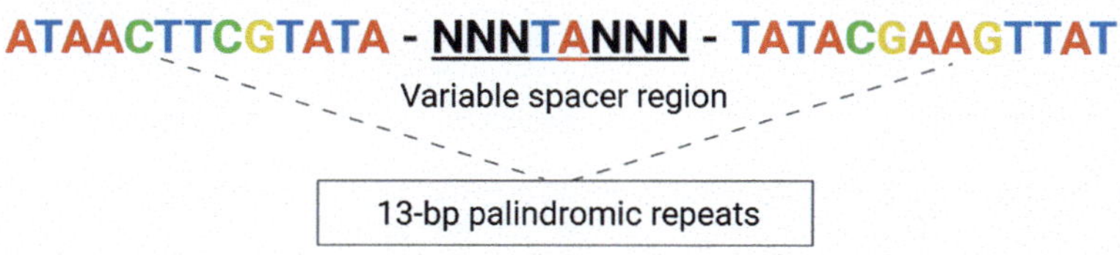

Fig. 1.3 A 34-bp *loxP* sequence indicating 13-bp palindromic repeats with spacer region. "*N*" indicates the variable bases in the spacer region sequence

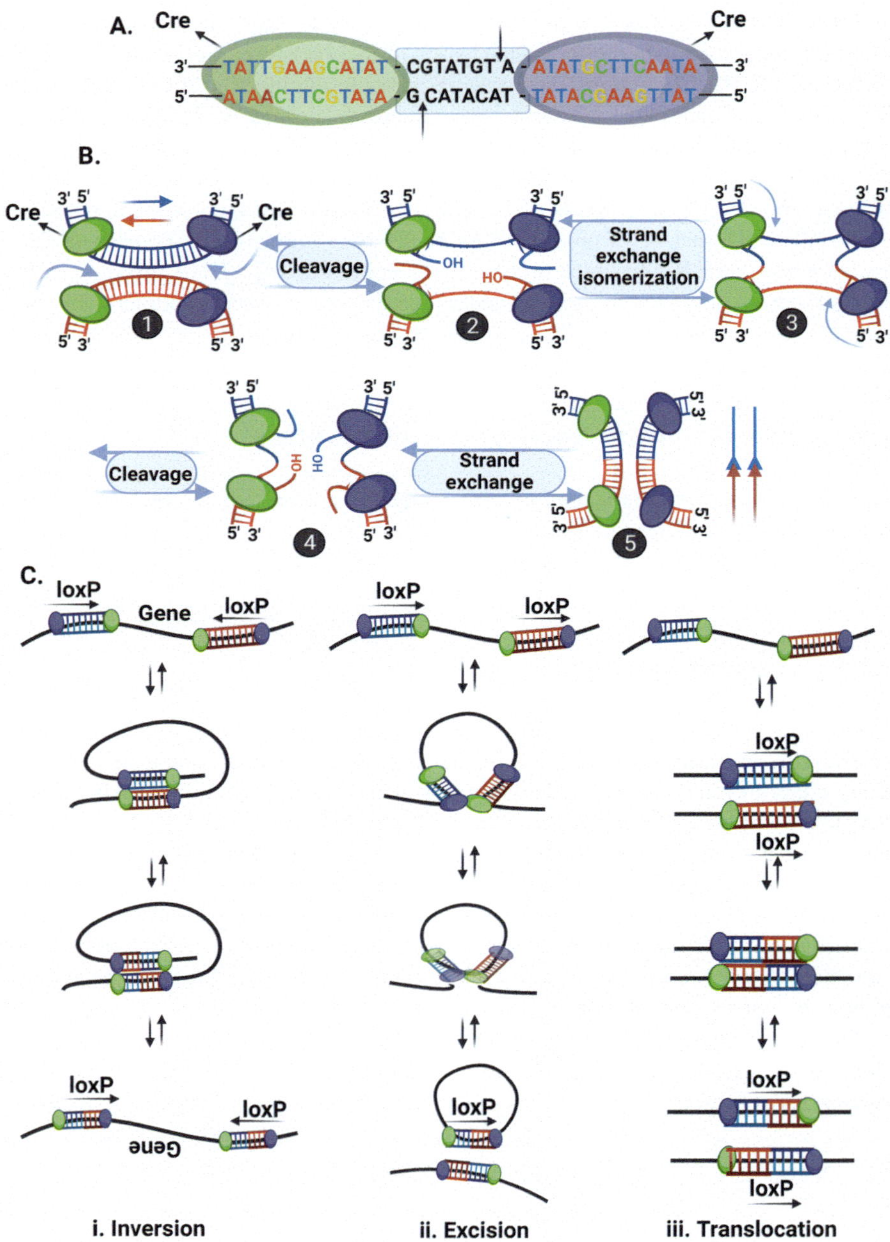

Fig. 1.4 Mechanism of action of Cre-loxP system: (**a**) The 13-bp Cre (recombinase) binding *loxP* sites with central variable region where cleavage and strand exchange take place. Top strand cleavage is between T and A, and bottom strand cleavage is between G and C base pairs. (**b**) Steps of reaction mechanism. (**c**) Reactions affected by the orientation of *loxP* site: (i) If *loxP* is in the opposite direction, Cre facilitates inversion of the DNA segment. (ii) Cre makes excision, if *loxP* is in the same direction. (iii) If *loxP* is on the different DNA and in the same direction, Cre mediates a translocation of the DNA segment

to form a tetramer, which brings the loxP sites together with opposing directionality. Cre protein cuts double-stranded DNA at both loxP sites and generates DSBs that are repaired by DNA ligase enzyme very efficiently (Fig. 1.4b). Meanwhile, a crossover event results in inversion, deletion, and translocation of the DNA sequence. The location and orientation of loxP sites determine three types of rearrangement of the genetic material (Fig. 1.4c):

- **Inversion**: A gene is inverted during a recombinant event if loxP is present on the same DNA strand but in the opposite direction. This is a reversible process, and genes can flip back and forth at anytime.
- **Excision**: A DNA sequence can be deleted if loxP sites are present on the same DNA strand in the same direction. A gene may be deleted from its original locus by recombinase, irreversibly so in this case.
- **Translocation**: In the presence of loxP sites on the different DNA strands, recombination results in a reversible translocation of the DNA sequence.

Because of its ease of manipulation, the Cre-loxP system has widely been used for gene editing in mammalian cells. It was used to develop transgenic mouse models in the 1990s (Zijlstra et al. 1989; De Chiara et al. 1990; Koller et al. 1990; Thomas and Capecchi 1990). Since Cre-loxP is exclusively present in bacteriophage and the loxP sequence does not exist in plant or animal cells, non-specific binding of Cre recombinase is minimal when used in mammalian cells. Moreover, the long 34-bp sequence of loxP reduces the chances of off-target binding of Cre protein. This technology has predominately been used for gene excision and to generate conditional knockout mice. The tetracycline-inducible Cre-loxP system, CreERT, has been used to regulate gene expression and is known as a potent tool for conditional gene manipulation (Nagy 2000; McLellan et al. 2017; Kim et al. 2018).

Cre-loxP technology has been used widely: in neurobiology (Tsien 2016); in immunology to understand viral pathogenesis (Sharma and Zhu 2014); in oncology (Kersten et al. 2017); and in behavioral and physiological studies in mice. While Cre-lox provided an opportunity to control gene expression more easily than HR, the efficiency of the system was low because of the large genetic distances between loxP sites (Zheng et al. 2000). Some websites provide information on Cre mouse lines for research purposes (Smedley et al. 2011; Chandras et al. 2012; Murray et al. 2012; Schofield et al. 2012; Sung et al. 2012), but no single, independent, worldwide platform is available to collect or share information with researchers easily. Moreover, development of transgenic systems by gene insertion with Cre-lox technology requires significant time and effort. There are newer, simple, and more efficient protein-based nuclease systems which have been recognized as genetic tools for quick and precise genome editing with wider applications in biological sciences.

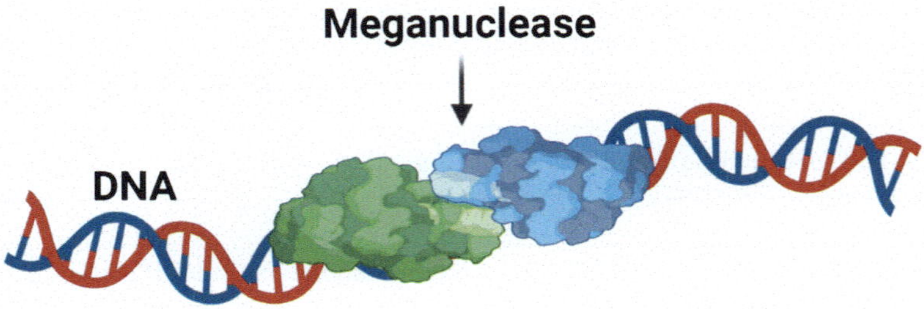

Fig. 1.5 Binding of meganuclease with target DNA

1.3.2 Meganucleases

Meganucleases or homing nucleases (HEs) are natural endonucleases which have been divided into five families on the basis of their sequence and structural motifs (Orlowski et al. 2007; Zhao et al. 2007). The five families are (1) LADGLIDADG; (2) HNH; (3) GIY-YIG; (4) His-Cys box; and (5) PD-(D/E)XK. The families differ in their catalytic mechanisms, biological distribution, and similarities with non-homing nucleases. Meganucleases are found in the biological domains of Archaea, Bacteria, and Eukarya and their respective viruses (Stoddard 2005) and are known as "selfish genetic elements" because of their unknown function in the cell. Among the five families, LADGLIDADG, which is also known as LHE family, is widely distributed and well characterized.

Because of their high-DNA-protein contact, high target specificity and flexibility, and ability to bind with long DNA sequences (22 bp or longer), members of LHE family are considered more suitable candidates for genome editing than members of other HE families. The LHE family endonucleases exist in both homodimeric and monomeric forms, and they recognize specific DNA sequences through protein-DNA interactions. These endonucleases bind with many DNA bases and the flanking phosphodiester backbone in the major groove through the β-sheet of the protein (Shen et al. 2016) (Fig. 1.5). Therefore, efforts were made to manipulate LAGLIDADG protein-DNA interactions by protein engineering techniques for novel targets. However, several studies have shown that related proteins can use different subsets of residues to recognize similar DNA sequences (Chevalier et al. 2003; Lucas et al. 2001).

The DNA binding and cleavage function of meganucleases are coupled with each other, with almost 50 amino acids involved in direct or indirect DNA contact, which creates a significant challenge in altering their specificity to the DNA sequence of interest (Takeuchi et al. 2014). Hence, it is difficult to reengineer the specificity of meganucleases. Nevertheless, several site-specific DNA cleavages by engineered meganucleases, including maize genomic sequence (Gao et al. 2010) and two human genes XPC and *RAG1* (Smith et al. 2006; Arnould et al. 2007), have been reported in the literature.

In addition to their natural origin, an advantage of meganucleases for genome editing is the comparatively small size (approximately 40 kDa) of these endonucleases, which makes their packing and delivery into cells easy for certain applications. A hybrid molecule known as "megaTAL," generated by the fusion of a TAL effector in the DNA domain with meganucleases, has been used for targeted gene modification in human cells (Boissel et al. 2014). In vitro compartmentalization (IVC) using iterative cycles for redesign of extensive protein-DNA interfaces of meganucleases has been employed as an efficient approach for reprogramming target specificity and cleavage of meganuclease (Takeuchi et al. 2014). Nevertheless, DNA methylation and chromatin structure, which affect the nuclease activity of meganucleases, limit the practical applications of these enzymes in epigenetics (Valton et al. 2012; Daboussi et al. 2012). Although an engineering pipeline for altering target specificities of meganucleases has been developed on an industrial scale (Arnould et al. 2007), specialized knowledge and expertise in protein engineering are required to utilize this technique in the laboratory. Moreover, the high probability of off-target binding is another limitation of the technique (Argast et al. 1998). The wider adoption of this technology still depends on the informal approaches for reengineering these nucleases for desired targets. Therefore, other genome editing techniques such as ZFNs and TALENs, which are more flexible in terms of reprogramming for novel target sequences, availability of engineering resources, time, and cost-effectiveness, are preferred over meganucleases.

1.3.3 Zinc Finger Nucleases (ZFNs)

A versatile and effective DNA manipulating tool is a ZFN that consists of two components: (i) a programmable DNA-binding domain and (ii) a DNA cleavage domain. The DNA-binding domain is DNA-binding zinc finger (ZnF) proteins, which were discovered by Klug and co-workers in transcription factor IIIa (TFIIIa) from *Xenopus* oocytes in 1985 (Miller et al. 1985). Zinc fingers are the most abundant group of proteins in humans, and they interact with a specific part of DNA—thus they play a crucial role in transcriptional regulation. Studies reveal that each ZnF contains 30-amino-acid repeating regions folded into unique two β-sheets and one α-helix structure. This unique ββα structure is maintained by a zinc ion that binds with two invariant cysteines and histidine pairs in different combinations (Pavletich and Pabo 1991). The crystal structure of ZnFs suggest that each ZnF recognizes a specific 3–4-bp DNA sequence and binds with DNA by inserting α-helix into the major groove of DNA (Buck-Koehntop et al. 2012; Fairall et al. 1993). Interestingly, six amino acids in the α-helix at positions −1, +1, +2, +3, +5, and + 6 are modifiable and can be used to generate new DNA specificities in ZnFs, while other conserved amino acids serve as the ZnF's backbone (Chandrasegaran and Carroll 2016). Therefore, ZnFs offer an ideal framework for a reprogrammable DNA-binding domain by changing amino acids present in α-helix according to the target DNA sequence.

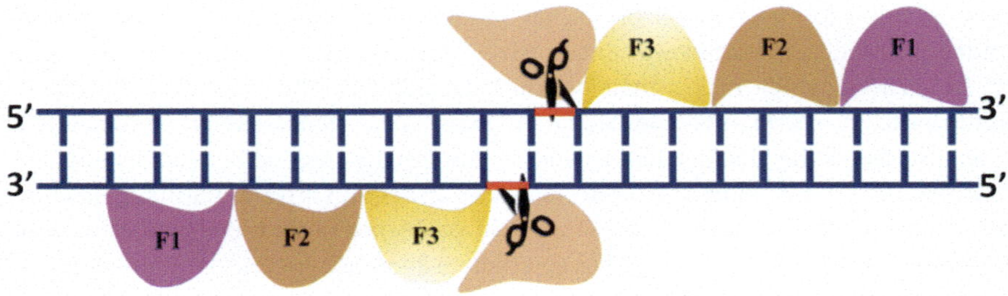

Fig. 1.6 Illustration of a zinc finger nuclease (ZFN) binding and cleavage of target DNA. A set of two ZFNs comprises three to six zinc finger proteins (ZFPs) (the DNA-binding domain) and the FokI restriction enzyme (the DNA cleavage domain) that bind with both strands of target DNA and aid the FokI dimerization required for the nuclease activity of FokI

After recognition of target DNA and binding, cleavage of double-stranded DNA is required to provoke internal DNA repair systems, which then may lead to modifications at that specific site during repair of the DSBs. Therefore, a DNA cleavage domain is required to attach with a DNA-binding domain to make DSBs for precise genome editing. It has been observed that type II restriction enzymes such as *BamHI* and *EcoRI* are not suitable as genome editing tools because of an overlap between their DNA recognition and DNA restriction function. Modification of DNA-binding amino acids to alter the target specificity of type II restriction enzymes change the catalytic site of these enzymes, producing catalytically inactive enzyme (Chandrasegaran and Carroll 2016). However, type IIS FokI restriction enzyme not only recognizes the non-palindromic five DNA bases 5′-GGATG-3′ to 5′-CATCC-3′ in duplex but also contains two separable domains: (1) the 41-kDa N-terminal DNA-binding domain and (2) the 25-kDa C-terminal DNA cleavage domain (Li et al. 1992). Normally, FokI endonuclease cleaves 9/13 nucleotides downstream of recognition sequence, but the addition of 4 amino acid linkers between 2 domains moves the cutting site 1 bp further downstream of the original cut site, i.e., to 10/14 bp rather than 9/13 bp in natural enzyme (Li and Chandrasegaran 1993). Therefore, a chimeric nuclease was constructed by linking the cleavage domain of FokI with DNA-binding zinc finger proteins to establish a novel-sequence specific nuclease to cut target DNA (Kim et al. 1996, 1998). Three to six zinc fingers are normally linked together in tandem repeats to generate a ZnF protein (ZFP), which binds to a 9–18 bp DNA target site. The fusion of tailor-made ZFPs with FokI chimeric nuclease creates designer nuclease that cuts DNA at specific target sites. Since it was observed that FokI requires dimerization for its nuclease activity (Bitinaite et al. 1998; Vanamee et al. 2001), double-stranded DNA cleavage requires two copies of sequence-specific ZFPs to bind on the opposite strands of DNA in an inverted tail-to-tail orientation that promotes dimerization of FokI to produce DSBs in DNA (Fig. 1.6) (Smith et al. 2000; Bibikova et al. 2001, 2003). When a pair of ZFPs binds with specific target DNA in forward and reverse strand, FokI dimerizes to make double-stranded DNA breaks that invoke endogenous DNA repair mechanism. Therefore, small base mutations, deletions, or gene silencing could occur by non-homologous end joining, or a gene could be inserted by HDR in the presence of a donor template.

The fusion of ZFPs with FokI nuclease domain neither affects the binding capabilities of ZFPs to target site nor destroys the restriction activity of FokI nuclease domain (Smith et al. 1999). Thus, this technique has been widely accepted for genome editing in animals and plants (Townsend et al. 2009; Shukla et al. 2009; Urnov et al. 2010; Mashimo et al. 2010; Kim et al. 2009). Zinc finger activators (ZFAs) and zinc finger repressors (ZFRs) have also been created by fusing ZFPs with an activator or repressor domains (Choo et al. 1994; Segal et al. 1999; Kim and Pabo 1997) that bind to a single recognition site of target DNA to activate or repress specific gene (Kim et al. 1997; Beerli et al. 1998; Xu and Bestor 1997).

Design and selection of highly specific ZFPs for target DNA sequence in complex genome is laborious and time-consuming (Carroll et al. 2006). The target recognition of ZnFs is highly influenced by neighboring zinc finger motifs (Ramirez et al. 2008). The presence of aspartate residue at position +2 in α-helix structure of a ZnF motif changes ZnF recognition from 3 to 4 bp, which influences the specificity of the neighboring ZnF aligned in the array. Thus, the design and selection of ZnF arrays is time-consuming and challenging with unpredictable target specificity in the final array (Carroll et al. 2006). Moreover, the toxicity of ZFNs due to off-target cleavage is another serious issue which has restricted the application of this technology in human therapeutics.

Efforts have been made to improve the technology. Off-target binding of ZFNs has been observed to be concentration-dependent; this could be lowered by using controllable expression vector. Moreover, the number of ZnFs could be increased in ZFNs to improve the specificity of the target. Additionally, highly-specific ZFPs may be explored to achieve high affinity with the target DNA and lower the toxicity of ZFNs in vivo (Chandrasegaran and Carroll 2016). Nevertheless, once limitations of the technology became apparent, researchers started to look for alternative strategies, which resulted in the discovery of the robust genome editing technologies now known as TALENs and CRISPR/Cas 9, described below.

1.3.4 TALENs

TALEN is a reprogrammable genome editing tool based on a novel DNA-binding module known as transcription activator-like effector (TALE) protein, fused with catalytic domain of FokI endonuclease. TALE was found in plant pathogenic bacteria of *Xanthomonas* species. The endogenous function of TALE proteins is to bind with the host's genomic DNA and alter transcription of their genes to facilitate colonization of pathogenic bacteria in the host (Boch and Bonas 2010). TALE protein consists of 33–35 highly conserved amino acid repeating motifs with two hypervariable amino acids at positions 12 and 13 in each repeat. These two variable amino acids, referred to as the repeat variable di-residues (RVDs), recognize specific DNA bases, for example, RVDs such as Asn and Ile (NI), Asn and Gly (NG), two Asn (NN), and His and Asp (HD) bind with nucleotides A, T, G, and C, respectively

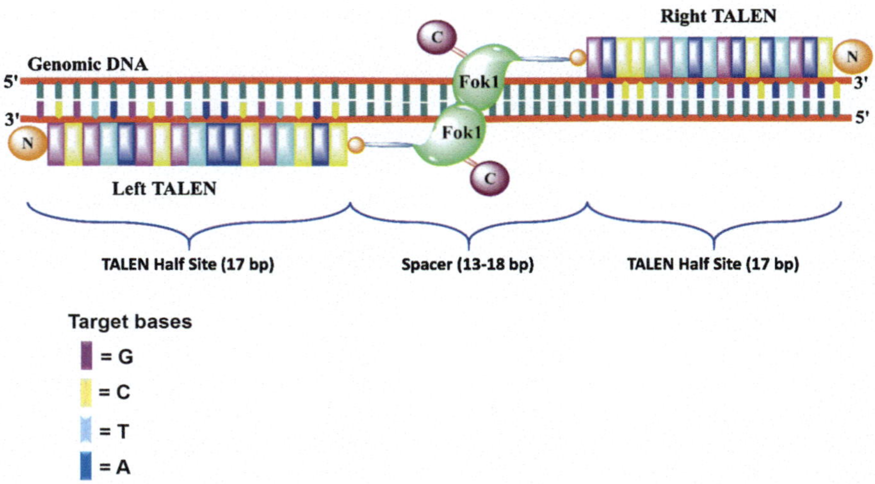

Fig. 1.7 Binding and mechanism of action of transcription activator-like effector nuclease (TALEN). A TALEN consists of left and right monomers of TALE proteins (DNA-binding domain) and the FokI restriction enzyme (DNA cleavage domain), which cleaves DNA when a dimer is formed. Each TALE recognizes a single base pair of target DNA sequence

(Mak et al. 2012; Boch et al. 2009). Unlike zinc fingers, which bind with three nucleotides, a single TALE motif recognizes one nucleotide (Fig. 1.7). The crystal structure of TALE bound to its DNA target suggested that di-variable residues in each repeat occupy a position in the major groove of DNA, and the amino acid at position 13 makes nucleotide-specific contact with DNA, while the amino acid at position 12 gives structural stability by making a contact with the amino acid at position 8 in the domain (Deng et al. 2012). A change in the RVDs alters the specificity of each repeat to its target nucleotide so an array of TALEs can bind with a longer DNA sequence of 30–40 bp. The DNA-binding specificity of each TALE motif is independent and does not affect the binding specificity of neighboring TALEs, providing one-to-one correspondence between TALE repeats and the target DNA sequence (Moscou and Bogdanove 2009). The number of repeats in an array depends on the length of the target sequence. The simple DNA recognition mechanism of TALEs makes them an ideal candidate to construct a custom nuclease. Fusion of the customizable DNA-binding module of TALE with FokI nucleases represented a genome editing technique with broad importance in the biological sciences and in 2011 was described as "method of the year" by *Nature Methods* (Baker 2012).

The general mechanism of action of TALENs is similar to ZFNs. The target-specific TALE domain recognizes the DNA target in the genome, and non-specific FokI nuclease domain cleaves DNA after dimerization and introduces targeted DSBs (Fig. 1.7), which is essential for genome editing. As target recognition by TALENs is restricted to one nucleotide, TALENs are more target-specific than ZFNs, where specificity is highly influenced by a neighboring ZF motif.

1.3.5 CRISPR/Cas System

Organisms in the life domains of Bacteria, Archaea, and Eukarya are constantly exposed to foreign genetic material from various sources, including bacteriophages, transposons, and plasmids. Consequently, they naturally adapt defense mechanisms to protect themselves from invaders. One such type of a natural immune system adapted by bacteria and archaea is CRISPR/Cas. Research indicates that CRISPR/Cas is present in almost 40% of bacterial and 90% of archaeal immune systems, respectively (Mojica et al. 2000). This natural defense system among prokaryotes works uniquely in combination with short RNA molecules (crRNAs) and the specific protein, Cas. The trans-activating crRNA recognizes foreign DNA and guides the Cas protein to bind with the invader's DNA, which may then be cleaved by Cas nuclease activity. Both crRNA and different Cas protein family members make an effector complex that not only identify foreign DNA but also introduce breaks into a DNA-specific site to degrade it (Karginov and Hannon 2010). Recognition of foreign DNA through CRISPR/Cas requires the presence of a specific sequence of that DNA in crRNA—this sequence is termed a "spacer." Spacer is normally acquired from a previous encounter of the host with that foreign DNA. With a recurrent infection, CRISPR crRNA recognizes the matching sequence of the invader, known as "protospacer," and binds to it through complementary base pairing and allows Cas protein to bind and break double-stranded DNA via its nuclease activity and ultimately inactivates the invader (Gasiunas et al. 2012). Gaining an understanding of CRISPR/Cas' mechanism of action in various microorganisms allowed scientists to use it as a technique to manipulate DNA (Swarts et al. 2012). Distinct classes of the CRISPR/Cas system based on structural variations and the organizational style of *cas* genes have been identified (Makarova et al. 2011, 2015), and these developments have helped in turning this system into an indispensable genome editing tool.

1.3.5.1 Mechanism of Action of CRISPR/Cas

The mechanism of action of the CRISPR/Cas system has been divided generally into three phases (Hille and Charpentier 2016).

Phase I is an immunization phase, also known as the "adaptation" or "spacer acquisition phase." During this phase, a bacterial cell adapts the invader's DNA by capturing a small DNA sequence of its genome and integrating it into its CRISPR locus as spacer. Every new spacer is inserted at the beginning of the CRISPR array, next to an AT-rich leader sequence. Generally, the CRISPR/Cas genomic locus consists of a set of *cas* genes, a leader sequence, and short repetitive elements (repeats) separated by unique spacers (Fig. 1.8). The spacer serves as a genetic record of prior infections that develops an immunological memory in the host to recognize the attacker in the future. Experimental evidences indicate that spacer

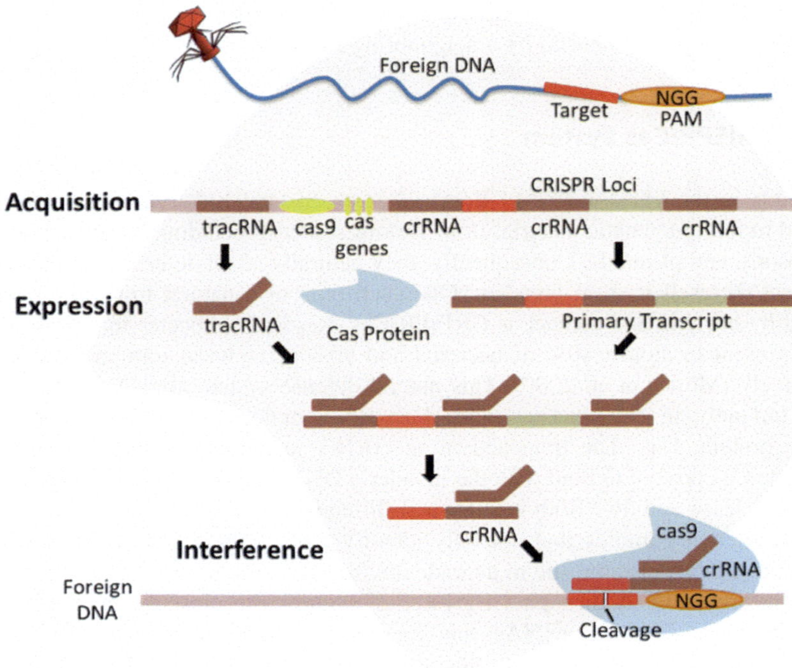

Fig. 1.8 Three phases of mechanism of action of natural CRISPR/Cas immunity system. **Acquisition** indicates the adaptation of invader's DNA piece called spacer and insert it into CRISPR locus of the host's genome. **Expression** is the transcription of a CRISPR array into a long precursor RNA (pre-crRNA) and its maturation into guide RNA having space sequence. **Interference** is the recognition of target DNA in the pathogen, in case of reinfection, using the specific PAM sequence upstream or downstream of the protospacer

adaptation and integration require Cas1 and Cas2 proteins (Yosef et al. 2012). It has been observed that the removal of the Cas1 and Cas2 proteins from the system stops the process of adaptation without affecting the CRISPR immune response, which indicates the involvement of Cas1 and Cas2 proteins in a "spacer acquisition" mechanism (Amitai and Sorek 2016).

Phase II is an **immunity phase** known as "expression and maturation." "Expression" indicates the expression or transcription of a CRISPR array into a long precursor RNA (pre-crRNA), while "maturation" is subsequent processing of pre-crRNA to make a mature guide crRNA containing spacer sequences. The processing of pre-crRNA in a type II CRIPSR system requires the binding of pre-crRNA with a short non-coding *trans*-activated crRNA (tracer RNA) through anti-repeat sequences of tracer RNA, which create an RNA duplex. This RNA duplex is stabilized by Cas9 protein and can be recognized by endogenous RNA III to produce an intermediate crRNA, which undergoes further maturation by a yet unknown detailed process to produce a small mature guide crRNA

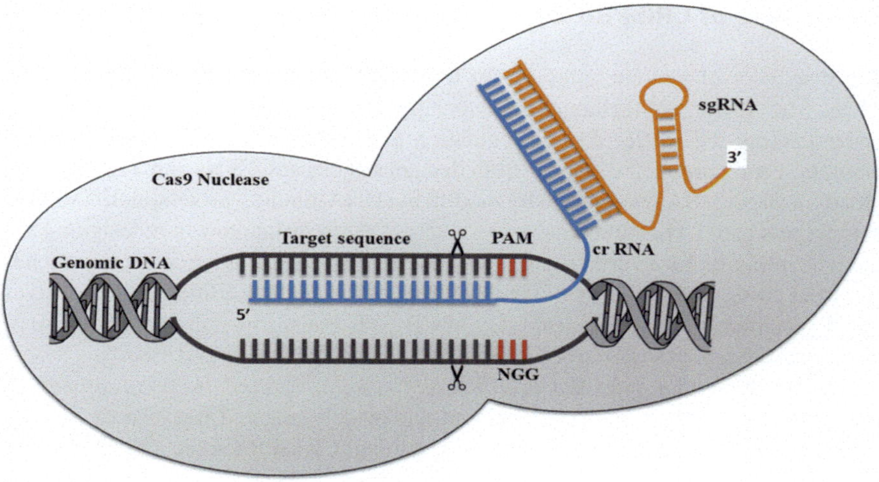

Fig. 1.9 An illustration of the general assembly and target recognition of the CRISPR/Cas genome editing tool, showing a single-guide RNA (sgRNA), consisting of a crRNA sequence specific to the DNA target, and a tracer RNA sequence that interacts with the Cas9 protein, which has DNA endonuclease activity. The resulting complex causes the target-specific double-stranded DNA cleavage that is required for genome editing

(gRNA) (Hatoum-Aslan et al. 2011). The mature crRNA contains a short repetitive element (repeat) portion at the 3′ end and a full spacer sequence at the 5′ end and which remains attached with tracer RNA and Cas9 to form an active Cas-crRNA complex (Fig. 1.8).

Phase III of CRISPR/Cas immunity is referred to as "interference." The Cas-crRNA complex scans the cell for foreign DNA targets and interferes with invading pathogens by recognizing a complementary sequence of spacer in the invader's genome, i.e., the protospacer. Once the protospacer in target DNA is identified by spacer through complementary base pairing, Cas-crRNA complex cleaves the protospacer by nuclease activity resulting in the destruction of the foreign DNA. Here it is important to note that target recognition and cleavage of target DNA by an active Cas-crRNA complex depends on the presence of a short protospacer adjacent motif (PAM) next to the protospacer sequence (Fig. 1.9). PAM is a 2–5-bp conserved sequence that is recognized by Cas protein which plays an important role in target recognition and discrimination between self and non-self-DNA for avoiding autoimmunity (Gleditzsch et al. 2019).

Advances in understanding the CRISPR/Cas mechanism (Figs. 1.8 and 1.9) have led scientists to build this system into a highly efficient and precise genome editing tool.

1.4 Rise of CRISPR/Cas

The discovery of genome engineering tools such as meganucleases, followed by ZFNs and TALENs, facilitated site-specific DNA manipulation and helped in understanding the mode of gene regulation and the relationship between genetic disruption and particular diseases. Both designer nucleases, ZFNs and TALENs, are based on target DNA recognition by modifiable DNA-binding protein motifs such as zinc fingers and TALE domains, respectively. Due to difficulties in redesigning the modular proteins for every target DNA, researchers sought alternative approaches that offer ease and modularity of reengineering of the genome editing tool, as well as target specificity. The recent discovery of CRISPR transformed the field because of its several advantages (described in Sect. 1.5), over ZFNs and TALENs.

The most popular type II CRISPR/Cas9 system, derived from *Streptococcus pyogenes*, has been extensively explored and widely adapted because of its easy design and manipulation for target genome editing. CRISPR/Cas acts as a "memory storage system" to limit reinfection by pathogens in prokaryotes (Jinek et al. 2012). Key findings in recent years have converted the CRISPR immune system into a powerful reprogrammable CRISPR gene editing tool. The presence of CRISPR repeats was initially noticed by Atsuo Nakata's group in the DNA of *Escherichia coli* in 1987 (Ishino et al. 1987) long before the name "CRISPR" was coined (Jansen et al. 2002). Subsequently, recognition of the nature and origin of spacer sequence through computational analysis of complete genome sequences of phages and other organisms led researchers to notice that spacer sequences belong to bacteriophages and other mobile genetic elements (Mojica et al. 2005; Bolotin et al. 2005; Pourcel et al. 2005). Moreover, the presence of multiple, highly conserved CRISPR-associated Cas protein-related genes boosted the interest of researchers to explore the potential function and mechanism of action of the whole CRISPR cascade.

The functional significance of CRISPR system was brought to light by Horvath et al. (2008), which provided experimental evidence of a correlation between CRISPR activity and adaptive immune response in prokaryotes. They observed the presence of a new spacer sequence in the CRIPSR loci of *Streptococcus thermophilus* after a viral challenge. Interestingly, the new spacer sequence adapted by bacterium belonged to a phage genomic sequence, resulting in a phage-resistant *Streptococcus thermophilus* strain. The acquisition of the spacer sequence was prerequisite to the direct targeting specificity of Cas nucleases that provide a unique defense system against the phage (Barrangou et al. 2007). After this key discovery, researchers further recognized that short CRISPR RNAs transcribed from spacer sequences are required to guide Cas nucleases to their target (Brouns et al. 2008). Another significant finding was the recognition of PAMs as a strict requirement for the action of the CRISPR system (Deveau et al. 2008). An important observation by Sapranauskas et al. (2011) was the ability of the CRISPR system to transfer from one bacterium to other bacterial species. This stimulated researchers to further characterize individual components of the CRISPR system biochemically and understand its molecular mechanism in greater detail. A crucial finding was that, among several Cas proteins, only Cas 9 from *S. thermophilus* had DNA catalytic activity; moreover, it required two short RNAs for its function (Garneau et al. 2010; Deltcheva et al. 2011).

These findings, together with the observation that Cas9 is reprogrammable in accordance with the target DNA sequence (Jinek et al. 2012; Gasiunas et al. 2012), marked the beginning of CRISPR/Cas as a reprogrammable genome editing tool. The construction of a chimeric sgRNA by the fusion of two short-guide RNAs (tracer RNA and crRNA) to guide Cas9 to its target simplified the designing of CRISPR/Cas technology for specific targets (Jinek et al. 2012). This was followed by the ground-breaking adaptation of this tool for genome editing in eukaryotic cells with high efficiency, specificity, and flexibility (Jinek et al. 2013; Cong et al. 2013; Mali et al. 2013). Since the inauguration of CRISPR/Cas as a genome editing tool, researchers from diverse fields have employed the technology for precise remodeling of prokaryotic and eukaryotic genomes and have demonstrated its potential applications in all fields of biotechnology, including antimicrobial, therapeutic, and diagnostic applications.

1.5 Comparison of CRISPR/Cas with Other Genome Editing Tools

Genome editing tools with programmable nucleases gained close attention because of their potential applications in the fields of genetics, biotechnology, and therapeutics. Continuous effort and investigation have allowed researchers to discover new genome editing candidates and improve the properties of existing tools. Current attention is mainly focused on the development of simple, cost-effective, and target-specific genome editing techniques with wide applications in basic and biomedical sciences. All four genome editing techniques introduced so far (meganucleases, ZFNs, TALENs, and CRISPR/Cas9) have been used effectively to modify targeted genome in model organisms, plants, and human cell lines. However, the important question over which reprogrammable technology is best for all-purpose genome editing is difficult to answer. The application of any genome editing tool depends on its efficiency, target specificity, simplicity of design, cost, and reliability. A comparison of programmable nucleases is shown in Table 1.1.

While effective, the design of ZFNs and TALENs requires protein engineering expertise to construct a specific architecture for the target DNA by modifying the DNA-binding domain of ZnFs or TALE. Protein modification in 3–9 fingers for a target site, or the engineering of multiple TALE repeats to design a target-specific DNA-binding domain, is challenging, highly time-consuming, and expensive. CRISPR/Cas9 offers several advantages over ZFNs and TALENs because of its simple design, high target specificity, reduced off-target toxicity, multiple gene target ability, low cost, and ease of delivery to cells.

However, the use of CRISPR/Cas for therapeutic purposes requires further investigation to confirm its safety because of remaining off-target effects (Cox et al. 2015; Doudna and Gersbach 2015). High-frequency off-target effects are a challenging obstacle that remain to be solved for higher and more precise specificity

Table 1.1 Comparison of commercially available genome editing techniques

Features		Meganucleases	Zinc finger nucleases (ZFNs)	TALENs	CRISPR/Cas9
Origin and year of introduction		Microbial mobile genetic elements (1990)	*Xenopus* oocytes (1996)	*Xanthomonas* species (2010)	Immune system of bacteria and archaea (2012)
Natural function		Mobile genetic elements	DNA-binding motifs in eukaryotes	Plant pathogenic protein	Prokaryotic defense protein
Mechanism of action	Target recognition	Protein-DNA	Protein-DNA (one to triplet)	Protein-DNA (one to one)	RNA-DNA (one to one)
	Target cleavage	Homing nucleases	Fok1 nuclease	Fok1 nuclease	Cas nucleases
Target specificity	Target site/ length	12–24 bp	~ 9–36 nt	~ 12–50 nt	~ 20–23 nt
	Target efficiency	Binds with DNA bases in major groove and phosphodiester backbone	Binds to a triplet of DNA bases	Needs T base at 5′	Needs PAM region (5′NGG)
	Off-target constraints	Low	Low	Very few	High
Engineering constraints	Design feasibility	Engineering of protein required for novel targets	Engineering of DNA-binding domain required for every target	Engineering of DNA-binding domain required for every target DNA	RNA complementary to target DNA required
	Assembly	No assembly required	Difficult	Technical but easy	Easy
	Time	Time-consuming	Highly time-consuming	Less time-consuming	Less time consuming
	Cost	Expensive	Expensive	Not cost-effective	Cost effective

Commercial tools		Not independent database available	Available	Available	Readily available
Uses	Genome editing possibilities	Gene addition, gene deletion	Gene disruption, gene deletion, gene correction, gene addition, tag ligation, ObLiGaRe	Gene activation, gene repression, gene disruption, gene deletion, gene correction, gene addition, tag ligation, ObLiGaRe	Gene disruption, gene deletion, gene correction, gene addition

of this technique than is currently possible (Sander and Joung 2014; Tsai and Joung 2016). The following strategies have been adopted by different researchers to prevent the off-target effects of CRISPR/Cas:

- Direct injection of Cas9 nucleases in the target cells instead of recombinant expression of the protein to reduce nuclease activity (Ramakrishna et al. 2014; Kim et al. 2014).
- Use of two sgRNAs to target both DNA strands of the target sequence in combination with a DNA-nicking variant of Cas9. This double-nicking strategy has shown significant reduction in off-target effects (Ran et al. 2013).
- Truncation of 2–3 nucleotides of sgRNA has exhibited high target specificity of CRISPR/Cas (Fu et al. 2014).
- Cho et al. (2014) have demonstrated a reduction in off-target effects by the addition of two guanine nucleotides at the 5′ end next to the target complementary region of sgRNA.

In addition to its off-target effects, another obstacle for CRISPR/Cas is overall delivery of this system into eukaryotic cells. Several viral and non-viral vector systems have proved promising for delivering Cas9 and sgRNA in vivo (Savić and Schwank 2016; Gori et al. 2015). An ex vivo strategy where patient-driven cells are modified by CRISPR/Cas in vitro and re-transplanted back into the patient gives a significant advantage in terms of selection of correctly modified cells and subsequent re-transplant without undesired off-target mutations. This approach has been used successfully to modify B cells to improve immunity and for T cells in cancer treatment (Ren and Zhao 2017). Efforts to use CRISPR/Cas for diagnostic purposes have been very successful (Freije et al. 2019). With continuing efforts to improve known CRISPR/Cas systems and discover new types of CRISPR systems, it seems that this technology will be safe and effective for applications ranging from biotechnology to therapeutics in the very near future.

References

Amitai G, Sorek R (2016) CRISPR–Cas adaptation: insights into the mechanism of action. Nat Rev Microbiol 14(2):67

Argast GM, Stephens KM, Emond MJ, Monnat RJ Jr (1998) I-PpoI and I-CreI homing site sequence degeneracy determined by random mutagenesis and sequential in vitro enrichment. J Mol Biol 280(3):345–353

Arnould S, Perez C, Cabaniols JP, Smith J, Gouble A, Grizot S, Pâques F (2007) Engineered I-CreI derivatives cleaving sequences from the human XPC gene can induce highly efficient gene correction in mammalian cells. J Mol Biol 371(1):49–65

Baker M (2012) Gene-editing nucleases. Nat Methods 9:23–26

Barrangou R, Fremaux C, Deveau H, Richards M, Boyaval P, Moineau S, Horvath P (2007) CRISPR provides acquired resistance against viruses in prokaryotes. Science 315 (5819):1709–1712

Beerli RR, Segal DJ, Dreier B, Barbas CF (1998) Toward controlling gene expression at will: specific regulation of the erbB-2/HER-2 promoter by using polydactyl zinc finger proteins constructed from modular building blocks. Proc Natl Acad Sci U S A 95(25):14628–14633

Bibikova M, Carroll D, Segal DJ, Trautman JK, Smith J, Kim YG, Chandrasegaran S (2001) Stimulation of homologous recombination through targeted cleavage by chimeric nucleases. Mol Cell Biol 21(1):289–297

Bibikova M, Beumer K, Trautman JK, Carroll D (2003) Enhancing gene targeting with designed zinc finger nucleases. Science 300(5620):764–764

Bitinaite J, Wah DA, Aggarwal AK, Schildkraut I (1998) FokI dimerization is required for DNA cleavage. Proc Natl Acad Sci U S A 95(18):10570–10575

Boch J, Bonas U (2010) Xanthomonas AvrBs3 family-type III effectors: discovery and function. Annu Rev Phytopathol 48

Boch J, Scholze H, Schornack S, Landgraf A, Hahn S, Kay S, Bonas U (2009) Breaking the code of DNA binding specificity of TAL-type III effectors. Science 326(5959):1509–1512

Boissel S, Jarjour J, Astrakhan A, Adey A, Gouble A, Duchateau P, Scharenberg AM (2014) megaTALs: a rare-cleaving nuclease architecture for therapeutic genome engineering. Nucleic Acids Res 42(4):2591–2601

Bolotin A, Quinquis B, Sorokin A, Ehrlich SD (2005) Clustered regularly interspaced short palindrome repeats (CRISPRs) have spacers of extra chromosomal origin. Microbiology 151 (8):2551–2561

Brouns SJ, Jore MM, Lundgren M, Westra ER, Slijkhuis RJ, Snijders AP, Van Der Oost J (2008) Small CRISPR RNAs guide antiviral defense in prokaryotes. Science 321(5891):960–964

Buck-Koehntop BA, Stanfield RL, Ekiert DC, Martinez-Yamout MA, Dyson HJ, Wilson IA, Wright PE (2012) Molecular basis for recognition of methylated and specific DNA sequences by the zinc finger protein Kaiso. Proc Natl Acad Sci U S A 109(38):15229–15234

Capecchi MR (1989) Altering the genome by homologous recombination. Science 244 (4910):1288–1292

Carroll D, Morton JJ, Beumer KJ, Segal DJ (2006) Design, construction and in vitro testing of zinc finger nucleases. Nat Protoc 1(3):1329–1341

Chandras C, Zouberakis M, Salimova E, Smedley D, Rosenthal N, Aidinis V (2012) CreZOO—the European virtual repository of Cre and other targeted conditional driver strains. Database 2012

Chandrasegaran S, Carroll D (2016) Origins of programmable nucleases for genome engineering. J Mol Biol 428(5):963–989

Chang HH, Pannunzio NR, Adachi N, Lieber MR (2017) Non-homologous DNA end joining and alternative pathways to double-strand break repair. Nat Rev Mol Cell Biol 18(8):495

Chevalier B, Turmel M, Lemieux C, Monnat RJ Jr, Stoddard BL (2003) Flexible DNA target site recognition by divergent homing endonuclease isoschizomers I-CreI and I-MsoI. J Mol Biol 329 (2):253–269

Cho SW, Kim S, Kim Y, Kweon J, Kim HS, Bae S, Kim JS (2014) Analysis of off-target effects of CRISPR/Cas-derived RNA-guided endonucleases and nickases. Genome Res 24(1):132–141

Choo Y, Sánchez García I, Klug A (1994) In vivo repression by a site-specific DNA-binding protein designed against an oncogenic sequence

Cong L, Ran FA, Cox D, Lin S, Barretto R, Habib N, Zhang F (2013) Multiplex genome engineering using CRISPR/Cas systems. Science 339(6121):819–823

Cox DBT, Platt RJ, Zhang F (2015) Therapeutic genome editing: prospects and challenges. Nat Med 21(2):121–131

Daboussi F, Zaslavskiy M, Poirot L, Loperfido M, Gouble A, Guyot V, Perez C (2012) Chromosomal context and epigenetic mechanisms control the efficacy of genome editing by rare-cutting designer endonucleases. Nucleic Acids Res 40(13):6367–6379

Danna K, Nathans D (1971) Specific cleavage of simian virus 40 DNA by restriction endonuclease of Hemophilus influenzae. Proc Natl Acad Sci U S A 68(12):2913–2917

De Chiara TM, Efstratiadis A, Robertsen EJ (1990) A growth-deficiency phenotype in heterozygous mice carrying an insulin-like growth factor II gene disrupted by targeting. Nature 345 (6270):78–80

Deltcheva E, Chylinski K, Sharma CM, Gonzales K, Chao Y, Pirzada ZA, Charpentier E (2011) CRISPR RNA maturation by trans-encoded small RNA and host factor RNase III. Nature 471 (7340):602–607

Deng D, Yan C, Pan X, Mahfouz M, Wang J, Zhu JK, Yan N (2012) Structural basis for sequence-specific recognition of DNA by TAL effectors. Science 335(6069):720–723

Deveau H, Barrangou R, Garneau JE, Labonté J, Fremaux C, Boyaval P, Moineau S (2008) Phage response to CRISPR-encoded resistance in Streptococcus thermophilus. J Bacteriol 190 (4):1390–1400

Doudna JA, Gersbach CA (2015) Genome editing: the end of the beginning. Genome Biol 16 (1):292

Fairall L, Schwabe JW, Chapman L, Finch JT, Rhodes D (1993) The crystal structure of a two zinc-finger peptide reveals an extension to the rules for zinc-finger/DNA recognition. Nature 366 (6454):483–487

Freije CA, Myhrvold C, Boehm CK, Lin AE, Welch NL, Carter A, Yozwiak NL (2019) Programmable inhibition and detection of RNA viruses using Cas13. Mol Cell 76(5):826–837

Fu Y, Sander JD, Reyon D, Cascio VM, Joung JK (2014) Improving CRISPR-Cas nuclease specificity using truncated guide RNAs. Nat Biotechnol 32(3):279–284

Gao H, Smith J, Yang M, Jones S, Djukanovic V, Nicholson GM, West A, Bidney S, Falco SC, Jantz D, Lyznik LA (2010) Heritable targeted mutagenesis in maize using a designed endonuclease. Plant J 61(1):176–187

Garneau JE, Dupuis MÈ, Villion M, Romero DA, Barrangou R, Boyaval P, Moineau S (2010) The CRISPR/Cas bacterial immune system cleaves bacteriophage and plasmid DNA. Nature 468 (7320):67–71

Gasiunas G, Barrangou R, Horvath P, Siksnys V (2012) Cas9–crRNA ribonucleoprotein complex mediates specific DNA cleavage for adaptive immunity in bacteria. Proc Natl Acad Sci U S A 109(39):E2579–E2586

Gleditzsch D, Pausch P, Müller-Esparza H, Özcan A, Guo X, Bange G, Randau L (2019) PAM identification by CRISPR-Cas effector complexes: diversified mechanisms and structures. RNA Biol 16(4):504–517

Gori JL, Hsu PD, Maeder ML, Shen S, Welstead GG, Bumcrot D (2015) Delivery and specificity of CRISPR/Cas9 genome editing technologies for human gene therapy. Hum Gene Ther 26 (7):443–451

Hatoum-Aslan A, Maniv I, Marraffini LA (2011) Mature clustered, regularly interspaced, short palindromic repeats RNA (crRNA) length is measured by a ruler mechanism anchored at the precursor processing site. Proc Natl Acad Sci U S A 108(52):21218–21222

Hille F, Charpentier E (2016) CRISPR-Cas: biology, mechanisms and relevance. Philos Trans R Soc B Biol Sci 371(1707):20150496

Horvath P, Romero DA, Coûté-Monvoisin AC, Richards M, Deveau H, Moineau S, Barrangou R (2008) Diversity, activity, and evolution of CRISPR loci in Streptococcus thermophilus. J Bacteriol 190(4):1401–1412

Ishino Y, Shinagawa H, Makino K, Amemura M, Nakata A (1987) Nucleotide sequence of the iap gene, responsible for alkaline phosphatase isozyme conversion in Escherichia coli, and identification of the gene product. J Bacteriol 169(12):5429–5433

Jacquier A, Dujon B (1985) An intron-encoded protein is active in a gene conversion process that spreads an intron into a mitochondrial gene. Cell 41(2):383–394

Jansen R, Embden JDV, Gaastra W, Schouls LM (2002) Identification of genes that are associated with DNA repeats in prokaryotes. Mol Microbiol 43(6):1565–1575

Jinek M, Chylinski K, Fonfara I, Hauer M, Doudna JA, Charpentier E (2012) A programmable dual-RNA–guided DNA endonuclease in adaptive bacterial immunity. Science 337 (6096):816–821

Jinek M, East A, Cheng A, Lin S, Ma E, Doudna J (2013) RNA-programmed genome editing in human cells. elife 2:e00471
Karginov FV, Hannon GJ (2010) The CRISPR system: small RNA-guided defense in bacteria and archaea. Mol Cell 37(1):7–19
Kelly TJ Jr, Smith HO (1970) A restriction enzyme from Hemophilus influenzae: II. Base sequence of the recognition site. J Mol Biol 51(2):393–409
Kersten K, de Visser KE, van Miltenburg MH, Jonkers J (2017) Genetically engineered mouse models in oncology research and cancer medicine. EMBO Mol Med 9(2):137–153
Kim H, Kim JS (2014) A guide to genome engineering with programmable nucleases. Nat Rev Genet 15(5):321–334
Kim JS, Pabo CO (1997) Transcriptional repression by zinc finger peptides exploring the potential for applications in gene therapy. J Biol Chem 272(47):29795–29800
Kim YG, Cha J, Chandrasegaran S (1996) Hybrid restriction enzymes: zinc finger fusions to Fok I cleavage domain. Proc Natl Acad Sci U S A 93(3):1156–1160
Kim JS, Kim J, Cepek KL, Sharp PA, Pabo CO (1997) Design of TATA box-binding protein/zinc finger fusions for targeted regulation of gene expression. Proc Natl Acad Sci U S A 94 (8):3616–3620
Kim YG, Smith J, Durgesha M, Chandrasegaran S (1998) Chimeric restriction enzyme: Gal4 fusion to Fokl cleavage domain. Biol Chem 379(4–5):489–496
Kim HJ, Lee HJ, Kim H, Cho SW, Kim JS (2009) Targeted genome editing in human cells with zinc finger nucleases constructed via modular assembly. Genome Res 19(7):1279–1288
Kim S, Kim D, Cho SW, Kim J, Kim JS (2014) Highly efficient RNA-guided genome editing in human cells via delivery of purified Cas9 ribonucleoproteins. Genome Res 24(6):1012–1019
Kim H, Kim M, Im SK, Fang S (2018) Mouse Cre-LoxP system: general principles to determine tissue-specific roles of target genes. Lab Anim Res 34(4):147–159
Koller BH, Marrack P, Kappler JW, Smithies O (1990) Normal development of mice deficient in β2M, MHC class I, proteins, and CD8+ T cells. Science 248:1227–1230
Kosicki M, Tomberg K, Bradley A (2018) Repair of double-strand breaks induced by CRISPR–Cas9 leads to large deletions and complex rearrangements. Nat Biotechnol 36(8):765–771
Li L, Chandrasegaran S (1993) Alteration of the cleavage distance of Fok I restriction endonuclease by insertion mutagenesis. Proc Natl Acad Sci U S A 90(7):2764–2768
Li L, Wu LP, Chandrasegaran S (1992) Functional domains in Fok I restriction endonuclease. Proc Natl Acad Sci U S A 89(10):4275–4279
Lucas P, Otis C, Mercier JP, Turmel M, Lemieux C (2001) Rapid evolution of the DNA-binding site in LAGLIDADG homing endonucleases. Nucleic Acids Res 29(4):960–969
Mak ANS, Bradley P, Cernadas RA, Bogdanove AJ, Stoddard BL (2012) The crystal structure of TAL effector PthXo1 bound to its DNA target. Science 335(6069):716–719
Makarova KS, Haft DH, Barrangou R, Brouns SJ, Charpentier E, Horvath P, Van Der Oost J (2011) Evolution and classification of the CRISPR–Cas systems. Nat Rev Microbiol 9(6):467–477
Makarova KS, Wolf YI, Alkhnbashi OS, Costa F, Shah SA, Saunders SJ, Horvath P (2015) An updated evolutionary classification of CRISPR–Cas systems. Nat Rev Microbiol 13 (11):722–736
Mali P, Yang L, Esvelt KM, Aach J, Guell M, DiCarlo JE, Church GM (2013) RNA-guided human genome engineering via Cas9. Science 339(6121):823–826
Mashimo T, Takizawa A, Voigt B, Yoshimi K, Hiai H, Kuramoto T, Serikawa T (2010) Generation of knockout rats with X-linked severe combined immunodeficiency (X-SCID) using zinc-finger nucleases. PLoS One 5(1):e8870
McLellan MA, Rosenthal NA, Pinto AR (2017) Cre-loxP-mediated recombination: general principles and experimental considerations. Curr Protoc Mouse Biol 7(1):1–12
Miller J, McLachlan AD, Klug A (1985) Repetitive zinc-binding domains in the protein transcription factor IIIA from Xenopus oocytes. EMBO J 4(6):1609–1614

Mojica FJ, Díez-Villaseñor C, Soria E, Juez G (2000) Biological significance of a family of regularly spaced repeats in the genomes of archaea, bacteria and mitochondria. Mol Microbiol 36(1):244–246

Mojica FJ, García-Martínez J, Soria E (2005) Intervening sequences of regularly spaced prokaryotic repeats derive from foreign genetic elements. J Mol Evol 60(2):174–182

Moscou MJ, Bogdanove AJ (2009) A simple cipher governs DNA recognition by TAL effectors. Science 326(5959):1501–1501

Murray SA, Eppig JT, Smedley D, Simpson EM, Rosenthal N (2012) Beyond knockouts: cre resources for conditional mutagenesis. Mamm Genome 23(9–10):587–599

Nagy A (2000) Cre recombinase: the universal reagent for genome tailoring. Genesis 26(2):99–109

Orban PC, Chui D, Marth JD (1992) Tissue-and site-specific DNA recombination in transgenic mice. Proc Natl Acad Sci U S A 89(15):6861–6865

Orlowski J, Boniecki M, Bujnicki JM (2007) I-Ssp6803I: the first homing endonuclease from the PD-(D/E) XK superfamily exhibits an unusual mode of DNA recognition. Bioinformatics 23 (5):527–530

Pavletich NP, Pabo CO (1991) Zinc finger-DNA recognition: crystal structure of a Zif268-DNA complex at 2.1 A. Science 252(5007):809–817

Pourcel C, Salvignol G, Vergnaud G (2005) CRISPR elements in Yersinia pestis acquire new repeats by preferential uptake of bacteriophage DNA, and provide additional tools for evolutionary studies. Microbiology 151(3):653–663

Puchta H (2005) The repair of double-strand breaks in plants: mechanisms and consequences for genome evolution. J Exp Bot 56(409):1–14

Ramakrishna S, Dad ABK, Beloor J, Gopalappa R, Lee SK, Kim H (2014) Gene disruption by cell-penetrating peptide-mediated delivery of Cas9 protein and guide RNA. Genome Res 24 (6):1020–1027

Ramirez CL, Foley JE, Wright DA, Müller-Lerch F, Rahman SH, Cornu TI, Cathomen T (2008) Unexpected failure rates for modular assembly of engineered zinc fingers. Nat Methods 5 (5):374–375

Ran FA, Hsu PD, Lin CY, Gootenberg JS, Konermann S, Trevino AE, Zhang F (2013) Double nicking by RNA-guided CRISPR Cas9 for enhanced genome editing specificity. Cell 154 (6):1380–1389

Ren J, Zhao Y (2017) Advancing chimeric antigen receptor T cell therapy with CRISPR/Cas9. Protein Cell 8(9):634–643

Rosen LE, Morrison HA, Masri S, Brown MJ, Springstubb B, Sussman D, Seligman LM (2006) Homing endonuclease I-CreI derivatives with novel DNA target specificities. Nucleic Acids Res 34(17):4791–4800

Rothstein RJ (1983) One-step gene disruption in yeast. In: Methods in enzymology, vol 101. Academic Press, pp 202–211

Rouet P, Smih F, Jasin M (1994) Expression of a site-specific endonuclease stimulates homologous recombination in mammalian cells. Proc Natl Acad Sci U S A 91(13):6064–6068

Sander JD, Joung JK (2014) CRISPR-Cas systems for editing, regulating and targeting genomes. Nat Biotechnol 32(4):347–355

Sapranauskas R, Gasiunas G, Fremaux C, Barrangou R, Horvath P, Siksnys V (2011) The *Streptococcus thermophilus* CRISPR/Cas system provides immunity in *Escherichia coli*. Nucleic Acids Res 39(21):9275–9282

Savić N, Schwank G (2016) Advances in therapeutic CRISPR/Cas9 genome editing. Transl Res 168:15–21

Schofield PN, Hoehndorf R, Gkoutos GV (2012) Mouse genetic and phenotypic resources for human genetics. Hum Mutat 33(5):826–836

Segal DJ, Meckler JF (2013) Genome engineering at the dawn of the golden age. Annu Rev Genomics Hum Genet 14:135–158

Segal DJ, Dreier B, Beerli RR, Barbas CF (1999) Toward controlling gene expression at will: selection and design of zinc finger domains recognizing each of the 5′-GNN-3′ DNA target sequences. Proc Natl Acad Sci U S A 96(6):2758–2763

Seligman LM, Chisholm KM, Chevalier BS, Chadsey MS, Edwards ST, Savage JH, Veillet AL (2002) Mutations altering the cleavage specificity of a homing endonuclease. Nucleic Acids Res 30(17):3870–3879

Shalem O, Sanjana NE, Zhang F (2015) High-throughput functional genomics using CRISPR–Cas9. Nat Rev Genet 16(5):299–311

Sharma S, Zhu J (2014) Immunologic applications of conditional gene modification technology in the mouse. Curr Protoc Immunol 105(1):10–34

Shen BW, Lambert A, Walker BC, Stoddard BL, Kaiser BK (2016) The structural basis of asymmetry in DNA binding and cleavage as exhibited by the I-SmaMI LAGLIDADG meganuclease. J Mol Biol 428(1):206–220

Shukla VK, Doyon Y, Miller JC, DeKelver RC, Moehle EA, Worden SE, Choi VM (2009) Precise genome modification in the crop species Zea mays using zinc-finger nucleases. Nature 459 (7245):437–441

Smedley D, Salimova E, Rosenthal N (2011) Cre recombinase resources for conditional mouse mutagenesis. Methods 53(4):411–416

Smith HO, Wilcox KW (1970) A restriction enzyme from Hemophilus influenzae: I. Purification and general properties. J Mol Biol 51(2):379–391

Smith J, Berg JM, Chandrasegaran S (1999) A detailed study of the substrate specificity of a chimeric restriction enzyme. Nucleic Acids Res 27(2):674–681

Smith J, Bibikova M, Whitby FG, Reddy AR, Chandrasegaran S, Carroll D (2000) Requirements for double-strand cleavage by chimeric restriction enzymes with zinc finger DNA-recognition domains. Nucleic Acids Res 28(17):3361–3369

Smith J, Grizot S, Arnould S, Duclert A, Epinat JC, Chames P, Prieto J, Redondo P, Blanco FJ, Bravo J, Montoya G, Pâques F, Duchateau P (2006) A combinatorial approach to create artificial homing endonucleases cleaving chosen sequences. Nucleic Acids Res 34(22):e149

Stoddard BL (2005) Homing endonuclease structure and function. Q Rev Biophys 38(1):49

Sung YH, Baek IJ, Seong JK, Kim JS, Lee HW (2012) Mouse genetics: catalogue and scissors. BMB Rep 45(12):686

Sussman D, Chadsey M, Fauce S, Engel A, Bruett A, Monnat R Jr, Seligman LM (2004) Isolation and characterization of new homing endonuclease specificities at individual target site positions. J Mol Biol 342(1):31–41

Swarts DC, Mosterd C, Van Passel MW, Brouns SJ (2012) CRISPR interference directs strand specific spacer acquisition. PLoS One 7(4):e35888

Takeuchi R, Choi M, Stoddard LB (2014) Redesign of extensive protein–DNA interfaces of meganucleases using iterative cycles of in vitro compartmentalization. PNAS 111 (11):4061–4066

Thomas KR, Capecchi MR (1990) Targeted disruption of the murine int-1 proto oncogene resulting in severe abnormalities in midbrain and cerebellar development. Nature 346(6287):847–850

Townsend JA, Wright DA, Winfrey RJ, Fu F, Maeder ML, Joung JK, Voytas DF (2009) High-frequency modification of plant genes using engineered zinc-finger nucleases. Nature 459 (7245):442–445

Tsai SQ, Joung JK (2016) Defining and improving the genome-wide specificities of CRISPR–Cas9 nucleases. Nat Rev Genet 17(5):300–312

Tsien JZ (2016) Cre-Lox neurogenetics: 20 years of versatile applications in brain research and counting. Front Genet 7:19

Urnov FD, Rebar EJ, Holmes MC, Zhang HS, Gregory PD (2010) Genome editing with engineered zinc finger nucleases. Nat Rev Genet 11(9):636–646

Valton J, Daboussi F, Leduc S, Molina R, Redondo P, Macmaster R, Duchateau P (2012) 5′-Cytosine-phosphoguanine (CpG) methylation impacts the activity of natural and engineered meganucleases. J Biol Chem 287(36):30139–30150

Vanamee ÉS, Santagata S, Aggarwal AK (2001) FokI requires two specific DNA sites for cleavage. J Mol Biol 309(1):69–78

Verma P, Greenberg RA (2016) Noncanonical views of homology-directed DNA repair. Genes Dev 30(10):1138–1154

Xu GL, Bestor TH (1997) Cytosine methylation targetted to pre-determined sequences. Nat Genet 17(4):376–378

Yosef I, Goren MG, Qimron U (2012) Proteins and DNA elements essential for the CRISPR adaptation process in *Escherichia coli*. Nucleic Acids Res 40(12):5569–5576

Zhao L, Bonocora RP, Shub DA, Stoddard BL (2007) The restriction fold turns to the dark side: a bacterial homing endonuclease with a PD-(D/E)-XK motif. EMBO J 26(9):2432–2442

Zheng B, Sage M, Sheppeard EA, Jurecic V, Bradley A (2000) Engineering mouse chromosomes with Cre-loxP: range, efficiency, and somatic applications. Mol Cell Biol 20(2):648–655

Zijlstra M, Li E, Sajjadi F, Subramani S, Jaenisch R (1989) Germ-line transmission of a disrupted β 2 microglobulin gene produced by homologous recombination in embryonic stem cells. Nature 342(6248):435–438

History and Classification of CRISPR/Cas System

2

Zulqurnain Khan, Zulfiqar Ali, Asif Ali Khan, Tahmina Sattar, Ali Zeshan, Tahira Saboor, and Barkha Binyamin

Contents

Abstract

CRISPR/Cas usually depends upon the archaeal and bacterial adaptive immune system and works by using transposable genetic elements called spacers of plasmids and viruses. It is used to cleavage targeted RNA/DNA at a specific site and provides an effective platform for genome engineering of species from all three kingdoms. In 1987, CRISPR/Cas system was discovered after the identification of similar DNA sequences in the genome of *Escherichia coli* while studying genes that are helpful in phosphate metabolism. These DNA sequences

Z. Khan (✉) · Z. Ali · A. A. Khan · T. Sattar · T. Saboor · B. Binyamin
Institute of Plant Breeding and Biotechnology, Muhammad Nawaz Sharif University of Agriculture, Multan, Pakistan
e-mail: zulqurnain.khan@mnsuam.edu.pk

A. Zeshan
Center for Advanced Studies in Agriculture and Food Security, University of Agriculture, Faisalabad, Pakistan

A. Ahmad et al. (eds.), *The CRISPR/Cas Tool Kit for Genome Editing*,
https://doi.org/10.1007/978-981-16-6305-5_2

are known as the CRISPR (clustered regularly interspaced short palindromic repeats). Later on, these sequences have been identified in other bacterial genomes including halophilic archaea; these sequences play an important role in the evolutionary relationship of an organism. The functional characterization of CRISPR/Cas depends on the Cas or Cas9 associated proteins. In hyperthermophilic archaea, it is hypothesized that Cas protein is involved in the DNA repairing mechanism. Researchers in the fields of advanced biochemistry, comparative or structural genomics have been working hard to explore the CRISPR/Cas-based genome editing tools. Further investigations about the origin and evolution of these systems from genetic elements known as casposons are in progress. CRISPR/Cas structure is categorized into two major classes, and these classes are further divided into six subclasses (Types I–VI). This classification depends on variation present in interference complexes. Class 1 has Types I, III, and IV, while Class 2 has Types II, V, and VI. Emmanuelle Charpentier and Jennifer A. Doudna got the Nobel Prize in 2020 for developing the CRISPR/Cas9 gene editing tool that has been used for targeting and modifying the human genome. In this chapter, we are studying about the various aspects of the origin of CRISPR/Cas system. Moreover, types/classes of CRISPR/Cas system will also be discussed along with their applications for different purposes.

Keywords

CRISPR/Cas System · Classification · Types · Variants · Applications

Abbreviations

ADAR	Adenosine deaminase acting on RNA
CARF	CRISPR-associated Rossmann fold
Cas	CRISPR-associated
CASCADE	CRISPR-associated antiviral defense complex
COA	Cyclic oligonucleotide
CRISPR	Clustered regularly interspaced short palindromic repeats
dCas9	Deactivated Cas9
DSB	Double-stranded break
dsDNA	Double-stranded DNA
dsRNA	Double-stranded RNA
E. coli	*Escherichia coli*
gRNA	Guided RNA
HD	Histidine
HDR	Homology-directed repair
indels	Deletion
NHEJ	Non-homologous end-joining
PAM	Protospacer adjacent motif
PFS	Protospacer flanking site
pre-crRNA	Pre-CRISPR RNA

RGN	RNA-guided Cas9 nuclease
RNP	Ribonucleoprotein
sgRNA	Single-guide RNA
ssDNA	Single-stranded DNA
ssRNA	Single-stranded RNA
TALENs	Transcription activator-like effector nucleases
tracrRNA	Trans-activating crRNAs
ZFNs	Zinc finger nucleases

2.1 Introduction of CRISPR/Cas System

The targeted genome editing has been rising in contrast to traditional plant breeding and transgenic methods for the improvement of different living organisms which guarantee efficient gene manipulation at pre-defined target sites. Different types of nucleases such as transcription activator-like effector nucleases (TALENs) and zinc finger nucleases (ZFNs) have been used for mutagenizing the genome at a specific locus; however, these methods require two distinctive DNA-binding proteins to target a sequence of interest, each with a C-terminal FokI nuclease module. Consequently, the plant research area has not broadly received these techniques. Recently, another new strategy, CRISPR/Cas (clustered regularly interspaced short palindromic repeats/CRISPR-associated protein), has been efficiently used for identifying the helicase as well as nuclease activity (Chylinski et al. 2014). It is the kind of Type II prokaryotic immune system that was present in bacteria and archaea (Sorek et al. 2013; Ishino et al. 1987; Grissa et al. 2007) and has risen as a technique for editing a genome (Jinek et al. 2012) along with its various remedial uses in eukaryotic systems. The bacterial cells have also been engineered through the artificial CRISPR/Cas system for industrial purposes. Classification of this system is done as Class 1 and Class 2, which comprises a large Cas protein and a complex of different Cas proteins, correspondingly, on the basis of their signature proteins and complexity; it is further divided into various subtypes (Makarova et al. 2015). Because of the entire comprehension of the response system against the foreign DNA, the specific RNA-guided CRISPR/Cas9 system has quickly progressed in engineering a genome in different cell systems (Carroll 2011; Boch et al. 2009). This system allows the efficient modification of targeted sequence by replacing the 20-nucleotide (nt) sequence of a single-guide RNA (sgRNA). sgRNA is the sequence that is opposite (complementary) to the targeted region. Also, the deactivated Cas9 (dCas9) is used to exhibit the gene expression and regulation which is further guided by sgRNA and has the DNA binding capacity (luo et al. 2015; Senturk et al. 2017). The adaptive immune system is relayed on three stages known as adaptation, expression, and interference (Jackson et al. 2017; Mohanraju et al. 2016). In case of adaptation, the unique spacer sequences have been used for coordinating the CRISPR/Cas array (Amitai and Sorek 2016). The transcription of

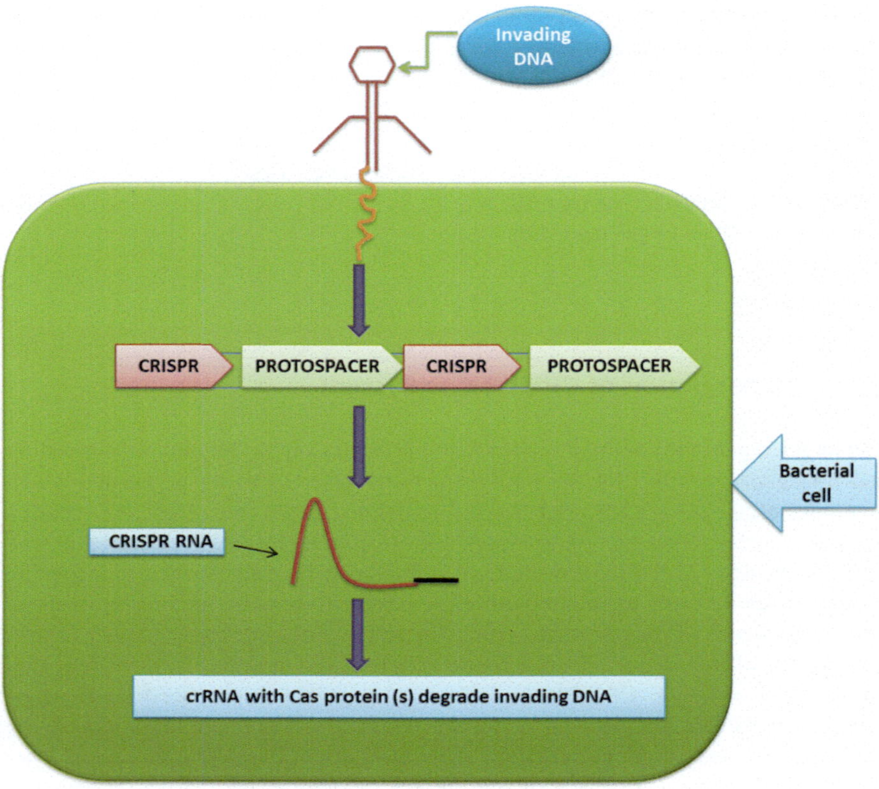

Fig. 2.1 The system of CRISPR/Cas is configured in three steps; the invading DNA is inserted into the genome of host, interspersed with CRISPR repeats. crRNA has been used to express in the second level. In the third stage, the complex of ribonucleoprotein based on crRNA and Cas nucleases cleaves the invading DNA

repaired spacer sequence occurs, and after that it is adapted into the genome. The proteins of Cas1, Cas2, and Cas4 are responsible for controlling adaptation of known CRISPR/Cas system which adapted the spacer sequence into CRISPR array (Jackson et al. 2017). Multiple lab experiments based on using CRISPR for gene editing in plants and microbes are in progress in our lab. We are using different platforms and reagents provided by Addgene, VectorBuilder, GeneCopoeia, Home – Nootropics Frontline, etc. The mechanism of CRISPR/Cas system in the bacterial cell is provided in Fig. 2.1.

In the expression stage, the numerous sequences contained by CRISPR arrangement are transcribed into the pre-CRISPR RNA (pre-crRNA). CRISPR/Cas arrangement has been homologous to particular targeted regions (protospacers) (Shmakov et al. 2017; Mohanraju et al. 2016). Consequently, homologous bonds are formed between pre-crRNA and smaller trans-activating crRNAs (tracrRNAs) (Shmakov et al. 2017). When this complex is assembled, it is attached to a protein of Cas9, and then the pre-crRNAs are cut by RNase III. In the process of interference, the crRNAs/tracrRNAs are used to guide the Cas9 to move towards the targeted sequence, and then crRNA is able to bind with the targeted region which is present near the sequence of protospacer adjacent motif (PAM). PAM sequences are defined as the small sequence of DNA (seed sequence) that is used for the targeting of gRNA. At this stage, the targeted sequence is unwound, and the domains of Cas9 proteins (RuvC and HNH) are used to cleave the targeted sequence (Mohanraju et al. 2016). Interference occurs by targeting invading viral or plasmid DNA and creates the double-stranded DNA breaks that are further used to detach/deactivate the complex of Cas9. The system can be harnessed to repair damaged DNA by homology-directed repair (HDR) or through Non-homologous end-joining (NHEJ) method. Previously, during an infection, these processes make some bacteria able to integrate the genome of any virus into their own genome; therefore, bacteria contain an immune response which is more effective during an upcoming infection (Makarova et al. 2015). Recently, for genetic engineering and molecular biology, these procedures have been enhanced to work as a more advanced tool.

2.2 History of CRISPR/Cas

In the 1980s, some researchers had paid attention to *Escherichia coli* alkaline phosphatase and had identified something unusual which was explained as a peculiar genomic topology based on a chain of 32 unique nucleotide sequence, flanking a small invariable palindromic repeat of the phosphatase gene at the 3′ end (Ishino et al. 1987). They found CRISPR array is the first known description of the unusual genomic architecture. It would be 15 years before further work on these new sites had been completed. More research will show that several genes of protein coding near to the CRISPR arrays are strongly or greatly conserved among archaea and bacteria (Jansen et al. 2002). In 2005, the recognition of this ubiquitous and unusual loci showed that the special spacer sequence present in CRISPR arrays was mapped to the genome of phage indicating that CRISPR has the ability to respond against RNA-guided infection of phage (Bolotin et al. 2005; Mojica et al. 2005; Pourcel et al. 2005). In 2012, the molecular strategy of this immune response was explained; few research papers demonstrated that the transcription of CRISPR arrays occurs into RNA and is then cut and loaded into Cas9. The complex of RNA protein is necessary for nuclease operation of the RNA-guided dsDNA (Jinek et al. 2012; Gasiunas et al. 2012). Additionally, it was demonstrated that in vitro Cas9 could be reprogrammed. And it was transcribed into sgRNA to target novel sequences (Jinek et al. 2012). They verified that a nuclease domain might be nonfunctional when the two amino acids in Cas9 would be changed (Jinek et al. 2012), a term used by a new group to build new tools to control the expression of a gene. Scientists have been looking for a method with focused mode of transformation for several years. Although few advancements had been prepared with meganucleases and ZFNs and also for TALENs, but some drawbacks are associated with these techniques. Each one was both costly and time-consuming because the mechanisms of targeting were all focused on the interaction of protein and nucleic acid and required a designed protein for site-specific editing.

While, in case of eukaryotes, the CRISPR approaches for RNA-guided nuclease activity has led many groups to immediately identify the potential of this innovation to bring double-stranded breaks (DSB) in a targeted manner which could just be achieved with great complexity. The DSBs can also be created by previously accessible techniques; currently, CRISPR-based methods are repaired through the low fidelity of DNA repair mechanism that is basically observed after the development of insertion-deletion (indels) mutations, mutations characterized by the spontaneous nucleotide deletion and insertion at the site of DSB. The indels in a coding region can either be due to frame shift or de novo. Insertion of a premature stop codon can lead to a shortened product or transcribed mRNA's introduction of nonsense-mediated decay itself when the targeted gene is expressed. DSB development is also utilized to facilitate the efficient knock-in of new genetic material by flanking the new element with homologous sequence resultant from the target gene and through co-delivering the flanked new elements along with Cas9 and sgRNA. In 2013, the foremost description of RNA-guided mutation in eukaryotic cells was done (Mali et al. 2013; Cong et al. 2013). In case of reprogramming the sgRNAs, the sgRNAs are not considered as an advanced discovery, but Jennifer Doudna and colleagues identified that Cas9 reprogramming could be easily attained and in vitro cleavage of DNA could take place. These findings provide ease to the scientists to work efficiently using these tools in the laboratories. Laterally, with the publication of these articles, several laboratories may acquire CRISPR builds, buy pairs of oligonucleotides, react simply with cloning, and rapidly produce cell knock-in and knockout in animals with a number of extra tools (Mali et al. 2013; Cong et al. 2013; Wang et al. 2013). The dCas9 (Wiedenheft et al. 2012) has the ability to bind with the target sequence and distract the initiation or elongation of transcription through steric hindrance, thus repressing the expression of gene with no DSBs into the DNA or a genome (Qi et al. 2013). Several groups utilized these additional components after some development in the fusion protein of Cas9 that could modify the expression of gene, epigenetic state (Gilbert et al. 2013; Kearns et al. 2015; Hilton et al. 2015), and fluorescent genome imaging in the live cell (Chen et al. 2013). Additionally, via exploiting the benefit of the presence of sgRNAs or reducing the cost of oligonucleotide production, a few libraries of screening were created (Wang et al. 2014a, b; Shalem et al. 2014; Sanjana et al. 2014; Konermann et al. 2015; Joung et al. 2016; Doench et al. 2016). CRISPR innovations have captured the imagination of biologists with all those potential applications. However, some limitations are associated with this new technique. Historical timeline of the gene editing system such as CRISPR/Cas has been provided in Fig. 2.2.

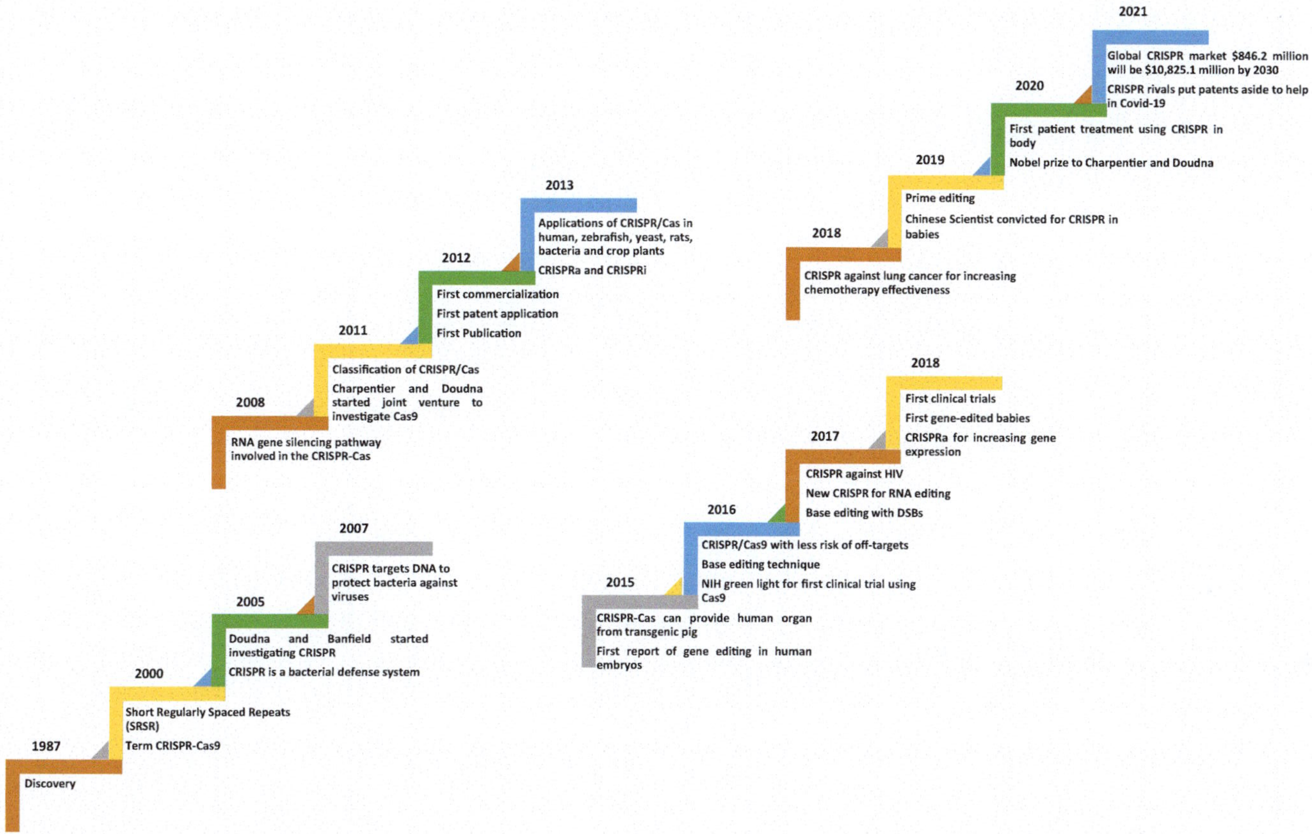

Fig. 2.2 Historical timeline of CRISPR/Cas gene editing system

2.3 Classification of CRISPR/Cas System

The CRISPR/Cas system has been categorized into two major classes that are further divided into other six subtypes. This classification depends on the variation present in the interference complexes. Types I, III, and IV belong to Class 1, while the remaining belong to the other class (Mohanraju et al. 2016; Makarova et al. 2015). Multiple subunits have been utilized by Class 1 system, whereas in the ribonucleoprotein (RNP) complex, the crRNA and one single protein are utilized against invaders in Class 2 (Shmakov et al. 2015). Almost, all the already described systems of CRISPR/Cas (excluding Type III) identify short sequence of nucleotide called protospacer or PAM at 5′ end and interact with hereditary components (Yamano et al. 2017; Wang et al. 2015a, b; Li et al. 2014; Anders et al. 2014; Westra et al. 2013; Peng et al. 2013). The multi-subunit RNP complex is utilized by Type I system identified as CRISPR-associated protein complex that is used for the identification and cutting of targeted DNA (Brouns et al. 2008). In nearly all the Cas5d for subtype I-C in Type I systems, the Cas6 (endoribonucleases) recognized or manipulated the CRISPR repeats of RNA (Sefcikova et al. 2017; Garside et al. 2012; Carte et al. 2008). PAM sequence is used to guide the cascade to recognize the DNA sequence and activates the Cas3 endonucleases (Hayes et al. 2016; Mulepati et al. 2014). Type III system which resembles with the complexes of Type I utilizes crRNA that is able to bind with the multi-protein complex for developing the defense against virus, attacking either DNA or RNA substrates. Seed motifs are used for targeting the rear-end subunit Cmr1 to target the RNA that was present in the 3′ end of crRNA (Pan et al. 2019; Li et al. 2017), subsequently, in the backbone of targeting transcript of Cas10 subunit (Liu et al. 2017b; Han et al. 2017; Kazlauskiene et al. 2016; Goldberg et al. 2014; Estrella et al. 2016; Samai et al. 2015; Elmore et al. 2016). The cleavage of DNA/RNA is performed by the nuclear cleavage domain of HD (histidine) of Cas10 subunit. Cyclic oligonucleotides (COA) are based on the synthase activity which catalyze using Cas10 domain, triggered by the similar targeted RNA, and signaling pathway of COA systematizes the RNP complex and CARF (CRISPR-associated Rossmann fold) ribonuclease (Csm6/Csx1 families) domain for the prevention against viral diseases (You et al. 2019; Wang et al. 2019; Han et al. 2018; Niewoehner et al. 2017; Kazlauskiene et al. 2017). Type IV is additionally utilized in Class 1, and few subunits in the harbor yet need interference nucleases like cas3 or cas10 and adaptation module like cas1 or cas2 (Koonin et al. 2017). The limited information about maturation of CrRNA and RNP complex formation is present contrariwise; the activities of Type IV are still not satisfactory (Özcan et al. 2019). Although, in case of Class 2, Types II, V, and VI using the crRNA and protein, forms complex of RNP that acts as defense against the invaders. For CRISPR RNA maturation, type Cas12a and Cas13 may process the crRNA itself (Fonfara et al. 2016; East-Seletsky et al. 2016; Dong et al. 2016; Abudayyeh et al. 2016), while tracrRNA and RNase III are actively used for the development of crRNA (Deltcheva et al. 2011). In contrast to Type II Cas9, which produces a blunt DSBs using various non-targeted stranded PAMs (Nishimasu et al. 2014; Fonfara et al. 2014), the activity of Cas12a and Cas12b depends on the identification of

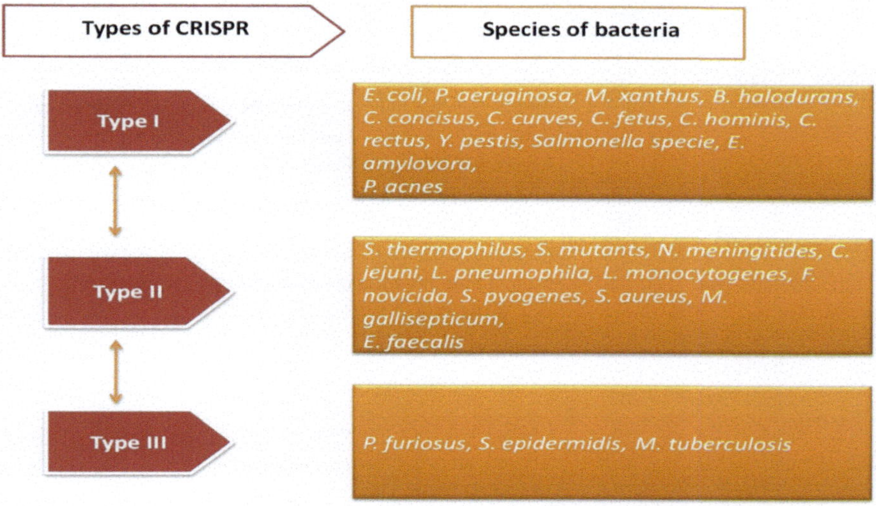

Fig. 2.3 CRISPR/Cas system and its types according to the bacterial species

T-rich PAM sequence, and then cleavage of DSB occurs (Yang et al. 2016; Gao et al. 2016; Fonfara et al. 2016). The Cas13 protein is encoded by Type VI system, which cleavages the ssRNA and degrades them and further targets the RNA that is opposite to the crRNA (Liu et al. 2017a; Smargon et al. 2017; East-Seletsky et al. 2016; Abudayyeh et al. 2016; Shmakov et al. 2015). For the identification of pathogen, Cas13a protein has been used. In addition to this, RNP complexes of Cas12a and Cas12b may as well catalyzed the cleavage of non-specific ssDNA when bound as an activator through a complementary target of ssDNA (Li et al. 2018; Chen et al. 2018). Various kinds of CRISPR system and their presence in the bacterial species are given in Fig. 2.3.

2.4 Types of CRISPR

CRISPR/Cas system is further categorized into three basic subtypes, namely, Type I, II, or III. This distinction is based on genes contained by each type of signature (Javed et al. 2018). Type I, for example, contains Cas3 protein, and Types II and III contain Cas9 protein or Cas10. It is noteworthy to remember that all CRISPR system subtypes and its forms have Cas1 protein and Cas2 proteins play an important role in the spacer (Shabbir et al. 2016a, b). Classes of CRISPR/Cas system are given in Table 2.1.

Table 2.1 Classes of CRISPR/Cas

Major classes	Signature proteins	Cas type	Role
1	Cas3	I	ATP-dependent helicase and ssDNA nuclease (HD domain)
	Cas5; Cas8b	IB	Subunit of interference module. Recognizing the PAM sequence to target the foreign DNA
	Cas8a	IA	
	Cas8c	IC	
	Cse2, Cse1	IE	Exhibit a domain which is homologous to the palm domain of nucleic acid polymerase and nucleotide cyclase
	Cas10d	ID	
	Csy1, Csy2, Csy3	IF	Not determined
	GSU0054	IU	
	Cas10	III	Cas10d and Cse1 homology
	Cmr5	IIIB	Not determined
	Csm2	IIIA	Not determined
	Cas10, Csx11	IIIC	
	Csx10	IIID	
	Csf1	IV	
		IVA	
		IVB	
2	Cas9	II	Nucleases HNH and RuvC separately can create single-strand breaks and in combination can create DSBs. During adaptation. It ensures the acquisition of functional spacers.
	Csn2	IIA	DNA-binding protein of ring shaped. In Type II. It has been used in process of adaptation
	Cas4	IIB	Non-determined
		IIC	Characterize by the absence of either Cas4 or Csn2
	C2c1, C2c3, Cpf1	V	Lacks HNH. Nuclease RuvC
	Cas13a	VI	RNA-guided RNase

2.4.1 CRISPR/Cas Type I

Type I system is present mostly in the archaeal and bacterial genome (Makarova et al. 2011). This system is further processed into six (encoding cas3 gene) subtypes (IA to IF). Cas3 is a helicase and nuclease-activated multidomain protein (Sinkunas et al. 2011). Cas3 protein contains two domains: a DNA cleavage N-terminal of HD phosphohydrolase and a C-terminal of DExH helicase domain for unwinding the double stranded DNA (Shabbir et al. 2016a). These domains combine to degrade the foreign hereditary material. On the other hand, Cas3 not only recognizes the DNA of the invader but also defends the cells against infection (Jore et al. 2011; Brouns et al.

2008; Shabbir et al. 2019). Each subtype of the Type I, having different Cas proteins subtype, accumulates to structure a complex called crRNA-guided antiviral defense complex (CASCADE) or surveillance complex related with CRISPR. Complementary to crRNA spacer, these complexes have an important role in defining and binding of targeted sequence. In *E. coli* K12 (Type I-E), the surveillance complex directed by crRNA was firstly described. It is a mixture of five Cas proteins. The Cas6e (earlier referred to as CasEor Cse3) assists in crRNA maturing. Mature crRNA always remains connected with the CASCADE which is basically used for the recognition of the targeted DNA. Similar complex type was documented in *S. solfataricus* (Lintner et al. 2011). Additionally, *Pseudomonas aeruginosa* of Type I-F and *Bacillus halodurans* of Type I-C have described crRNA-guided surveillance complexes (Wiedenheft et al. 2011; Nam et al. 2012).

2.4.2 CRISPR/Cas Type II

This mechanism only exists in bacteria (Shabbir et al. 2019; Makarova et al. 2011). In contrast with other subtypes, Type II is the simplest one (Bhaya et al. 2011). The system of Type II is based on four genes known as *Cas1*, *Cas9*, *Cas2*, and *Cas4*. Type II protein is characteristic of the Cas9, which is also utilized in the invader DNA cleavage (Deltcheva et al. 2011) and the biogenesis of crRNA. Cas9 gene has two major domains, i.e., HNH and RuvC (Jinek et al. 2012). The HNH domain assists in DNA cleavage, after finding the complementary sequence of crRNA, although the RuvC domain cuts the non-complement strand, while, in the case of Type II, a trans-activating crRNA or tracrRNA has been successfully used for the biogenesis of crRNA. TracrRNA encoding in *Streptococcus pyogenes* occurs at the reverse CRISPR/Cas locus strand (Shabbir et al. 2016a). Hybridization between tracRNA and crRNA repeats results in the creation of RNA known as double-stranded RNA (dsRNA), which further used the RNase III (non-Cas cellular enzyme) for identifying and cutting the target site. The biogenesis of crRNA happens due to the complete removal of Cas9; but its function is still unclear in the biogenesis of crRNA. Jinek and his colleagues have clearly shown that the enzyme of Cas9 required both types of RNA either crRNA or tracrRNA for site-specific cleavage of DNA. Remarkably, multiple domains are used which are essential for cutting the specific site of DNA that further combines with a protein called Cas9 and makes Type II a more efficient and perfect genome editing method of the CRISPR (Cong et al. 2013).

2.4.3 CRISPR/Cas Type III

Type III was divided into two subtypes known as Type III-A and Type III-B (Makarova et al. 2011; Shabbir et al. 2019). It is prevalent in archaea, but Type III-B form is only present in combination with other forms of CRISPR.

Type III of CRISPR/Cas encoded the genes in both Cas10 and Cas6. Cas10 is theoretically involved in DNA cleavage and the maturation of crRNA; it is known as the repeat-associated mysterious protein (RAMP) (Anantharaman et al. 2010). The Cas6 endoribonuclease is not connected to complex CAS-CADE and functions independently (Zhang et al. 2012). CRISPR Type III CASCADE complex binds to mature crRNA and then cleaves the invading RNA (Wang et al. 2011). Additionally, Cas6 may be shared with Type I-A or I-B (Deng et al. 2013) in some archaea and bacteria which have the Type III system. Although there are similarities between these two, Type III systems and its subtypes have been used for targeting chemically diverse substrates. For example, in Type III-A CRISPR system, *S. epidermidis* targeted the DNA sequence, whereas in Type III-B *S. solfataricus* and *Pyrococcus furiosus* cut the RNA sequence (Shabbir et al. 2016a). Various forms of CRISPR systems are shown in Fig. 2.2.

2.4.4 Other Variants

In many bacterial genomes, the CRISPR/Cas system of Type IV is not fully understood. This system includes the proteins Cas5, Cas7, and Csf1 (Makarova et al. 2015). Type V of this system has been introduced recently which consists of Cpf1 protein. In the absence of extra RNA, this protein is performing the role of crRNA and forms complexes that later cleave the DNA, so it simplifies the process of editing (Luo et al. 2016). The role of CRISPR/Cas in prokaryotic evolution and ecology is given in Fig. 2.4.

2.5 CRISPR/Cas System for DNA Editing

After the development of the central dogma, researchers made efforts for the advancements to control and edit the DNA or a genome. The regulation of genetic information and precise editing play an important role to understand the function of any gene. In the previous decade, new developments and technologies have been created for the significant and simpler regulation and editing of a genome. One latest development has adjusted the CRISPR/Cas which is based on bacterial immune system as an easy and simple tool based on gRNA for exceptionally specific and efficient DNA regulation and editing. CRISPR/Cas is a progressive tool used for treating hereditary diseases and efficiently used in medical research (Mali et al. 2013; Cong et al. 2013; Jinek et al. 2013; Hsu et al. 2014). The advances in Cas 9 can provide the specific site cleavage of the targeted sequence, and it was also known as a precise genome editing tool in the living cell. The system is based on the combined utilization of a sgRNA and Cas9 endonuclease to delete and insert substitute DNA sequences in genome at a specific location (Hsu et al. 2014). Other than these applications, the CRISPR/Cas9 has been used for manipulating the DNA, and this manipulation was done in three stages: (1) for identifying the complementary sequence of sgRNA, genome scanning is done through the RNA-guided Cas9

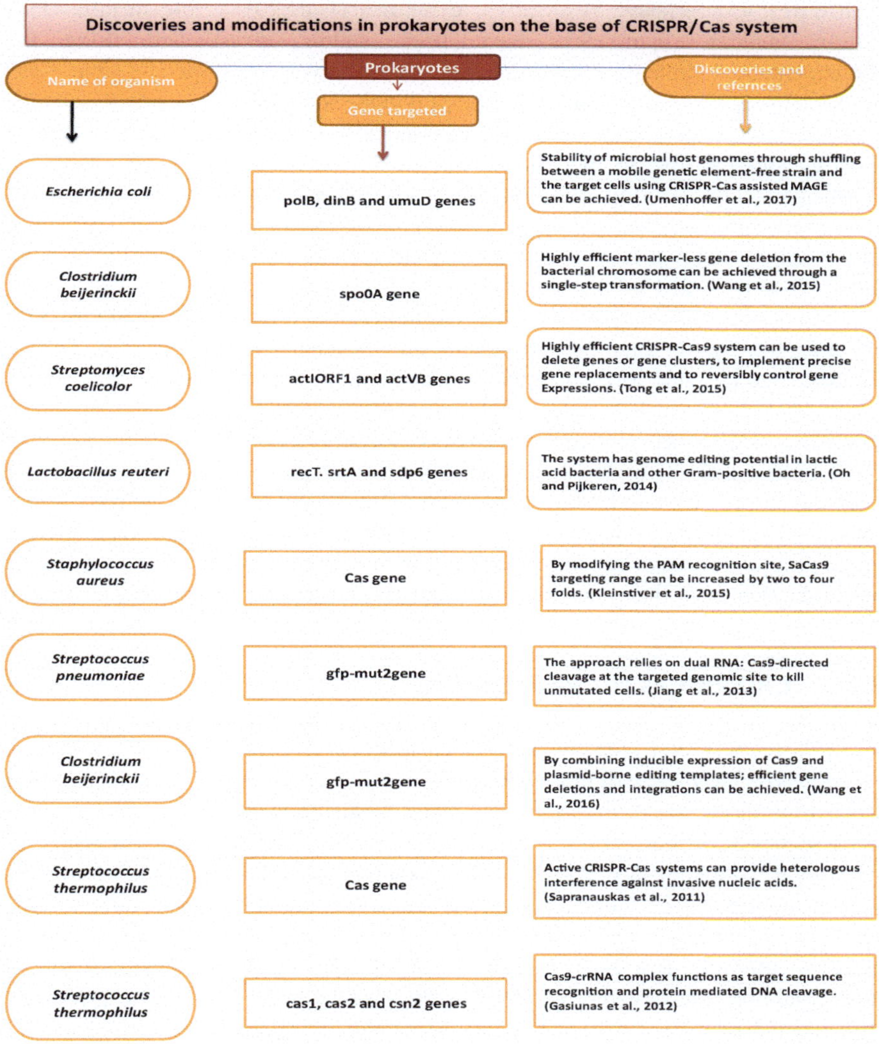

Fig. 2.4 Role of CRISPR in prokaryotic evolution

nuclease (RGN); (2) production of DSB in DNA through the Cas9; and (3) using the endogenous DNA repair machinery to repair the cleaved region (Hsu et al. 2014). Both the precision and proficiency of procedures engaged in each step clearly influence the result of CRISPR-based genome editing and the usefulness in an innovation. To oprimize the CRISPR as a tool for manipulating the DNA, many researchers have made improvements in CRISPR/Cas system and have devised various strategies (Horlbeck et al. 2016; van Overbeek et al. 2016; Henser-Brownhill et al. 2017; Uusi-Mäkelä et al. 2018; Brinkman et al. 2018). Site-specific DNA editing nuclease approaches have been widely investigated using the abovementioned strategy. For understanding the functions of a specific gene, Cas9 has been used as an editing tool that is helpful in the study of reverse genetics, which exhibits new restorative plans in various models of hereditary and transferable diseases and for disease modeling (Xiao-Jie et al. 2015; Hsu et al. 2014). In expansion to DNA editing, dCas9 has been utilized to control the entire profiles of gene expression by either inhibiting (CRISPRi) expression or activating (CRISPRa) expression to target at least one or more gene at the same time (Dominguez et al. 2016; Pham et al. 2016; Farasat and Salis 2016; Xue et al. 2016).

The important application of CRISPR system is that it can be efficiently used for targeting the human genome and developing resistance against diseases. Due to CRISPR, we are able to modify the mutant genes and develop resistance against some lethal disease such as cancer, HIV, and other diseases (Zhang et al. 2016; Savić and Schwank 2016; Kleinstiver et al. 2015; White and Khalili 2016; Slaymaker et al. 2016; Wojtal et al. 2016). On the other hand, essential questions concerning how the mammalian proteins and DNA interact with sgRNAs and Cas9 and the response of cells during DNA damage caused by CRISPR array remained unanswered. Expanding our insight into the CRISPR/Cas based systems and considering the risk and potential of the system is very important to use it for benefit of humankind.

2.6 CRISPR/Cas for RNA Editing

In the last few years, biomedical research testified that the figure of non-coding RNAs/DNAs has increased tremendously (Calin and Croce 2006; Munker and Calin 2011; Ling et al. 2015). Precisely, the future medicine is created by changing the RNA sequence (Mullin 2017). It could be an excellent and attractive alternative of targeting RNA in the place of DNA. When targeting RNA, the result will be tunable, and nucleotide change will be reversible. Numerous attractive nucleotide changes are distant or useless at the level of genome, for instance, when a gene loss is either deadly or promptly redressed (El-Brolosy and Stainier 2017; Rossi et al. 2015). In such condition, the targeting of RNA at the point signaling would be especially attractive (El-Brolosy and Stainier 2017). Moreover, the revelation of novel layers of control in epitranscriptome stimulates the need to investigate RNA targeting techniques (Wang et al. 2014a, b; Schwartz et al. 2014).

Importantly, not all the CRISPR-associated systems exclusively target DNA; few of them target RNA and could transform gene expression without persuading cytotoxicity. So far, to target RNA, Types I, II, III, and VI of CRISPR have been demonstrated (Strutt et al. 2018; Rousseau et al. 2018; Dugar et al. 2018; Li et al. 2016; Abudayyeh et al. 2016; Samai et al. 2015; O'Connell et al. 2014; Hale et al. 2012). If RNA is targeted instead of DNA, the autoimmunity from self-targeting spacers would not be attained, but the expression pattern of a gene is altered through the degradation of mRNA. Type III-B in *Myxococcus xanthus* contains latent model that degrades the mRNA, and this mechanism is established completely (Wallace et al. 2014). Point mutations can create substitution of single amino acid that promotes to produce altered types of proteins and connect with various diseases like hemophilia A (Higuchi et al. 1991), cystic fibrosis (Ferec et al. 1992), brain cancer (Leisegang et al. 2016), and other diseases (Bolscher et al. 1991). Point mutation is modified at the RNA/DNA level (Rees and Liu 2018). At post-transcriptional level, RNA editing happens with a practical result fundamentally upon various types of proteins (Maas and Rich 2000); through substitution and insertion, the changes in the RNA structure can happen (Kim and Kim 2014; Nishikura 2010). In mammalian cell, during RNA editing process, two fundamental groups of enzymes are liable. In these enzymes adenosine is changed over to inosine (guanosine-like complementarity), and cytosine is changed over to uracil-AID-APOBEC compounds (Nishikura 2016; Slotkin and Nishikura 2013). The abovementioned modifications can occur in various types of RNA and in various regions (Rosenberg et al. 2011). ADAR can embrace a large range of RNAs, incorporating essential transcripts of microRNA with impacts leading to gene silencing. But the activity of AID-APOBEC mainly occurs in transcript 3′-UTRs (Blow et al. 2006; Kawahara et al. 2007). Though, when correcting nucleotide changes, the downstream impacts can fundamentally change the phenotype of the cell. Substitutions in target region can impact in the signaling pathways, post-transcriptional modifications, and furthermore catalysis and disturb the open reading frames, splicing signals, and furthermore microRNA seed sequences (Azad et al. 2017). The advancement of an exact and adaptable RNA base editing utilized the CRISPR-mediated RNA-guided RNase Cas13 (Type VI) (Smargon et al. 2017; Shmakov et al. 2015, 2017; Abudayyeh et al. 2016). Cas13 contains two nucleotide binding (HEPN) endoRNase domains of prokaryotes and eukaryotes. These domains mediate the exact cleavage of RNA with an inclination of targets with protospacer flanking site (PFS) domains, which is identified in microbes (Shmakov et al. 2015; Abudayyeh et al. 2016). There are three protein families of Cas13 that have been distinguished to date: Cas13(a) (recently known as C2c2), Cas13(b), and Cas13(c) (Shmakov et al. 2017; Smargon et al. 2017). Cas13a can be adjusted as a tool for recognition of DNA/RNA (Gootenberg et al. 2017), just as plant cell and mammalian RNA knockdown and tracking of transcript. The programmable idea of Cas13 proteins is the beginning stage to create tools for RNA binding (Abudayyeh et al. 2017). Müller-Esparza and Randau demonstrated that CRISPR systems in *P. aeruginosa* UCBPP-PA14 require5'-GGN-3' follow by nine/twelve complementary nucleotides for crRNA and mRNA interactions. They distinguish approximately 189 putative target mRNAs, proposing extra necessities, for example, to target mRNA, auxiliary structure of mRNA is required. Accordingly, for mRNA degradation further investigations are important to explain Type I-F necessities in *P. aeruginosa* and many other different living beings (Müller-Esparza and Randau 2017).

The nuclease activity of Cas9 is generally observed for targeting DNA except a few examples of Cas9 being used for targeting the RNA (Strutt et al. 2018; O'Connell et al. 2014; Rousseau et al. 2018). In *Streptococcus pyogenes* unique investigations about Cas9 proposed that among RNA and DNA, the activity of effector protein could be different (Gasiunas et al. 2012), where targeting RNA must be accomplished by P hybridization introducing oligonucleotide (PAMmer) that has the RNA sequence (Nelles et al. 2016; O'Connell et al. 2014). Afterward, it was demonstrated that several Cas9 enzymes can cut RNA even lacking PAMmer. In Type II-C in *N. meningitides*, Cas9 has been appeared for targeting the RNA in vitro, whereas Type II-C and Type II-A in *C. jejuni* or in *S. aureus* were appeared to cleavage sequence of RNA in vivo/in vitro (Strutt et al. 2018; Rousseau et al. 2018; Dugar et al. 2018). In this way, targeting of DNA/RNA didn't require a flanked sequence called motif. For example, in *C. jejuni*, the cutting and the binding of RNAs by the spacers naturally occur. Dugar and collaborators didn't clearly discover a phenotype that is related to targeting the RNA through the endogenous Cas9, despite the fact that (Strutt et al. 2018; Dugar et al. 2018) from *S. aureus* the Cas9 could hinder the expression of a gene by programmable targeting of RNA in *E. coli* deprived of cell death. Previously, few examples demonstrated that some systems of DNA targeting can also be used for RNA targeting, having the potential to degrade RNA and alter the expression of gene in the selected host, but it is considered that the CRISPR will be better than other methodologies of RNA targeting, as its one-of-a-kind property to liquefy the DNA is not required. Various CRISPR tools are available and may be used by the researchers based on the mode of application.

2.7 Future Prospects and Final Remarks

CRISPR/Cas system is a well-organized tool of genome modification over the past few years, giving great stimulus to the advancement in life sciences. In addition to CRISPR systems of Class 2 that are successfully used for modifying the genomic sequences with single-RNP complex, more precise endogenous Cas9 can be useful for the genome editing of other species of archaea and bacteria, particularly those species that contain heterogeneous Cas proteins. After being repurposed as antimicrobial, CRISPR/Cas systems have effectively interfered in precise chromosome sites (Yosef et al. 2015; Li et al. 2016; Selle and Barrangou 2015). There are some barriers that need to be tackled to the successful implementation of this technology in other new species. A major challenge is to select an appropriate combination of lytic and temperate phages that are unique for sensitized pathogens. Therefore, there is an urgent need to establish a standardized means for effective DNA transmission to all pathogens (Shabbir et al. 2019). Selection of a suitable tool for a specific DNA modification is critical along with the method of delivery and type of vectors for a successful genome editing study. Moreover, the selection of an appropriate CRISPR tool for a specific type of application is the major decision which will result in effective and efficient genome editing. We have given a number of types of the system with potential applications which may help the researchers to select a tool of choice for desirable gene modifications.

References

Abudayyeh OO, Gootenberg JS, Konermann S, Joung J, Slaymaker IM, Cox DB, Shmakov S, Makarova KS, Semenova E, Minakhin L (2016) C2c2 is a single-component programmable RNA-guided RNA-targeting CRISPR effector. Science 353(6299):aaf5573

Abudayyeh OO, Gootenberg JS, Essletzbichler P, Han S, Joung J, Belanto JJ, Verdine V, Cox DB, Kellner MJ, Regev A (2017) RNA targeting with CRISPR–Cas13. Nature 550(7675):280–284

Amitai G, Sorek R (2016) CRISPR–Cas adaptation: insights into the mechanism of action. Nat Rev Microbiol 14(2):67

Anantharaman V, Iyer LM, Aravind L (2010) Presence of a classical RRM-fold palm domain in Thg1-type 3′-5′ nucleic acid polymerases and the origin of the GGDEF and CRISPR polymerase domains. Biol Direct 5(1):43

Anders C, Niewoehner O, Duerst A, Jinek M (2014) Structural basis of PAM-dependent target DNA recognition by the Cas9 endonuclease. Nature 513(7519):569–573

Azad MT, Bhakta S, Tsukahara T (2017) Site-directed RNA editing by adenosine deaminase acting on RNA for correction of the genetic code in gene therapy. Gene Ther 24(12):779–786

Bhaya D, Davison M, Barrangou R (2011) CRISPR-Cas systems in bacteria and archaea: versatile small RNAs for adaptive defense and regulation. Annu Rev Genet 45:273–297

Blow MJ, Grocock RJ, van Dongen S, Enright AJ, Dicks E, Futreal PA, Wooster R, Stratton MR (2006) RNA editing of human microRNAs. Genome Biol 7(4):R27

Boch J, Scholze H, Schornack S, Landgraf A, Hahn S, Kay S, Lahaye T, Nickstadt A, Bonas U (2009) Breaking the code of DNA binding specificity of TAL-type III effectors. Science 326(5959):1509–1512

Bolotin A, Quinquis B, Sorokin A, Ehrlich SD (2005) Clustered regularly interspaced short palindrome repeats (CRISPRs) have spacers of extrachromosomal origin. Microbiology 151(8):2551–2561

Bolscher B, De Boer M, De Klein A, Weening R, Roos D (1991) Point mutations in the beta-subunit of cytochrome b558 leading to X-linked chronic granulomatous disease. Blood 77:2482

Brinkman EK, Chen T, de Haas M, Holland HA, Akhtar W, van Steensel B (2018) Kinetics and fidelity of the repair of Cas9-induced double-strand DNA breaks. Mol Cell 70(5):801–813.e806

Brouns SJ, Jore MM, Lundgren M, Westra ER, Slijkhuis RJ, Snijders AP, Dickman MJ, Makarova KS, Koonin EV, Van Der Oost J (2008) Small CRISPR RNAs guide antiviral defense in prokaryotes. Science 321(5891):960–964

Calin GA, Croce CM (2006) Genomics of chronic lymphocytic leukemia microRNAs as new players with clinical significance. In: Seminars in oncology, vol 2. Elsevier, pp 167–173

Carroll D (2011) Genome engineering with zinc-finger nucleases. Genetics 188(4):773–782

Carte J, Wang R, Li H, Terns RM, Terns MP (2008) Cas6 is an endoribonuclease that generates guide RNAs for invader defense in prokaryotes. Genes Dev 22(24):3489–3496

Chen B, Gilbert LA, Cimini BA, Schnitzbauer J, Zhang W, Li G-W, Park J, Blackburn EH, Weissman JS, Qi LS (2013) Dynamic imaging of genomic loci in living human cells by an optimized CRISPR/Cas system. Cell 155(7):1479–1491

Chen JS, Ma E, Harrington LB, Da Costa M, Tian X, Palefsky JM, Doudna JA (2018) CRISPR-Cas12a target binding unleashes indiscriminate single-stranded DNase activity. Science 360(6387):436–439

Chylinski K, Makarova KS, Charpentier E, Koonin EV (2014) Classification and evolution of type II CRISPR-Cas systems. Nucleic Acids Res 42(10):6091–6105

Cong L, Ran FA, Cox D, Lin S, Barretto R, Habib N, Hsu PD, Wu X, Jiang W, Marraffini LA (2013) Multiplex genome engineering using CRISPR/Cas systems. Science 339(6121):819–823

Deltcheva E, Chylinski K, Sharma CM, Gonzales K, Chao Y, Pirzada ZA, Eckert MR, Vogel J, Charpentier E (2011) CRISPR RNA maturation by trans-encoded small RNA and host factor RNase III. Nature 471(7340):602–607

Deng L, Garrett RA, Shah SA, Peng X, She Q (2013) A novel interference mechanism by a type IIIB CRISPR-Cmr module in Sulfolobus. Mol Microbiol 87(5):1088–1099

Doench JG, Fusi N, Sullender M, Hegde M, Vaimberg EW, Donovan KF, Smith I, Tothova Z, Wilen C, Orchard R (2016) Optimized sgRNA design to maximize activity and minimize off-target effects of CRISPR-Cas9. Nat Biotechnol 34(2):184

Dominguez AA, Lim WA, Qi LS (2016) Beyond editing: repurposing CRISPR–Cas9 for precision genome regulation and interrogation. Nat Rev Mol Cell Biol 17(1):5

Dong D, Ren K, Qiu X, Zheng J, Guo M, Guan X, Liu H, Li N, Zhang B, Yang D (2016) The crystal structure of Cpf1 in complex with CRISPR RNA. Nature 532(7600):522–526

Dugar G, Leenay RT, Eisenbart SK, Bischler T, Aul BU, Beisel CL, Sharma CM (2018) CRISPR RNA-dependent binding and cleavage of endogenous RNAs by the Campylobacter jejuni Cas9. Mol Cell 69(5):893–905.e897

East-Seletsky A, O'Connell MR, Knight SC, Burstein D, Cate JH, Tjian R, Doudna JA (2016) Two distinct RNase activities of CRISPR-C2c2 enable guide-RNA processing and RNA detection. Nature 538(7624):270–273

El-Brolosy MA, Stainier DY (2017) Genetic compensation: a phenomenon in search of mechanisms. PLoS Genet 13(7):e1006780

Elmore JR, Sheppard NF, Ramia N, Deighan T, Li H, Terns RM, Terns MP (2016) Bipartite recognition of target RNAs activates DNA cleavage by the Type III-B CRISPR–Cas system. Genes Dev 30(4):447–459

Estrella MA, Kuo F-T, Bailey S (2016) RNA-activated DNA cleavage by the type III-B CRISPR–Cas effector complex. Genes Dev 30(4):460–470

Farasat I, Salis HM (2016) A biophysical model of CRISPR/Cas9 activity for rational design of genome editing and gene regulation. PLoS Comput Biol 12(1):e1004724

Ferec C, Audrezet M, Mercier B, Guillermit H, Moullier P, Quere I, Verlingue C (1992) Detection of over 98% cystic fibrosis mutations in a Celtic population. Nat Genet 1(3):188–191

Fonfara I, Le Rhun A, Chylinski K, Makarova KS, Lecrivain A-L, Bzdrenga J, Koonin EV, Charpentier E (2014) Phylogeny of Cas9 determines functional exchangeability of dual-RNA and Cas9 among orthologous type II CRISPR-Cas systems. Nucleic Acids Res 42(4):2577–2590

Fonfara I, Richter H, Bratovič M, Le Rhun A, Charpentier E (2016) The CRISPR-associated DNA-cleaving enzyme Cpf1 also processes precursor CRISPR RNA. Nature 532(7600):517–521

Gao P, Yang H, Rajashankar KR, Huang Z, Patel DJ (2016) Type V CRISPR-Cas Cpf1 endonuclease employs a unique mechanism for crRNA-mediated target DNA recognition. Cell Res 26(8):901–913

Garside EL, Schellenberg MJ, Gesner EM, Bonanno JB, Sauder JM, Burley SK, Almo SC, Mehta G, MacMillan AM (2012) Cas5d processes pre-crRNA and is a member of a larger family of CRISPR RNA endonucleases. RNA 18(11):2020–2028

Gasiunas G, Barrangou R, Horvath P, Siksnys V (2012) Cas9–crRNA ribonucleoprotein complex mediates specific DNA cleavage for adaptive immunity in bacteria. Proc Natl Acad Sci 109(39): E2579–E2586

Gilbert LA, Larson MH, Morsut L, Liu Z, Brar GA, Torres SE, Stern-Ginossar N, Brandman O, Whitehead EH, Doudna JA (2013) CRISPR-mediated modular RNA-guided regulation of transcription in eukaryotes. Cell 154(2):442–451

Goldberg GW, Jiang W, Bikard D, Marraffini LA (2014) Conditional tolerance of temperate phages via transcription-dependent CRISPR-Cas targeting. Nature 514(7524):633–637

Gootenberg JS, Abudayyeh OO, Lee JW, Essletzbichler P, Dy AJ, Joung J, Verdine V, Donghia N, Daringer NM, Freije CA (2017) Nucleic acid detection with CRISPR-Cas13a/C2c2. Science 356(6336):438–442

Grissa I, Vergnaud G, Pourcel C (2007) CRISPRFinder: a web tool to identify clustered regularly interspaced short palindromic repeats. Nucleic Acids Res 35(Suppl 2):W52–W57

Hale CR, Majumdar S, Elmore J, Pfister N, Compton M, Olson S, Resch AM, Glover CV III, Graveley BR, Terns RM (2012) Essential features and rational design of CRISPR RNAs that function with the Cas RAMP module complex to cleave RNAs. Mol Cell 45(3):292–302

Han W, Li Y, Deng L, Feng M, Peng W, Hallstrøm S, Zhang J, Peng N, Liang YX, White MF (2017) A type III-B CRISPR-Cas effector complex mediating massive target DNA destruction. Nucleic Acids Res 45(4):1983–1993

Han W, Stella S, Zhang Y, Guo T, Sulek K, Peng-Lundgren L, Montoya G, She Q (2018) A Type III-B Cmr effector complex catalyzes the synthesis of cyclic oligoadenylate second messengers by cooperative substrate binding. Nucleic Acids Res 46(19):10319–10330

Hayes RP, Xiao Y, Ding F, Van Erp PB, Rajashankar K, Bailey S, Wiedenheft B, Ke A (2016) Structural basis for promiscuous PAM recognition in type I–E Cascade from E. coli. Nature 530(7591):499–503

Henser-Brownhill T, Monserrat J, Scaffidi P (2017) Generation of an arrayed CRISPR-Cas9 library targeting epigenetic regulators: from high-content screens to in vivo assays. Epigenetics 12(12):1065–1075

Higuchi M, Kazazian HH, Kasch L, Warren TC, McGinniss MJ, Phillips JA, Kasper C, Janco R, Antonarakis SE (1991) Molecular characterization of severe hemophilia A suggests that about half the mutations are not within the coding regions and splice junctions of the factor VIII gene. Proc Natl Acad Sci 88(16):7405–7409

Hilton IB, D'ippolito AM, Vockley CM, Thakore PI, Crawford GE, Reddy TE, Gersbach CA (2015) Epigenome editing by a CRISPR-Cas9-based acetyltransferase activates genes from promoters and enhancers. Nat Biotechnol 33(5):510–517

Horlbeck MA, Witkowsky LB, Guglielmi B, Replogle JM, Gilbert LA, Villalta JE, Torigoe SE, Tjian R, Weissman JS (2016) Nucleosomes impede Cas9 access to DNA in vivo and in vitro. elife 5:e12677

Hsu PD, Lander ES, Zhang F (2014) Development and applications of CRISPR-Cas9 for genome engineering. Cell 157(6):1262–1278

Ishino Y, Shinagawa H, Makino K, Amemura M, Nakata A (1987) Nucleotide sequence of the iap gene, responsible for alkaline phosphatase isozyme conversion in Escherichia coli, and identification of the gene product. J Bacteriol 169(12):5429–5433

Jackson SA, McKenzie RE, Fagerlund RD, Kieper SN, Fineran PC, Brouns SJ (2017) CRISPR-Cas: adapting to change. Science 356(6333):eaal5056

Jansen R, Embden JDV, Gaastra W, Schouls LM (2002) Identification of genes that are associated with DNA repeats in prokaryotes. Mol Microbiol 43(6):1565–1575

Javed MR, Sadaf M, Ahmed T, Jamil A, Nawaz M, Abbas H, Ijaz A (2018) CRISPR-Cas system: history and prospects as a genome editing tool in microorganisms. Curr Microbiol 75(12):1675–1683

Jiang W, Bikard D, Cox D, Zhang F, Marraffini LA (2013) RNA-guided editing of bacterial genomes using CRISPRCas systems. Nature Biotechnol 31(3):233–239

Jinek M, Chylinski K, Fonfara I, Hauer M, Doudna JA, Charpentier E (2012) A programmable dual-RNA–guided DNA endonuclease in adaptive bacterial immunity. Science 337(6096):816–821

Jinek M, East A, Cheng A, Lin S, Ma E, Doudna J (2013) RNA-programmed genome editing in human cells. elife 2:e00471

Jore MM, Lundgren M, Van Duijn E, Bultema JB, Westra ER, Waghmare SP, Wiedenheft B, Pul Ü, Wurm R, Wagner R (2011) Structural basis for CRISPR RNA-guided DNA recognition by Cascade. Nat Struct Mol Biol 18(5):529

Joung J, Konermann S, Gootenberg JS, Abudayyeh OO, Platt RJ, Brigham MD, Sanjana NE, Zhang F (2016) Protocol: genome-scale CRISPR-Cas9 knockout and transcriptional activation screening. BioRxiv:059626

Kawahara Y, Zinshteyn B, Sethupathy P, Iizasa H, Hatzigeorgiou AG, Nishikura K (2007) Redirection of silencing targets by adenosine-to-inosine editing of miRNAs. Science 315(5815):1137–1140

Kazlauskiene M, Tamulaitis G, Kostiuk G, Venclovas Č, Siksnys V (2016) Spatiotemporal control of type III-A CRISPR-Cas immunity: coupling DNA degradation with the target RNA recognition. Mol Cell 62(2):295–306

Kazlauskiene M, Kostiuk G, Venclovas Č, Tamulaitis G, Siksnys V (2017) A cyclic oligonucleotide signaling pathway in type III CRISPR-Cas systems. Science 357(6351):605–609

Kearns NA, Pham H, Tabak B, Genga RM, Silverstein NJ, Garber M, Maehr R (2015) Functional annotation of native enhancers with a Cas9–histone demethylase fusion. Nat Methods 12(5):401–403

Kim H, Kim J-S (2014) A guide to genome engineering with programmable nucleases. Nat Rev Genet 15(5):321–334

Kleinstiver BP, Prew MS, Tsai SQ, Nguyen NT, Topkar VV, Zheng Z, Joung JK (2015) Broadening Staphylococcus aureus Cas9 targeting range by modifying PAM recognition. Nat Biotechnol 33(12):1293

Konermann S, Brigham MD, Trevino AE, Joung J, Abudayyeh OO, Barcena C, Hsu PD, Habib N, Gootenberg JS, Nishimasu H (2015) Genome-scale transcriptional activation by an engineered CRISPR-Cas9 complex. Nature 517(7536):583

Koonin EV, Makarova KS, Zhang F (2017) Diversity, classification and evolution of CRISPR-Cas systems. Curr Opin Microbiol 37:67–78

Leisegang M, Engels B, Schreiber K, Yew PY, Kiyotani K, Idel C, Arina A, Duraiswamy J, Weichselbaum RR, Uckert W (2016) Eradication of large solid tumors by gene therapy with a T-cell receptor targeting a single cancer-specific point mutation. Clin Cancer Res 22(11):2734–2743

Li M, Wang R, Xiang H (2014) Haloarculahispanica CRISPR authenticates PAM of a target sequence to prime discriminative adaptation. Nucleic Acids Res 42(11):7226–7235

Li R, Fang L, Tan S, Yu M, Li X, He S, Wei Y, Li G, Jiang J, Wu M (2016) Type I CRISPR-Cas targets endogenous genes and regulates virulence to evade mammalian host immunity. Cell Res 26(12):1273–1287

Li Y, Zhang Y, Lin J, Pan S, Han W, Peng N, Liang YX, She Q (2017) Cmr1 enables efficient RNA and DNA interference of a III-B CRISPR–Cas system by binding to target RNA and crRNA. Nucleic Acids Res 45(19):11305–11314

Li J, Shen S, Chen J, Liu W, Li X, Zhu Q, Wang B, Chen X, Wu L, Wang M (2018) Accurate annotation of accessible chromatin in mouse and human primordial germ cells. Cell Res 28(11):1077–1089

Ling H, Vincent K, Pichler M, Fodde R, Berindan-Neagoe I, Slack FJ, Calin GA (2015) Junk DNA and the long non-coding RNA twist in cancer genetics. Oncogene 34(39):5003–5011

Lintner NG, Kerou M, Brumfield SK, Graham S, Liu H, Naismith JH, Sdano M, Peng N, She Q, Copié V (2011) Structural and functional characterization of an archaeal clustered regularly interspaced short palindromic repeat (CRISPR)-associated complex for antiviral defense (CASCADE). J Biol Chem 286(24):21643–21656

Liu L, Li X, Wang J, Wang M, Chen P, Yin M, Li J, Sheng G, Wang Y (2017a) Two distant catalytic sites are responsible for C2c2 RNase activities. Cell 168(1–2):121–134.e112

Liu TY, Iavarone AT, Doudna JA (2017b) RNA and DNA targeting by a reconstituted Thermus thermophilus type III-A CRISPR-Cas system. PLoS One 12(1):e0170552

Luo ML, Mullis AS, Leenay RT, Beisel CL (2015) Repurposing endogenous type I CRISPR-Cas systems for programmable gene repression. Nucleic Acids Res 43(1):674–681

Luo ML, Leenay RT, Beisel CL (2016) Current and future prospects for CRISPR-based tools in bacteria. Biotechnol Bioeng 113(5):930–943

Maas S, Rich A (2000) Changing genetic information through RNA editing. BioEssays 22(9):790–802

Makarova KS, Haft DH, Barrangou R, Brouns SJ, Charpentier E, Horvath P, Moineau S, Mojica FJ, Wolf YI, Yakunin AF (2011) Evolution and classification of the CRISPR–Cas systems. Nat Rev Microbiol 9(6):467–477

Makarova KS, Wolf YI, Alkhnbashi OS, Costa F, Shah SA, Saunders SJ, Barrangou R, Brouns SJ, Charpentier E, Haft DH (2015) An updated evolutionary classification of CRISPR–Cas systems. Nat Rev Microbiol 13(11):722–736

Mali P, Yang L, Esvelt KM, Aach J, Guell M, DiCarlo JE, Norville JE, Church GM (2013) RNA-guided human genome engineering via Cas9. Science 339(6121):823–826
Mohanraju P, Makarova KS, Zetsche B, Zhang F, Koonin EV, Van der Oost J (2016) Diverse evolutionary roots and mechanistic variations of the CRISPR-Cas systems. Science 353(6299): aad5147
Mojica FJ, García-Martínez J, Soria E (2005) Intervening sequences of regularly spaced prokaryotic repeats derive from foreign genetic elements. J Mol Evol 60(2):174–182
Mulepati S, Héroux A, Bailey S (2014) Crystal structure of a CRISPR RNA–guided surveillance complex bound to a ssDNA target. Science 345(6203):1479–1484
Müller-Esparza H, Randau L (2017) Commentary: type I CRISPR-Cas targets endogenous genes and regulates virulence to evade mammalian host immunity. Front Microbiol 8:319
Mullin E (2017) Was the year of gene-therapy breakthroughs. MIT Technol Rev 2018
Munker R, Calin GA (2011) MicroRNA profiling in cancer. Clin Sci 121(4):141–158
Nam KH, Haitjema C, Liu X, Ding F, Wang H, DeLisa MP, Ke A (2012) Cas5d protein processes pre-crRNA and assembles into a cascade-like interference complex in subtype IC/ DvulgCRISPR-Cas system. Structure 20(9):1574–1584
Nelles DA, Fang MY, O'Connell MR, Xu JL, Markmiller SJ, Doudna JA, Yeo GW (2016) Programmable RNA tracking in live cells with CRISPR/Cas9. Cell 165(2):488–496
Niewoehner O, Garcia-Doval C, Rostøl JT, Berk C, Schwede F, Bigler L, Hall J, Marraffini LA, Jinek M (2017) Type III CRISPR–Cas systems produce cyclic oligoadenylate second messengers. Nature 548(7669):543–548
Nishikura K (2010) Functions and regulation of RNA editing by ADAR deaminases. Annu Rev Biochem 79:321–349
Nishikura K (2016) A-to-I editing of coding and non-coding RNAs by ADARs. Nat Rev Mol Cell Biol 17(2):83
Nishimasu H, Ran FA, Hsu PD, Konermann S, Shehata SI, Dohmae N, Ishitani R, Zhang F, Nureki O (2014) Crystal structure of Cas9 in complex with guide RNA and target DNA. Cell 156(5):935–949
O'Connell MR, Oakes BL, Sternberg SH, East-Seletsky A, Kaplan M, Doudna JA (2014) Programmable RNA recognition and cleavage by CRISPR/Cas9. Nature 516(7530):263–266
Oh JH, van Pijkeren JP (2014) CRISPR–Cas9-assisted recombineering in Lactobacillus reuteri. Nucleic Acids Res 42(17):e131
van Overbeek M, Capurso D, Carter MM, Thompson MS, Frias E, Russ C, Reece-Hoyes JS, Nye C, Gradia S, Vidal B (2016) DNA repair profiling reveals nonrandom outcomes at Cas9-mediated breaks. Mol Cell 63(4):633–646
Özcan A, Pausch P, Linden A, Wulf A, Schühle K, Heider J, Urlaub H, Heimerl T, Bange G, Randau L (2019) Type IV CRISPR RNA processing and effector complex formation in Aromatoleumaromaticum. Nat Microbiol 4(1):89–96
Pan S, Li Q, Deng L, Jiang S, Jin X, Peng N, Liang Y, She Q, Li Y (2019) A seed motif for target RNA capture enables efficient immune defence by a type III-B CRISPR-Cas system. RNA Biol 16(9):1166–1178
Peng W, Li H, Hallstrøm S, Peng N, Liang YX, She Q (2013) Genetic determinants of PAM-dependent DNA targeting and pre-crRNA processing in Sulfolobusislandicus. RNA Biol 10(5):738–748
Pham H, Kearns NA, Maehr R (2016) Transcriptional regulation with CRISPR/Cas9 effectors in mammalian cells. In: Post-transcriptional gene regulation. Springer, New York, NY, pp 43–57
Pourcel C, Salvignol G, Vergnaud G (2005) CRISPR elements in Yersinia pestis acquire new repeats by preferential uptake of bacteriophage DNA, and provide additional tools for evolutionary studies. Microbiology 151(3):653–663
Qi LS, Larson MH, Gilbert LA, Doudna JA, Weissman JS, Arkin AP, Lim WA (2013) Repurposing CRISPR as an RNA-guided platform for sequence-specific control of gene expression. Cell 152(5):1173–1183

Rees HA, Liu DR (2018) Publisher correction: base editing: precision chemistry on the genome and transcriptome of living cells. Nat Rev Genet 19(12):801–801

Rosenberg BR, Hamilton CE, Mwangi MM, Dewell S, Papavasiliou FN (2011) Transcriptome-wide sequencing reveals numerous APOBEC1 mRNA-editing targets in transcript 3′ UTRs. Nat Struct Mol Biol 18(2):230

Rossi A, Kontarakis Z, Gerri C, Nolte H, Hölper S, Krüger M, Stainier DY (2015) Genetic compensation induced by deleterious mutations but not gene knockdowns. Nature 524(7564):230–233

Rousseau BA, Hou Z, Gramelspacher MJ, Zhang Y (2018) Programmable RNA cleavage and recognition by a natural CRISPR-Cas9 system from Neisseria meningitidis. Mol Cell 69(5):906–914.e904

Samai P, Pyenson N, Jiang W, Goldberg GW, Hatoum-Aslan A, Marraffini LA (2015) Co-transcriptional DNA and RNA cleavage during type III CRISPR-Cas immunity. Cell 161(5):1164–1174

Sanjana NE, Shalem O, Zhang F (2014) Improved vectors and genome-wide libraries for CRISPR screening. Nat Methods 11(8):783

Savić N, Schwank G (2016) Advances in therapeutic CRISPR/Cas9 genome editing. Transl Res 168:15–21

Schwartz S, Bernstein DA, Mumbach MR, Jovanovic M, Herbst RH, León-Ricardo BX, Engreitz JM, Guttman M, Satija R, Lander ES (2014) Transcriptome-wide mapping reveals widespread dynamic-regulated pseudouridylation of ncRNA and mRNA. Cell 159(1):148–162

Sefcikova J, Roth M, Yu G, Li H (2017) Cas6 processes tight and relaxed repeat RNA via multiple mechanisms: a hypothesis. BioEssays 39(6):1700019

Selle K, Barrangou R (2015) Harnessing CRISPR–Cas systems for bacterial genome editing. Trends Microbiol 23(4):225–232

Senturk S, Shirole NH, Nowak DG, Corbo V, Pal D, Vaughan A, Tuveson DA, Trotman LC, Kinney JB, Sordella R (2017) Rapid and tunable method to temporally control gene editing based on conditional Cas9 stabilization. Nat Commun 8(1):1–10

Shabbir MA, Hao H, Shabbir MZ, Wu Q, Sattar A, Yuan Z (2016a) Bacteria vs. bacteriophages: parallel evolution of immune arsenals. Front Microbiol 7:1292

Shabbir MAB, Hao H, Shabbir MZ, Hussain HI, Iqbal Z, Ahmed S, Sattar A, Iqbal M, Li J, Yuan Z (2016b) Survival and evolution of CRiSPR–Cas system in prokaryotes and its applications. Front Immunol 7:375

Shabbir MAB, Shabbir MZ, Wu Q, Mahmood S, Sajid A, Maan MK, Ahmed S, Naveed U, Hao H, Yuan Z (2019) CRISPR-Cas system: biological function in microbes and its use to treat antimicrobial resistant pathogens. Ann Clin Microbiol Antimicrob 18(1):21

Shalem O, Sanjana NE, Hartenian E, Shi X, Scott DA, Mikkelsen TS, Heckl D, Ebert BL, Root DE, Doench JG (2014) Genome-scale CRISPR-Cas9 knockout screening in human cells. Science 343(6166):84–87

Shmakov S, Abudayyeh OO, Makarova KS, Wolf YI, Gootenberg JS, Semenova E, Minakhin L, Joung J, Konermann S, Severinov K (2015) Discovery and functional characterization of diverse class 2 CRISPR-Cas systems. Mol Cell 60(3):385–397

Shmakov S, Smargon A, Scott D, Cox D, Pyzocha N, Yan W, Abudayyeh OO, Gootenberg JS, Makarova KS, Wolf YI (2017) Diversity and evolution of class 2 CRISPR–Cas systems. Nat Rev Microbiol 15(3):169

Sinkunas T, Gasiunas G, Fremaux C, Barrangou R, Horvath P, Siksnys V (2011) Cas3 is a single-stranded DNA nuclease and ATP-dependent helicase in the CRISPR/Cas immune system. EMBO J 30(7):1335–1342

Slaymaker IM, Gao L, Zetsche B, Scott DA, Yan WX, Zhang F (2016) Rationally engineered Cas9 nucleases with improved specificity. Science 351(6268):84–88

Slotkin W, Nishikura K (2013) Adenosine-to-inosine RNA editing and human disease. Genome Med 5(11):105

Smargon AA, Cox DB, Pyzocha NK, Zheng K, Slaymaker IM, Gootenberg JS, Abudayyeh OA, Essletzbichler P, Shmakov S, Makarova KS (2017) Cas13b is a type VI-B CRISPR-associated RNA-guided RNase differentially regulated by accessory proteins Csx27 and Csx28. Mol Cell 65(4):618–630.e617

Sorek R, Lawrence CM, Wiedenheft B (2013) CRISPR-mediated adaptive immune systems in bacteria and archaea. Annu Rev Biochem 82:237–266

Strutt SC, Torrez RM, Kaya E, Negrete OA, Doudna JA (2018) RNA-dependent RNA targeting by CRISPR-Cas9. elife 7:e32724

Tong Y, Charusanti P, Zhang L, Weber T, Lee SY (2015) CRISPR-Cas9 based engineering of actinomycetal genomes. ACS Synth Biol 4(9):1020–1029

Umenhoffer K, Draskovits G, Nyerges A, Karcagi I, Bogos B, Tímář E, Csörgő B, Herczeg R, Nagy I, Feheŕ T, Pál C (2017) Genome-wide abolishment of mobile genetic elements using genome shuffling and CRISPR/Cas-assisted MAGE allows the efficient stabilization of a bacterial chassis. ACS Synth Biol 6(8):1471–1483

Uusi-Mäkelä MI, Barker HR, Bäuerlein CA, Häkkinen T, Nykter M, Rämet M (2018) Chromatin accessibility is associated with CRISPR-Cas9 efficiency in the zebrafish (Danio rerio). PLoS One 13(4):e0196238

Wallace RA, Black WP, Yang X, Yang Z (2014) A CRISPR with roles in Myxococcusxanthus development and exopolysaccharide production. J Bacteriol 196(23):4036–4043

Wang R, Preamplume G, Terns MP, Terns RM, Li H (2011) Interaction of the Cas6 riboendonuclease with CRISPR RNAs: recognition and cleavage. Structure 19(2):257–264

Wang H, Yang H, Shivalila CS, Dawlaty MM, Cheng AW, Zhang F, Jaenisch R (2013) One-step generation of mice carrying mutations in multiple genes by CRISPR/Cas-mediated genome engineering. Cell 153(4):910–918

Wang T, Wei JJ, Sabatini DM, Lander ES (2014a) Genetic screens in human cells using the CRISPR-Cas9 system. Science 343(6166):80–84

Wang X, Lu Z, Gomez A, Hon GC, Yue Y, Han D, Fu Y, Parisien M, Dai Q, Jia G (2014b) N 6-methyladenosine-dependent regulation of messenger RNA stability. Nature 505(7481):117–120

Wang J, Li J, Zhao H, Sheng G, Wang M, Yin M, Wang Y (2015a) Structural and mechanistic basis of PAM-dependent spacer acquisition in CRISPR-Cas systems. Cell 163(4):840–853

Wang Y, Zhang Z-T, Seo S-O, Choi K, Lu T, Jin Y-S, Blaschek HP (2015b) Markerless chromosomal gene deletion in Clostridium beijerinckii using CRISPR/Cas9 system. J Biotechnol 200: 1–5

Wang Y, Zhang Z-T, Seo S-O, Lynn P, Lu T, Jin Y-S, Blaschek HP (2016) Bacterial genome editing with CRISPR-Cas9: deletion, integration, single nucleotide modification, and desirable "clean" mutant selection in Clostridium beijerinckii as an example. ACS Synth Biol 5(7):721–732

Wang L, Mo CY, Wasserman MR, Rostøl JT, Marraffini LA, Liu S (2019) Dynamics of Cas10 govern discrimination between self and Non-self in type III CRISPR-Cas immunity. Mol Cell 73(2):278–290.e274

Westra ER, Semenova E, Datsenko KA, Jackson RN, Wiedenheft B, Severinov K, Brouns SJ (2013) Type IE CRISPR-Cas systems discriminate target from non-target DNA through base pairing-independent PAM recognition. PLoS Genet 9(9):e1003742

White MK, Khalili K (2016) CRISPR/Cas9 and cancer targets: future possibilities and present challenges. Oncotarget 7(11):12305

Wiedenheft B, van Duijn E, Bultema JB, Waghmare SP, Zhou K, Barendregt A, Westphal W, Heck AJ, Boekema EJ, Dickman MJ (2011) RNA-guided complex from a bacterial immune system enhances target recognition through seed sequence interactions. Proc Natl Acad Sci 108(25):10092–10097

Wiedenheft B, Sternberg SH, Doudna JA (2012) RNA-guided genetic silencing systems in bacteria and archaea. Nature 482(7385):331–338

Wojtal D, Kemaladewi DU, Malam Z, Abdullah S, Wong TW, Hyatt E, Baghestani Z, Pereira S, Stavropoulos J, Mouly V (2016) Spell checking nature: versatility of CRISPR/Cas9 for developing treatments for inherited disorders. Am J Hum Genet 98(1):90–101

Xiao-Jie L, Hui-Ying X, Zun-Ping K, Jin-Lian C, Li-Juan J (2015) CRISPR-Cas9: a new and promising player in gene therapy. J Med Genet 52(5):289–296

Xue H-Y, Ji L-J, Gao A-M, Liu P, He J-D, Lu X-J (2016) CRISPR-Cas9 for medical genetic screens: applications and future perspectives. J Med Genet 53(2):91–97

Yamano T, Zetsche B, Ishitani R, Zhang F, Nishimasu H, Nureki O (2017) Structural basis for the canonical and non-canonical PAM recognition by CRISPR-Cpf1. Mol Cell 67(4):633–645.e633

Yang H, Gao P, Rajashankar KR, Patel DJ (2016) PAM-dependent target DNA recognition and cleavage by C2c1 CRISPR-Cas endonuclease. Cell 167(7):1814–1828.e1812

Yosef I, Manor M, Kiro R, Qimron U (2015) Temperate and lytic bacteriophages programmed to sensitize and kill antibiotic-resistant bacteria. Proc Natl Acad Sci 112(23):7267–7272

You L, Ma J, Wang J, Artamonova D, Wang M, Liu L, Xiang H, Severinov K, Zhang X, Wang Y (2019) Structure studies of the CRISPR-Csm complex reveal mechanism of co-transcriptional interference. Cell 176(1–2):239–253.e216

Zhang J, Rouillon C, Kerou M, Reeks J, Brugger K, Graham S, Reimann J, Cannone G, Liu H, Albers S-V (2012) Structure and mechanism of the CMR complex for CRISPR-mediated antiviral immunity. Mol Cell 45(3):303–313

Zhang N, Zhi H, Curtis BR, Rao S, Jobaliya C, Poncz M, French DL, Newman PJ (2016) CRISPR/Cas9-mediated conversion of human platelet alloantigen allotypes. Blood 127(6):675–680

Bioinformatic Tools in CRISPR/Cas Platform

3

Aftab Ahmad, Sidra Ashraf, Humera Naz Majeed, Sabin Aslam, Muhammad Aamir Aslam, Muhammad Salman Mubarik, and Nayla Munawar

Contents

A. Ahmad
Department of Biochemistry/US-Pakistan Center for Advanced Studies in Agriculture and Food Security, University of Agriculture, Faisalabad, Pakistan

S. Ashraf
Department of Biochemistry, University of Agriculture, Faisalabad, Pakistan

H. N. Majeed
Department of Biochemistry, Government College Women University, Faisalabad, Pakistan

S. Aslam · M. S. Mubarik
Center for Advanced Studies in Agriculture and Food Security (CASAFS), University of Agriculture, Faisalabad, Pakistan

M. A. Aslam
Institute of Microbiology, University of Agriculture, Faisalabad, Pakistan

N. Munawar (✉)
Department of Chemistry, College of Science, United Arab Emirates University, Al-Ain, UAE
e-mail: nmunawar@uaeu.ac.ae

A. Ahmad et al. (eds.), *The CRISPR/Cas Tool Kit for Genome Editing*,
https://doi.org/10.1007/978-981-16-6305-5_3

Abstract

CRISPR/Cas has emerged as a game-changing technology for genome editing with widespread applications ranging from human therapeutics to engineering bacterial genomes for beneficial purposes to editing plant genomes for agricultural purposes. Successful genome editing through CRISPR/Cas relies on two components: an appropriate Cas endonuclease and a 20-base-pair (bp), single-guide RNA (sgRNA). CRISPR/Cas is currently favored as a genome editing technique due to its simple design rules and efficient editing capabilities that do not necessarily involve adding any foreign DNA at the target site. Cas endonucleases can be programmed to target any site in the genome by changing the gRNA sequence, highlighting the importance of gRNA design for increased specificity and efficiency, and reduced off-targeting in CRISPR/Cas genome editing. The rapid rise in CRISPR/Cas genome editing and associated applications has led to the development of numerous computational tools for effective sgRNA design. In this chapter, we discuss the essentials of gRNA design and provide an overview of the design process. In addition to summarizing factors which affect gRNA specificity and CRISPR cleavage efficiency, we discuss predictions of target efficiency and off-target detection algorithms.

Finally, we describe the application-specific (knockout, activation, repression, base editing, and RNA editing) requirements of gRNA design and different tools to facilitate gRNA design.

Keywords

CRISPR/Cas · Design tools · On-targeting efficiency · Off-targeting · Bioinformatic tools · Specificity of CRISPR/Cas

Abbreviations

ABE	Adenine base editor
BLESS	Breaks labeling, enrichment on streptavidin
BWA	Burrows-Wheeler aligner
CBE	Cytosine base editor
CFD	Cutting frequency determination
ChIP	Chromatin immunoprecipitation
Cmr	Cas module-RAMP
Cpf1	CRISPR from Prevotella and Francisella 1
CRISPR/Cas	Clustered regularly interspaced short palindromic repeats/CRISPR associated protein
CRISPRa	CRISPR activation
CRISPRi	CRISPR interference
CRISTA	CRISPR target assessment
crRNA	CRISPR RNA
DHS	DNase I hypersensitivity
DSB	Double-stranded break
GFP	Green fluorescent protein
gRNA	Guide RNA
HDR	Homology-directed repair
HEK293	Human embryonic kidney 293 cells
IDLV	Integrase-deficient lentiviral vectors
KI	Knock-in
KO	Knockout
KRAB	Krüppel-associated box
LSD1	Lysine-specific demethylase 1
MMEJ	Microhomology-mediated end joining
nCas9	Cas9 nickase
NGS	Next-generation sequencing
NHEJ	Non-homologous end joining
PAM	Protospacer adjacent motif
RAMP	Repeat-associated mysterious proteins
RNP	Ribonucleoprotein
sgRNA	Single-guide RNA
SHERLOCK	Specific high-sensitivity enzymatic reporter unlocking

TET1	Ten-eleven translocation gene protein 1
TSS	Transcription start site
WGE	Wellcome Sanger Institute genome editing

3.1 Introduction

CRISPR/Cas is an adaptive immune system of archaea and bacteria, providing a defense against invading plasmids and viruses (Garneau et al. 2010). Natural CRISPR/Cas systems consist of three core components:

- An array of repeats encompassing unique sequences called spacers
- A promoter sequence upstream of CRISPR arrays
- An operon encoding a set of effector Cas proteins, essential for processing information coded within arrays

Native CRISPR/Cas defense systems consist of three stages: adaptation or acquisition, expression or biogenesis, and interference. During acquisition, a foreign genetic element (a "protospacer") is cleaved and incorporated into the CRISPR locus as a new spacer. In biogenesis, these arrays are expressed as precursor CRISPR RNA (pre-crRNA) and subsequently processed into mature crRNA. Finally, in the interference stage, Cas endonucleases cleave the invading double-stranded DNA using crRNA as a guide sequence (Brazelton et al. 2015). Multiple studies have confirmed that the adoption and interference stages also require a protospacer adjacent motif (PAM) in the immediate vicinity of the protospacer (Fig. 3.1).

Based on effector Cas protein organization and non-coding RNA species architecture, CRISPR/Cas systems have been classified into two main classes and six types (Lino et al. 2018). Class 1 systems are defined as multi-Cas proteins acting in a cascade manner or Cas module-RAMP (repeat-associated mysterious proteins), i.e., Cmr complexes. In contrast, class 2 systems are compact and utilize a single effector Cas protein. For detailed classification of CRISPR/Cas systems, see Chap. 2. Due to their compact architecture and single effector Cas protein, class 2 systems have been adopted for genome editing applications in eukaryotes (Jinek et al. 2013; Makarova and Koonin 2015; Mali et al. 2013). Cas9 from *Streptococcus pyogenes* (SpCas9) requires a non-coding RNA known as transactivating crRNA (tracrRNA) in addition to crRNA. In today's genome editing applications, these two non-coding RNAs are synthetically fused into one sgRNA (Alkhnbashi et al. 2020). So, an sgRNA in an engineered CRISPR/Cas9 system consists of a permanent part and a programmable part. The programmable part can be tailored to target Cas9 anywhere in the genome. The target site in DNA consists of a 20-nucleotide (nt)-long region complementary to sgRNA plus a PAM sequence (NGG for SpCas9 and TTTV for Cpf1) (Table 3.1). If there is no PAM adjacent to the target site, Cas endonuclease will not cleave the target site. If the sgRNA pairs with the DNA target sequence followed by PAM, it

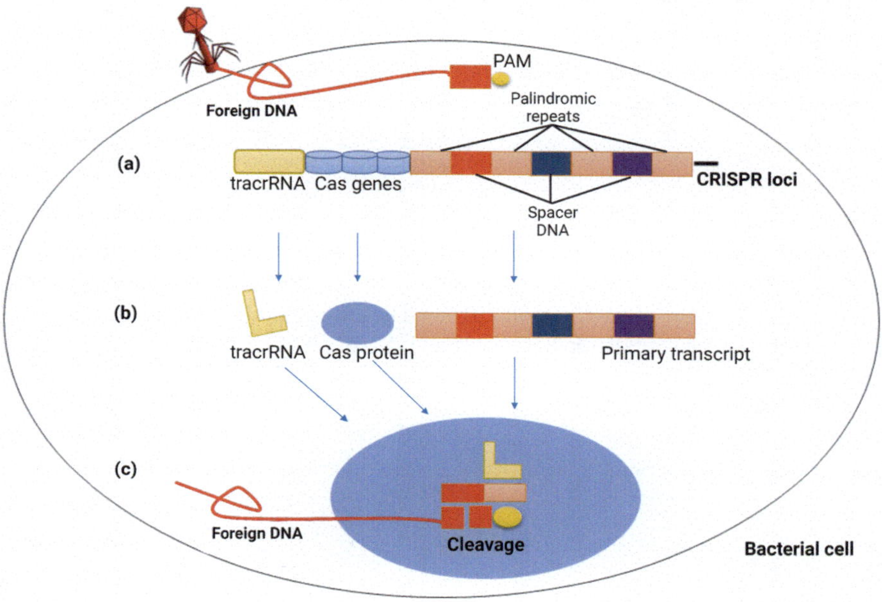

Fig. 3.1 Mechanism of a natural CRISPR/Cas9 system: (**a**) acquisition; (**b**) expression; and (**c**) interference

could create a double-stranded break (DSB) in the target site. The DSB will be repaired by either non-homologous end joining (NHEJ) or homology-directed repair (HDR) (Tian et al. 2017) (Fig. 3.2). The sgRNAs are not selected randomly; they must be associated with a PAM that is present in the target DNA but not included in the sgRNA. Bacteria use PAM to differentiate between self and non-self, thereby protecting their own DNA from cleavage because PAMs are only present in phage DNA (Fig. 3.3). With this simple and straightforward design, CRISPR/Cas can be programmed to any sequence in the genome. However, this simple, two-component (sgRNA and PAM) process also has disadvantages, as exactly similar or closely similar sgRNA sequences may occur at multiple locations and some of them could be tolerated by Cas endonuclease, leading to so-called off-targets (Cui et al. 2018). Cas endonuclease may also tolerate specific sequence changes in PAM. For example, while spCas9 specifically recognizes NGG (where N is any nucleotide base; G is guanine), it may also recognize NAG (where N is any nucleotide base; A is adenine; G is guanine), albeit less efficiently (Thomas et al. 2019). It is critical to reduce the number of potential off-target sites for improved CRISPR/Cas specificity, especially in human therapeutic applications, germline modifications, and genome editing for important agricultural purposes.

The rapid rise in CRISPR/Cas applications has prompted researchers to devise bioinformatic tools using different algorithms and design rules for effective sgRNA design, specific targeted modification, and low off-targets. Such tools facilitate gRNA design with maximum on-target efficiency in available genomes with

Table 3.1 PAM sequence, cutting site, and sgRNA length requirement for different Cas proteins

Nuclease	Species	PAM sequence	Cutting site	sgRNA length	Function	Reference
SpCas9-NG	*Streptococcus pyogenes*	5′-NTG, NAC, NTT, and NCG	5′ of PAM	–	Great potential for base editing, targeted genome editing, and genome regulation in plants	Ren et al. (2019)
SpCas9	*Streptococcus pyogenes*	5′-NGG-3′	5′ of PAM	20 bp	NGG sequence contains two G:C base pairs	Jinek et al. (2012)
SaCas9	*Staphylococcus aureus*	5′-NNGRRT-3′	5′ of PAM	21 bp	Facilitates efficient in vivo genome editing	Nishimasu et al. (2015)
St1Cas9	*Streptococcus thermophilus*	NNAGAAW	5′ of PAM	20 bp	Functions in human cells; also has a critical role in bacterial positive selection system	Kleinstiver et al. (2015)
St3Cas9	*Streptococcus thermophilus*	5′-NGGNG-3′	5′ of PAM	20 bp	Facilitates in vitro genome editing	Cong et al. (2013)
xCas9	–	GAA, NG, and GAT	–	–	Can identify various types of PAM	Wang et al. (2018)
FnCas9	*Francisella novicida*	5′-NGG-3′	5′ of PAM	20 bp	Creates variants that can identify more relaxed PAM	Hirano et al. (2016)
Nme Cas9	*Neisseria meningitidis*	5′-NNNNGATT-3′	5′ of PAM	20 and 24 bp	Requires longer PAM for site-specific cleavage	Lee et al. (2016)
AacC2c1	*Alicyclobacillus acidoterrestris*	T-rich PAM	5′ of PAM	20 bp	Endonuclease that contains NUC and REC lobe	Liu et al. (2017)
Cpf1 (AsCpf1)	*Acidaminococcus*	5′-TTTN-3′	3′ of PAM	24 bp	Identifies crRNA scaffold and PAM sequence. Can generate staggered DSB	Yamano et al. (2016)
Cpf1	*Francisella 1 and Prevotella*	TTTV	5′ of PAM	20 bp	Relies on T-rich PAM sequence	Moon et al. (2018)
Cas12a	*Acidaminococcus*	Thymine-rich PAM sequence	5′ of PAM	–	Generates DSB with staggered ends	Jeon et al. (2018)

Cas13	*Leptotrichia shahii*	Non-G nucleotide at 3′ of protospacer flanking site (PFS)	–	28 bp	Has trans-acting RNase activity	Garcia-Doval and Jinek (2017)
Cas14	*Archaea*	–	–	–	Can cut ssDNA without restrictive sequence requirement	Harrington et al. (2018)

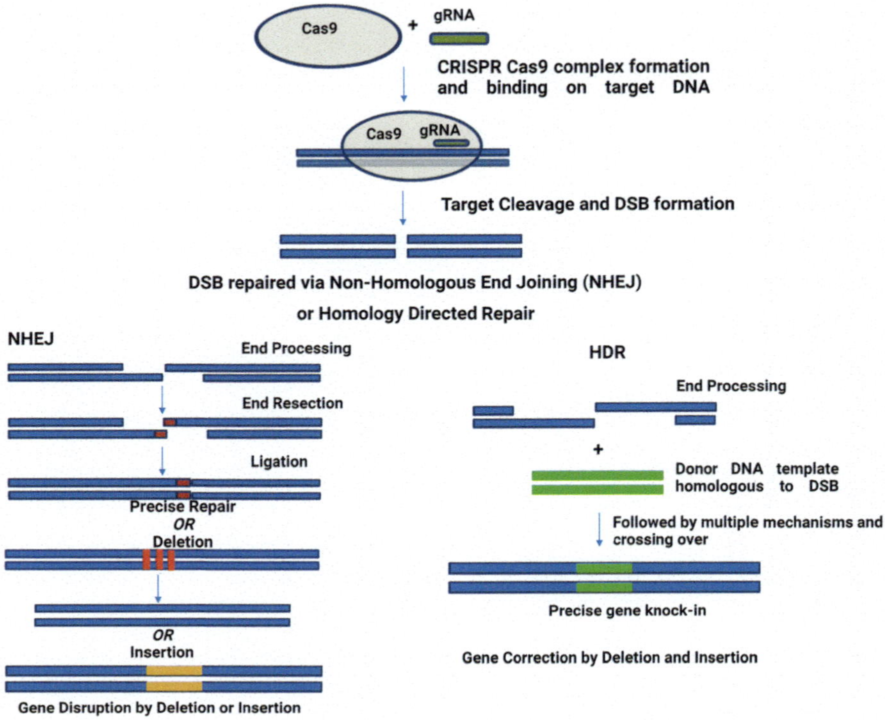

Fig. 3.2 Repair mechanisms for CRISPR/Cas-induced DSB. NHEJ and HDR are the two main DSB repair mechanisms

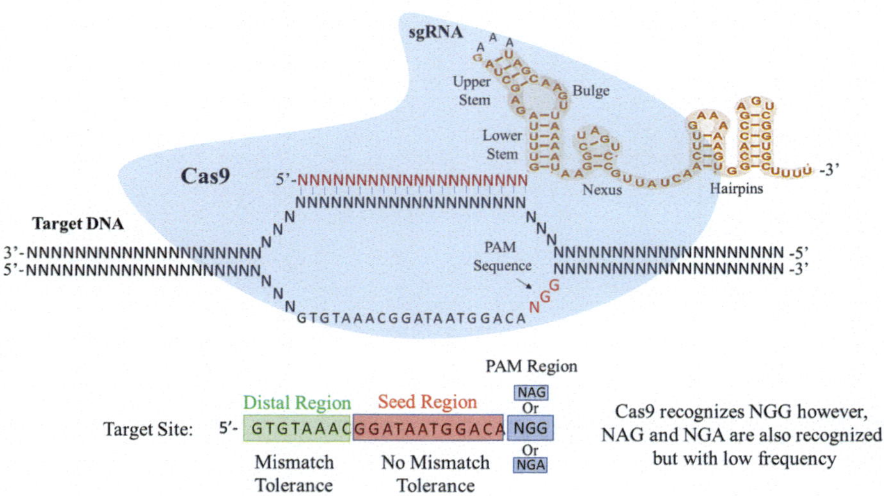

Fig. 3.3 CRISPR/Cas9 components. Seed region (consist of 12 nt region proximal to PAM, which is sensitive to mismatches) of sgRNA and PAM sequence for Cas9

user-defined PAM sequence and Cas endonuclease (Cui et al. 2018). Many design tools exist, but all have their own individual strengths and limitations. Most vary in terms of design parameters, specifications, available genomes, on-target efficiency score, off-target predictions, and so on. For example, design tools such as CRISPR-P (Li and Durbin 2009), E-CRISPR (Heigwer et al. 2014), CasOT (Xiao et al. 2014), and Cas-OFFinder (Bae et al. 2014) were mainly developed to predict off-targets in CRISPR/Cas experiments. However, in CRISPR/Cas applications such as CRISPR screening, cleavage efficiency is also important (Ma et al. 2016). Therefore, design tools such as sgRNA Designer, CRISPR-ERA (Liu et al. 2015), and Benchling predict on-targets as well as off-targets. Other genomic features such as sgRNA guanine-cytosine (GC) content, PAM flanking sequences, chromatin structure, methylation status, regulatory potential, and evolutionary conservation are also important in sgRNA design (Shi et al. 2015). Another critical factor in designing an efficient sgRNA is the application-specific (knockout (KO), knock-in (KI), CRISPR interference (CRISPRi), CRISPR activation (CRISPRa), and base editing) location of sgRNA in the genome. "WeReview: CRISPR Tools" is an online, live repository which helps researchers choose the best and latest tools for CRISPR/Cas applications (Torres-Perez et al. 2019). The current chapter aims to help researchers select the most useful tools for sgRNA design with maximum specificity and limited off-targets. This chapter also seeks to help users who are designing sgRNA with application-specific parameters in CRISPR/Cas.

3.2 Fundamentals of CRISPR/Cas Experiment and sgRNA Design

Engineered CRISPR/Cas system relies on sgRNA and PAM for genome modification in the target site of the genome. The prerequisites for designing an efficient sgRNA are:

1. Target gene and target region
2. Specific Cas endonuclease (e.g., Cas9, Cas9 nickase (nCas9), nuclease-dead Cas9 (dCas9), Cpf1) and an appropriate PAM for the Cas endonuclease
3. Promoter selection for in vivo or in vitro expression of sgRNA
4. Cloning strategy for sgRNA, e.g., sgRNA cloned in expression vector or used as template for RNA production
5. For multiple gRNA, whether expressed from a single promoter or individual promoters

Also important for sgRNA design are application-specific parameters (e.g., for KO, KI, CRISPRi, CRISPRa, and base editing) coupled with the intended DSB repair system. For example, in KO applications, off-targets on other chromosomes may be cleared by backcrossing. Moreover, the sgRNA position for CRISPRi and CRISPRa applications would be different to that for KO and KI applications. In addition, two or more sgRNAs are required in some applications, such as two sgRNAs with nCas9, a pair of sgRNAs in CRISPRa, and a pair of distal sgRNAs in KI applications (Mohr et al. 2016). Here we summarize the essentials of an effective sgRNA for different CRISPR/Cas systems.

3.2.1 Good Gene Annotation: An Essential Requirement

From a genome editing perspective, good gene annotation is a prerequisite for designing an appropriate sgRNA. Online databases and tools are available to help designers view sgRNA in a relevant genome browser, as successful editing in most CRISPR/Cas applications depends upon gRNA positioning relative to specific features of the gene. For example, in CRISPRa, the sgRNA must be located within 50–500 bp of the transcription start site (TSS), but in CRISPRi, the gRNA should be near TSS. For KO applications using NHEJ, appropriate target regions may include a common coding exon, while in KI, a specific coding exon, intron, or a region coding for a protein domain could be appropriate (Gilbert et al. 2014; Shalem et al. 2014; Wang et al. 2014; Shi et al. 2015). High-quality genome databases with regularly updated gene annotations based on experimental data are available for models such as drosophila, zebrafish, mouse, rat, and *Arabidopsis*. These databases assist in formed design of gRNA relative to the position of gene features. However, in non-model species, the lack of genome databases with appropriate gene annotations is a limiting factor on the design of specific gRNA (Mohr et al. 2016).

3.2.2 Different Guidelines for Different Applications

With rapid development in CRISPR/Cas systems has come the development of bioinformatic tools and algorithms to predict on-target efficiency, as well as off-targets. Off-target tools mostly focus on sequence similarity with on-target sites and use a defined cut-off for possible number of mismatches that can be tolerated. However, even for off-target sites with mismatches, creating a bulge or gap sometimes leads to a valid target site for a DSB. Although several tools can predict off-targets, it is not feasible to apply those rules for every gRNA and every application. Some rules for gRNA effectiveness are not relevant to all CRISPR/Cas applications or even the same application in different species (Mohr et al. 2016). For example, a CRISPRi application in *Escherichia coli* showed that gRNA must target the non-template strand (also called the coding strand or sense strand) (Qi et al. 2013), but similar studies in eukaryotes showed that gRNA binding to either strand is effective. Moreover, as compared with KO applications, off-target effects will be of less concern in CRISPRi and CRISPRa applications, because binding may not be within effective range of the promoter sequence (Mohr et al. 2016). A recent study showed that sgRNA effectiveness parameters for cleavage efficiency in CRISPRi were not valid for CRISPRa applications (Doench et al. 2016). This suggests that different applications require different design principles. However, it is not yet clear to what extent general design rules are relevant to various applications or to what extent optional parameters will be required for a particular species, tissue, or cell.

3.2.3 Best Design Linked with Availability of More Data

Improvements in CRISPR/Cas design require more data to be available. When designing sgRNA, researchers must be aware of the design tool's criteria for maximizing specificity and limiting off-targets. Researchers must also know the background of the design criteria: the study, species, delivery method, and specific applications from which a particular parameter was derived (Mohr et al. 2016). Sharing results and data from good designs and poor ones, along with species information and specific applications, will help researchers to continue improving the design and efficiency of CRISPR/Cas systems. In addition, information and data sharing will help researchers better understand the universal and application-specific factors that influence the effective design of sgRNA.

3.3 sgRNA Design Process: An Overview

The key aspect of sgRNA design is to define the target site in the genome. This can easily be done by locating the PAM sequence (NGG for spCas9 and TTTV for Cpf1) in the target region or gene. All PAM sequences recognized by different Cas endonucleases are listed in Table 3.1. Theoretically, if 5′-20 nt of the sgRNA pairs with a complementary target site in the genome, the sgRNA/Cas9 complex will create a DSB. However, several practical studies have suggested that cleavage efficiency varies significantly among different gRNAs. So, predictive models and algorithms are essential for selecting the best high-efficiency gRNA with limited off-targets. An additional challenge in CRISPR applications is off-target activity caused by both sgRNA and Cas9. Several studies have confirmed that CRISPR/Cas9 can tolerate several mismatches and cleave the DNA at sites other than the intended site of modification (target site) leading to off-target mutations. Although spCas9 systems recognize 5′-NGG-3′ as PAM, spCas9 can also recognize 5′-NAG-3′ and 5′-NGA-3′ albeit with low efficiency. Many models and computational tools are available to help researchers design an effective gRNA with high efficiency and specificity (Cui et al. 2018). In the following section, we present an overview of the design process in CRISPR/Cas applications.

3.3.1 Selection of Desired Genetic Modification

The first step in the design process is to define the desired genetic modification, e.g., KO, point mutation, transcriptional control, or KI. Because different modifications require different CRISPR/Cas reagents, a clear understanding of the desired genetic manipulation will narrow down the selection of appropriate CRISPR/Cas components (Thomas et al. 2019). However, although a broad range of CRISPR reagents and components exist, it is better to customize these components if perfect reagents do not exist for the chosen application.

3.3.2 Choice of Appropriate Expression System

To achieve the desired objective in a CRISPR/Cas experiment, Cas9 and gRNA must be expressed in the target cells or organism. Factors that can affect the desired modification, off-target numbers, and efficiency include the selected expression system (transient or stable), promoter choice (constitutive or tissue specific), reagents (plasmid, mRNA or RNPs), and delivery systems (viral, non-viral, or physical) (Graham and Root 2015). Standard protocols and reagents may suffice for CRISPR/Cas applications in easy-to-transfect cell lines, e.g., HEK293 (Banan 2020).

3.3.3 Selection of Appropriate Cas Endonuclease

Of the two classes of CRISPR/Cas systems described above, Class 1 systems use multiple Cas proteins, while Class 2 use a single effector Cas protein to create DSB in the target DNA. Choosing the right Cas endonuclease is essential. Cas9 and Cpf1 (Cas12a), the two most widely used Cas endonucleases, are both Class 2 CRISPR/Cas systems. Cas9 is a type II endonuclease that recognizes NGG as PAM sequence and creates DSB with blunt ends, three bp upstream of PAM site. Multiple engineered Cas9 variants have been generated, for example, nCas9, which produces single-stranded breaks (SSB), while dCas9 is used for site-specific binding of DNA. In contrast, Cpf1 is a type V endonuclease that recognizes the TTTN PAM sequence. Cpf1 cleaves 18–23 bp away from PAM and produces staggered ends with 5′ overhangs. Because it is smaller than spCas9, it is easy to pack into viral vectors for delivery. So, selection of expression system depends upon the desired modifications (Luo 2019).

3.3.4 Selection of Gene or Genetic Element

To manipulate a gene with a particular CRISPR application, a researcher must first identify the target gene's genomic sequence. Selection of target region (promoter, exons, or introns) in the gene will depend upon the desired genetic modification. For example, for KO applications, 5′ constitutive expressed exon is the best target. Alternatively, gRNA can be targeted to an exon that codes an essential protein domain. For HDR applications, the target sequence should be in close proximity (within 10 bp) to the desired edit site.

3.3.5 Searching of Target Site for Intended Gene Modification

Most CRISPR/Cas design tools search target regions using either a sequence-based or a genome-based approach. In sequence-based searching, the user must input the sequence to define the target site for gRNA design. The CRISPOR design tool searches on sequence and requires an input of <2000 nt for gRNA design and display. In a genome-based approach, the user must provide a gene name, ID, or similar input to display gRNA relative to the gene features. For example, the WGE (Wellcome Sanger Institute genome editing) tool requires a gene symbol in order to display sgRNA relative to the gene features (Thomas et al. 2019).

3.3.6 Sequencing of Target Site and Design of sgRNA

Once the desired manipulation, expression system, Cas endonuclease, and CRISPR reagents are decided, the next step is to confirm the site and design sgRNA. SgRNA design is a prime concern in CRISPR applications. Because features in the target DNA site affect the sgRNA efficiency, therefore, it is better to sequence the target region before designing gRNA, because variations in the target region and gRNA may occur and this can reduce cleavage efficiency. Most CRISPR/Cas applications require an efficient and specific sgRNA, but this task is quite challenging because there are many criteria to obey. So, to identify the most suitable gRNA with maximum efficiency, design criteria are very important. Various sequence features influence the efficiency of gRNA. For example, the presence of guanine (G) at 5′ end of sgRNA (GX19NGG) was crucial for expression from U6 promoter. G was also required on the first or second position adjacent to PAM, probably for loading of Cas9. The presence of cytosine (C) at this position was not favored. Thymine (T) at the fourth position closest to PAM is undesirable too, because the presence of multiple uracil (U) decreases sgRNA expression. Adenine (A) is suitable in the middle region of gRNA; G is preferred in the distal region of sgRNA. Overall, A and G make sgRNA more stable and more efficient. In addition to gRNA sequence features, novel features in PAM affect sgRNA reproducibility. For example, in the variable nucleotide N of NGG for spCas9, C is preferred, while T is not favored. Moreover, Cas9 preferences for particular sgRNA sequence features are quite different from those in a dCas9-mediated application. A 19-nucleotide sgRNA in dCas9-mediated CRISPRi and CRISPRa showed the highest efficiency compared with 20 nt or 17–18 nt truncated sgRNA for Cas9. Moreover, the seed region of sgRNA is of key importance in CRISPR/Cas9, while all sgRNA nucleotides contribute to gRNA efficiency in CRISPR/dCas9.

3.3.7 Selection of Suitable gRNA

A given target sequence or gene may have many potential gRNAs. It is important to select the most suitable gRNA with the highest efficiency for the intended modification. Suitability is assessed in terms of position relative to target site, high on-target activity, and low off-target activity. This can be achieved with tools such as WGE and CRISPOR using custom filters. Filtering for gRNAs with low off-targets will identify candidates with minimum off-targets. However, a gRNA with high on-target activity may have significantly low specificity leading to high off-targets. A gRNA with a high on-target score and high specificity would be an ideal sgRNA candidate for the desired CRISPR application (Thomas et al. 2019).

3.3.8 Design Criteria for Genome-Wide CRISPR Libraries

In contrast to individual gRNA design, CRISPR libraries are designed to screen mutations (or desired modifications) in many genes or across an entire genome. As a result, sgRNA design for genome-wide CRISPR libraries is entirely computer-based because it is impossible to evaluate each gRNA. Instead, multiple sgRNAs are designed for each gene in the genome at different locations. Users can design their own custom libraries or use libraries according to their chosen application (Thomas et al. 2019). Selected libraries and their applications are listed in Table 3.2.

3.4 Specificity in CRISPR/Cas

After selecting PAM and potential target sites, the next step is to identify the site most likely to result in efficient genome editing. In addition to choosing an sgRNA to match the target site, researchers try to select one with no additional binding sites in the genome. While the ideal sgRNA would have no homologous sites in the genome, in practice an sgRNA will have partial homology to many additional sites in the genome, i.e., off-targets (Duan et al. 2014). Off-target sites with mismatches near PAM will not be cleaved efficiently; such sgRNA would have lower off-targets effects and will be associated with the highest specificity as compared to those sgRNA in which mismatches are away from PAM in off-target sites. Off-target sites may be effectively minimized by predicting CRISPR/Cas specificity and designing a specific and optimal sgRNA. The two main approaches for predicting sgRNA specificity are based on either (1) alignment or (2) scoring. In the first method, sgRNA sequences are aligned to a given genome using conventional or specialized tools to discover all off-targets, and only frequency of the mismatches in the gRNA sequence is considered. In a scoring-based approach, sgRNA are scored and ranked after the initial alignment in order to select the most specific sgRNA for a given experiment. In this scoring-based approach, in addition to frequency of mismatches, positional weighing of each mismatch is calculated. Two scoring-based approaches are commonly used: (1) a learning-based method and (2) a hypothesis-driven method. Below we discuss alignment- and scoring-based methods in detail (Liu et al. 2020).

Table 3.2 Selected CRISPR libraries and their purposes

Library	Purpose	Reference
Activity-optimized genome-wide library	• Optimized for cleavage activity in order to maximize the likelihood of gene knockout	Wang et al. (2015)
Bassik Human CRISPR Knockout Library	• Lentiviral genome-scale CRISPR library targeting all ~20,500 protein-coding genes in the human genome	Morgens et al. (2017)
	• Includes safe-targeting controls, i.e., gRNAs that target the genome but are not expected to disrupt gene function	
Mouse CRISPR Knockout Pooled Library (Brie)	• Uses optimized metrics that combine improved on-target activity predictions with an off-target score	Doench et al. (2016)
Human CRISPR Activation Pooled Library (Calabrese p65-HSF)	• Activates over 18,000 human genes and is used for genome-wide activation screening	Sanson et al. (2018)
pC004—CRISPR Cas13a/C2c2 PFS Library	• Library of protospacer flanking sites (PFS) inserted at the 5′ end of the β-lactamase gene for screening PFS preference for Cas13a/C2c2	Abudayyeh et al. (2016)
Zhang E. coli Genome-wide Inhibition Library	• Genome-scale CRISPR inhibition library that targets ~4000 *E. coli* genes with ~15 gRNAs per gene	Wang et al. (2018)
	• Divided into five sub-libraries defined by biological categories, including a negative control (NC) library and an additional custom test library targeting 86 genes	
Human Genome-wide CRISPRa-v2 Libraries	• For activation of gene transcription	Ho et al. (2016)
Human CRISPR Knockout Pooled Library (GeCKO v2)	• Targets early consecutive exons for genome editing	Sanjana et al. (2014)
• Oxford Drosophila Genome-wide Knockout CRISPR Library	• 40,279 guides (three guides/gene) targeting 13,501 genes (98.8% of the *Drosophila* genome)	Bassett et al. (2013)
Yeo Lab RBP CRISPR Knockout Library	• Targets RNA-binding proteins with ten guide RNAs per target	Wheeler et al. (2020)

3.4.1 Alignment-Based Approach to Predict Specificity

Alignment-based methods for assessing sgRNA sequences involve aligning the sgRNA with a reference genome and identifying potential off-targets based on sequence homology. Bowtie (Langmead et al. 2009) and Burrows-Wheeler aligner (BWA) mapping tools are used to predict off-targets, but neither identify small PAM sequences. Because these tools allow a limited number of mismatches in the sgRNA seed region, they cannot identify all off-targets. CHOPCHOP and CCTOP design tools use Bowtie to find off-targets for a candidate sgRNA, while CRISPOR uses BWA. Alignment-based Cas-OFFinder and Cas-OT also predict off-targets (Liu et al. 2020). Cas-OFFinder is popular for finding off-targets with no mismatch limitations and can even predict off-targets with a 1-bp insertion or deletion (Thomas et al. 2019). Cas-OT can identify off-targets with 6-bp mismatches in the seed region and predict off-targets in coding exons of genes. Alignment-based CRISFlash and FlashFry use tree-based algorithms and user-defined data to optimize sgRNA. As well as off-target predictions, FlashFry provides additional information such as GC content and on-target score for sgRNA (Liu et al. 2020).

3.4.2 Specificity Prediction Through Scoring-Based Tools

3.4.2.1 Hypothesis-Driven Methods

Alignment-based methods can reliably predict off-targets. However, not all nucleotide positions with mismatches in sgRNA are equally effective in terms of off-target cleavage. In addition, alignment-based predictions for off-targets are sometimes false positives. One study found that only a few of the off-targets predicted by Cas-OFFinder and CC-Top were valid, and the tools also failed to predict some valid off-targets. So, there was a need to limit the features that contribute to the non-specific off-targets in CRISPR/Cas (Liu et al. 2020). These issues can be addressed in CRISPR/Cas systems by using the MIT specificity score (named after the institution) to evaluate off-targets (Hsu et al. 2013). Hsu et al. studied more than 700 sgRNAs and evaluated sgRNA/Cas9 sequence features such as contribution of position and numbers of mismatched nucleotide in the target site (Hsu et al. 2013). The MIT score is adopted to predict off-targets in design tools such as CHOPCHOP and CRISPOR (Haeussler et al. 2016; Labun et al. 2016). Cutting frequency determination (CFD) score is also popular for evaluating off-targets in CRISPR/Cas (Liu et al. 2020). In addition to recognizing NGG PAM, Cas9 recognizes non-canonical PAM sites such as NAG, NGA, and NCG, thus leading to off-targets. Doench et al. (2016) used PAM sequence features in their scoring matrix to predict off-targets. CFD score is considered a better performer better than MIT score and has been adopted by many design tools, such as GuideScan (Perez et al. 2017) and CRISPRscan (Moreno-Mateos et al. 2015). Other design tools use sgRNA/Cas9 structural features to predict off-targets. For example, CRISPR-OFF (Alkan et al. 2018) and uCRISPR (Zhang et al. 2019) use structural features because their off-target prediction accuracy is better than sequence features.

3.4.2.2 Learning-Based Methods

Compared to empirical algorithms, learning methods use multiple features (including PAM, GC contents, methylation state, and chromatic structure) to improve their off-target predictions. Most recent tools use machine learning with multiple features for predicting CRISPR/Cas system specificity and off-targets. For example, CRISPR target assessment (CRISTA), which uses machine learning to predict efficiency, was found to perform better than other tools (Liu et al. 2020). The computer platform DeepCRISPR, which incorporates sgRNA on-targets and off-targets into a single framework, has been found to perform better than other tools for predicting efficiency and off-targets (Chuai et al. 2018).

3.5 Factors Affecting Specificity

Numerous studies have revealed different factors that may affect CRISPR/Cas specificity. These factors can be classified into two categories: (a) an intrinsic specificity of Cas9 which recognize the importance of position of every sgRNA nucleotide to create DSB and (b) relative abundance of sgRNA/Cas9 for effective target cleavage. Factors that may contribute to CRISPR/Cas system specificity are discussed below.

3.5.1 Importance of PAM in CRISPR/Cas Specificity

To be recognized by an individual Cas9 domain, PAM must be next to the 3′ end of the genome target sequence (Wu et al. 2014b). Because PAM sequences vary across Cas endonucleases, users can select a different Cas endonuclease if a particular PAM (e.g., NGG for Cas9) does not exist in the target sequence. The most commonly used Cas endonuclease, Cas9, recognizes NGG for cleavage but can also recognize the canonical PAM sites NGA and NAG, thus increasing the number of off-targets. Some of these Cas proteins require a longer PAM sequence such as SaCas9 protein, derived from *Staphylococcus aureus*, which has "NNGRRT" PAM requirement. It is assumed that such Cas9 proteins which recognize a longer PAM will have less targetable sites in the genome and, therefore, will have fewer off-target sites in a given target DNA. PAM sequences with appropriate Cas endonucleases are listed in Table 3.1.

3.5.2 Seed Sequence of sgRNA

Recruiting Cas9 to the genome target site requires sgRNA. In vitro studies have shown that Cas9 can tolerate mismatches in the first seven nucleotides in the region distal to PAM. However, studies with bacteria and mammals have confirmed that mismatches in 10–12 bp PAM proximal region (also called seed region) of the gRNA will result in reduced cleavage or complete abolishment. Other studies suggest there is no clearly defined seed region, but have confirmed that mismatches in the PAM proximal region stop Cas9 cleavage of DNA (Cong et al. 2013). In contrast, genome-wide binding datasets have shown a clearly defined seed region, limited to five nucleotides proximal to PAM (Wu et al. 2014b). The differences in seed region might arise from factors such as concentration and time required for Cas9 binding and cleavage.

3.5.3 Effective Concentration of Cas9/sgRNA Complex

The effective concentration of Cas9/sgRNA influences the specificity of CRISPR/Cas systems. Studies have confirmed that cleavage becomes less specific at higher effective concentrations of Cas9/sgRNA. For example, an in vitro study found that higher concentrations of Cas9/sgRNA complex resulted in greater tolerance of mismatches, leading to cleavage of non-specific sites. Hsu and co-authors suggested that decreasing the amount of plasmid in transfected cells led to increasedCas9 specificity (Wu et al. 2014b; Hsu et al. 2013). Another study showed that a 2.6-fold increase in Cas9 concentration led to a similar increase in off-targets. When Cas9 level remained constant, the amount of sgRNA influenced off-target number (Wu et al. 2014a).

3.5.4 Importance of sgRNA Sequence

SgRNA sequence is the key to Cas9 specificity because it contributes to Cas9 loading and Cas9/sgRNA binding to the target site. Differences in sgRNA sequence influence Cas9 tolerance of mismatches at every position in 20 nucleotides. A possible underlying mechanism for this change in specificity is that different sgRNA sequences may influence effective concentration of sgRNA. For example, it has been reported that seed sequence mutations in sgRNA increase its transcription by U6 promoter. Changes in sgRNA sequence may also contribute to chromatin state, off-targets, and thermodynamic stability of sgRNA-DNA duplex (Wu et al. 2014b). We describe these effects in detail below.

3.5.4.1 Chromatin Accessibility and Epigenetic Features Affecting Binding of Cas

Chromatin state, i.e., whether packed or open, may influence Cas9's ability to access the target site. DNase I hypersensitivity (DHS) is a strong predictor of chromatin accessibility. DHS peaks for a number of accessible seed sequences and PAM have been found to accurately predict the number of chromatin immunoprecipitation (ChIP) peaks in vivo. Wu and colleagues have suggested that chromatin accessibility does not impact significantly on-target activity of sgRNA as compared to off-target binding (Wu et al. 2014a, b).

Methylation of CpG sites (where cytosine and guanine are adjacent, with guanine closer to 3′) is an epigenetic mechanism that has been found to be linked with chromatin silencing. A study confirmed that CpG methylation of target sites may restrict Cas9 binding to the target site. Target site methylation showed strong correlation with ChIP signal, and less binding was observed in highly methylated sites (Wu et al. 2014a, b). Hsu et al. showed that Cas9 can mutate highly methylated promoters in vivo. However, an in vitro study found that CpG methylation had no significant effect on Cas9 cleavage (Hsu et al. 2013). Taking these studies together suggests that CpG methylation may affect only off-target sites.

3.5.4.2 Numbers of Seed Sequence in the Genome

Depending on sgRNA seed sequence length (5–12 nt), a mammalian genome may contain hundreds of thousands of seed match sites followed by PAM. However, nucleotide preference in the seed regions may mean that specific seed match sequences could be dramatically low. For example, for Cas9, a mouse genome contains about one million AAGGA + NGG seed sites but less than 10,000 CGTCG + NGG sites (Wu et al. 2014a, b). The relative abundance of seed sites is an important factor in designing specific sgRNA, especially in dCas9 applications.

3.5.4.3 Length of Target Sequence Influences Specificity

Length of sgRNA is important for Cas9 specificity. A 20-bp gRNA is optimal for guiding Cas9 to a target site. Although one might speculate that specificity may increase with sgRNA length, Ran et al. found that when sgRNA length was increased by extending the 5′-end, the extended sequence at the 5′-end was degraded in vivo (Ran et al. 2013). In contrast, truncated sgRNA with 17–18 nt of length increased Cas9 specificity. While the underlying mechanism is not clear, it may be that the first two nucleotides do not contribute to Cas9 stability, but instead contribute to off-targets (Fu et al. 2014).

3.5.5 sgRNA Scaffold

The impact of modifications in the sgRNA scaffold region has not been studied in detail. However, it is known that truncation or extension at the 3′ end may contribute to Cas9 stability and specificity by changing sgRNA expression, in similar fashion to 5′-end modifications in sgRNA. Increasing the length of the hairpin bound by Cas9 has been found to increase sgRNA efficiency for imaging and transcriptional regulation, probably due to efficient loading of sgRNA, but the exact mechanism remains unclear (Hsu et al. 2013; Wu et al. 2014b).

3.5.6 Repair Outcomes of DSBs

In addition to the above factors, DNA repair outcomes and sequence variations are likely to influence the selection of specific sgRNA. Several studies have identified a bias in repair outcomes for KO applications. These studies have shown that nucleotide comparison of target site adjacent to the cleavage site is important for single-nucleotide insertion or deletion in NHEJ repair pathway (Mao et al. 2013). The presence of thymine (T) adjacent to the cleavage site was associated with precise insertion of a single homologous nucleotide at the cleavage site (T to TT). However, having a dinucleotide repeat adjacent to the cleavage site led to single-nucleotide deletion with removal of homologous base (CC to C). Moreover, microhomologies in sequences flanking the cleavage site resulted in deletion of 30 nucleotides through microhomology-mediated end joining (MMEJ) repair. These findings highlight a bias in repair outcomes linked to the presence of specific sequences in target sites and the competing roles of NHEJ and MMEJ. Based on these studies, computational tools such as Favored Outcomes of Repair Events at Cas9 Targets (FORECasT) and inDelphi have been developed to predict the most likely mutational outcomes of CRISPR/Cas experiments.

3.6 Efficiency of sgRNA

Initially it was believed that CRISPR/Cas9 could target any genome sequence that was followed by PAM (NGG). As a result, most of the early bioinformatic tools were constructed based on simple methods to locate target site and PAM to design sgRNAs. Some of these tools predicted sgRNA position relative to gene features. However, several later studies demonstrated that Cas9 cleavage efficiency varies significantly between different sites, i.e., not all sites are cleaved with the same efficiency (Cong et al. 2013; Jinek et al. 2012; Mali et al. 2013; Wang et al. 2014). For example, two sgRNAs can have 100% homology with their target sites but different cleavage efficiency, indicating that cleavage efficiency may also be affected by specific nucleotides and nucleotide composition. Subsequent studies identified additional factors such as sequence features (GC contents, specific nucleotide positions, and sequence composition), genetic and epigenetic factors (methylation and chromatin arrangement), and thermodynamic properties (sgRNA secondary structure, melting temperature (T_m), and free energy) that influence on-target cleavage efficiency.

Nucleotide position and composition in the target sequence is critical for CRISPR/Cas on-target efficiency (Wilson et al. 2018; Wong et al. 2015). CRISPR/Cas-based screening in mammals has shown that G is highly preferred at positions 1 and 2 upstream to PAM, while T is not favored at position 4 in close proximity to PAM. The GC content of positions 4–13 proximal to PAM is also important for Cas9 cleavage efficiency. Using sequence features such as GC content, preferred nucleotide position, and sgRNA position relative to gene features, predictive models have been developed to design efficient sgRNA for CRISPR/Cas applications. Several laboratories have used these models to develop individual design platforms such as E-CRISP, CHOPCHOP, CRISPR-FOCUS, and CCTOP for predicting sgRNA efficiency (Table 3.3).

Genetic and epigenetic features also contribute to target-site cleavage efficiency. Studies have shown that nucleosomes (sections of chromatin) may reduce Cas9 cleavage efficiency, and DNase I hypersensitivity (DHS) and epigenetic signatures may influence on-target efficiency. Predict-SGRNA is an R package (R is a free software environment) that uses epigenetic features to predict sgRNA cleavage efficiency (Liu et al. 2020). CRISPRpred and uCRISPR predict sgRNA efficiency using the energy properties of sgRNA, DNA, and Cas9 complex and sgRNA

Table 3.3 Bioinformatic tools for sgRNA activity

Tool name	gRNA activity prediction	Reference
CCTOP	Yes	Stemmer et al. (2015)
CRISPOR	Yes	Haeussler et al. (2016)
CRISPRscan	Yes	Naito et al. (2015)
GuideScan	Yes	Perez et al. (2017)
CHIOCHOP	No	Montague et al. (2014)
Cas-OFFinder	No	Bae et al. (2014)

secondary structure. Because not all sgRNAs are effective, even when using the best design tools, multiple sgRNAs are used for each target gene. Multiple sgRNAs are also required to distinguish on-target perturbation from any off-target effect of an individual sgRNA.

3.7 Off-Targeting in CRISPR/Cas

Off-targets are a major challenge for the CRISPR/Cas community because Cas9 can bind and create DSBs even when there is only partial complementarity between sgRNA and target site. Numerous studies have reported that CRISPR/Cas may produce substantial numbers of off-targets. For example, a study in human beings found that Cas9 can tolerate up to five mismatches between sgRNA and target site, leading to DNA cleavage frequencies even higher than the intended target site (Carroll 2013; Hsu et al. 2013; Xie et al. 2014). Off-targets are not random changes but are induced by the PAM and target site. Natural off-targets in a bacterial defense system may degrade hypervariable nucleic acids (i.e., those vary much more than their counterparts in other similar regions) or plasmids beneficial for archaea and bacteria. However, from a genome editing perspective, off-targets may lead to undesirable changes at random sites in the genome, thus compromising the benefits of genome modifications. Predicting and minimizing off-targets in advance is essential for safe use of CRISPR/Cas, especially in therapeutic applications and translational research. It is also important to identify all off-targets and confirm that a desired phenotype has arisen from on-target modification instead of off-targets.

Several sgRNA design tools have a special focus on limiting off-targets in CRISPR/Cas (Table 3.4). Most of these produce sgRNA with minimal off-targets and show predicted off-targets for a given sgRNA. Different tools use different

Table 3.4 Tools for evaluating off-targets in CRISPR/Cas system

Tool name	Off-target prediction	Reference
ZiFit (Zinc Finger Targeter)	Identifies potential off-target sites, their positions, and number of times each off-target site can occur	Uniyal et al. (2019)
Cas-OFFinder	Searches potential off-target sites	Bae et al. (2014)
CRISPR design	Identifies possible off-target sites	Uniyal et al. (2019), Hsu et al. (2013)
Cas-Designer	Identifies off-target sites in DNA and RNA bulges	Uniyal et al. (2019)
CHOPCHOP	Searches for off-targets using Bowtie	Uniyal et al. (2019)
E-CRISP	Searches for off-targets using Bowtie2	Uniyal et al. (2019)
sgRNAcas9	Identifies potential off-target sites in targeted genome	Uniyal et al. (2019)
Off-Spotter	Identifies potential off-targets	Uniyal et al. (2019), Pliatsika and Rigoutsos (2015)
CRISPR MultiTargeter	Uses Cas-OFFinder to predict off-targets	Uniyal et al. (2019)

scoring methods to predict off-targets. Most of these tools score off-targets either by using data from systematic mutation studies or by having user-provided input penalties such as mismatch number and positions. Others use binary criteria, e.g., defined proximal or distal region, or sites with less than a defined number of mismatches. SgRNA candidates are then ranked by off-target number or the weighted sum of all off-target scores (Wu et al. 2014b). Some tools give option of using alternate PAM site to predict off-targets, e.g., NAG or NGA for Cas9.

As with on-target prediction tools, most design tools for off-target prediction initially focused on Cas9 and predicted off-targets through alignment-based methods using seed sequence followed by NGG. However, the discovery that Cas9 also binds NAG or NGA PAM made it apparent that many off-targets were being missed. The early tools were superseded by tools that used sequence similarity or dCas9-mediated binding to confirm off-target sites, but these later approaches were biased and not comprehensive. Unbiased approaches were then developed based on high-throughput, next-generation sequencing (NGS). For example, DSBCapture used integrase-deficient lentiviral vectors (IDLV) and sequencing, while Digenome-seq, ChIP-NGS (whole genome binding), and direct in situ *b*reaks *l*abeling, *e*nrichment on *s*treptavidin, and next-generation *s*equencing (BLESS) were developed to detect off-targets in CRISPR/Cas applications. However, these approaches also had advantages and disadvantages. IDLV and BLESS could detect genome-wide off-targets, but they were less efficient because most off-target sites are transient. In addition, both approaches could generate false-positive off-targets because DSBs may arise from endogenous processes. Although whole genome sequencing is ideal and unbiased, it can miss perfectly repaired off-targets and binding sites without cleavage. Moreover, ChIP-NGS could be biased towards open chromatin and highly expressed genes. Guide-seq has good efficiency but does not work for DNA nicks (single-stranded cuts). Digested genome sequencing (Digenome-seq) does not consider other factors that affect cleavage. All things considered, the above approaches are all useful but need refinement because in vitro results can differ from in vivo (Peng et al. 2016).

Over the last few years, considerable effort has gone into limiting off-targets and improving specificity. Approaches have included lowering GC content, employing paired nickase enzymes, and using truncated sgRNA (17–18 bp). Lower GC content may reduce off-targets because higher GC content improves RNA/DNA duplex stability, thereby increasing the chance of tolerated mismatches. SgRNA and target site mismatches that produce bulges at the 5′ end, the 3′ end, or 7–12 nucleotides proximal to PAM must be avoided. The combined use of paired nickases and paired sgRNAs will generate two closely associated single-stranded breaks and eventually make a DSB.

3.8 Application-Specific Design of sgRNA

Although all CRISPR/Cas applications rely on sgRNA to guide Cas9 to the target sequence, DSBs are not always required. KO and KI applications always require DSB creation to delete or insert DNA at a precise location respectively. Large-scale deletions and insertions require more than one DSB. In KO applications, the NHEJ repair pathway will introduce a small indel into the coding framework, leading to a frameshift mutation and thus disruption of protein formation. However, for repair templates with suitable homology arms, DSBs will be repaired by HDR pathway, consequently leading to site-specific insertion of the repair template. Because NHEJ is the preferred pathway in cells, HDR efficiency must be improved for KI applications. In contrast to KO and KI applications, CRISPRi and CRISPRa use dCas9, which does not create a DSB, but instead recruits a transcriptional activator (VP64) or repressor (Krüppel-associated box (KRAB) domain proteins) to the promoter region of a gene (Graham and Root 2015). Similarly, sgRNA position in CRISPRi and CRISPRa varies significantly between KO and KI applications. However, despite differences in sgRNA position relative to gene features, the same basic principle underlies sgRNA design in all applications. Here we summarize application-specific sgRNA design in CRISPR/Cas.

3.8.1 sgRNA for KO Applications

Being able to KO an individual gene is a powerful tool for functional genomics. Knockout (KO) of single and multiple genes is often studied to evaluate phenotypic changes in cells, tissue, or organisms and by subsequently characterizing those genes for their potential roles in different functions. CRISPR/Cas has become the gold standard for producing KO models for functional characterization of genes (Graham and Root 2015). The KO of a gene or genetic element may be achieved by creating a DSB that is repaired through the NHEJ pathway. Exon size and relative position are important for generating KO alleles. For example, larger exons would have multiple choices of sgRNA, making it easier to select an efficient sgRNA. However, small exons are easy to delete with two DSBs. In addition, sgRNA position relative to the gene features may affect the outcomes of KO applications. Targeting sgRNA too close to a translation initiation codon ATG may reinitiate translation at a downstream ATG, leading to N-terminal truncated protein. Similarly, targeting sgRNA close to the 3′-end of a gene may result in insufficient disruption of protein functions. With sgRNA design for KO applications, selecting an optimal target such as a functional domain, active site, or the transmembrane helical domain of a protein (Fig. 3.4) can increase the likelihood of completely disrupting protein functions (Thomas et al. 2019). Using multiple sgRNAs can help ensure that the curated phenotype in a KO experiment has resulted from disrupting the respective gene instead of off-targets. For large-scale design, multiple sgRNAs per gene are also recommended for increased screen efficiency. In addition to generating KO for a single gene, multiplex

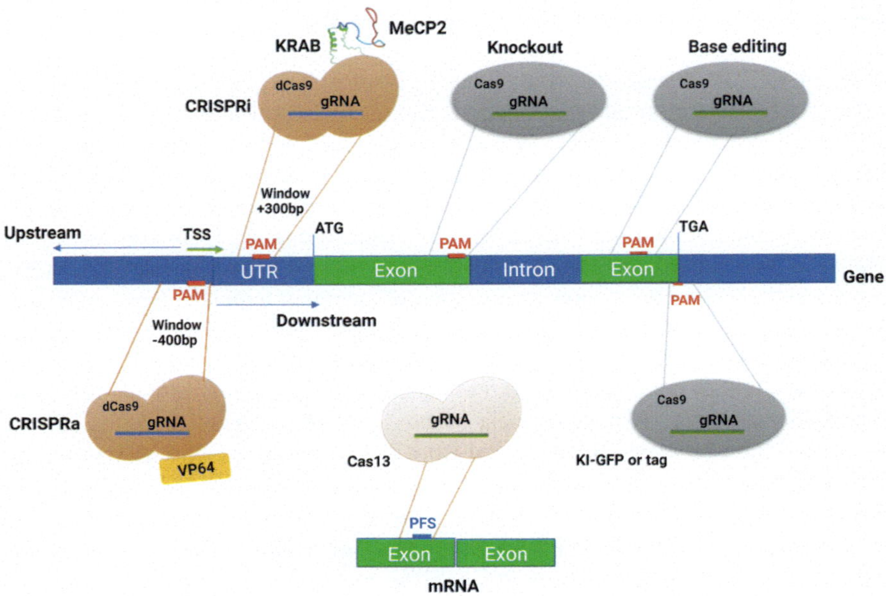

Fig. 3.4 Application-specific positioning of sgRNA in CRISPR/Cas systems

genomes using CRISPR/Cas can be used to simultaneously disrupt multiple genes in order to study their interactions and discover pathways.

3.8.2 Position of sgRNA for KI Applications

While the NHEJ pathway may lead to disruption of a gene, KI approaches using repair templates can use the HDR pathway to precisely insert a single nucleotide change or add a large template such as green fluorescent protein (GFP) (Wu et al. 2018), a tag (Chen et al. 2018), or a fluorophore. For the HDR-based repair pathway, the desired repair template must be introduced along with sgRNA and Cas9 or nCas9. The length and nature of the repair template depend on the size of the intended modification. For example, for a single-base replacement, ssDNA repair template with 50 bp homology arms on both sides of DSB could work efficiently. However, for larger insertions such as a GFP, tag, or fluorophore, a repair template with long homology arms (400–1000 bp) is desirable (Fig. 3.4). It is also advisable to exclude PAM site in the repair template. Moreover, mutating PAM site and sgRNA binding site with silent mutations would prevent subsequent binding and cleavage of target site after insertion of the repair template. These silent mutations may also assist genotyping following insertion of the desired repair template (Graham and Root 2015).

3.8.3 Designing sgRNA for CRISPRi and CRISPRa

In contrast to KO and KI applications that use Cas9 or nCas9, respectively, transcriptional regulation through CRISPR/Cas relies on dCas9, which does not create DSB but simply binds at a precise location in the genome. Binding dCas9 with an appropriate activator or repressor to a gene's promoter region may subsequently activate or repress the gene by blocking binding of RNA polymerase or transcriptional factors. SgRNA position relative to the transcription start site (TSS) may affect the efficiency of activation or repression. Accurate TSS identification is highly desirable for transcriptional regulation through CRISPR/Cas. Generally, the target site for sgRNA design in CRISPRi should be downstream (within a 300 bp window) of TSS, while for CRISPRa it should be upstream (within a 400 bp window) (Fig. 3.4). Designing multiple sgRNAs for a target region should assist in achieving the best results (Davis et al. 2018; Noguchi et al. 2017; Thomas et al. 2019).

3.8.4 SgRNA in Epigenetic Regulation

dCas9 can be used to alter gene expression by recruiting epigenetic modifiers such as lysine-specific demethylase 1 (LSD1), ten-eleven translocation gene protein 1 (TET1), DNA methyltransferase MQ1, and histone acetyltransferase p300 to modify the methylation state of cytosine in the promoter region by inducing demethylation or histone acetylation (Brocken et al. 2017). Epigenetic modifiers sometimes work better than CRISPRi or CRISPRa.

3.8.5 Design Criteria for Base Editing

In CRISPR/Cas system, base editing was initially achieved by providing a repair template using the HDR pathway, which has low efficiency. To overcome the low efficiency, researchers developed two CRISPR-mediated base editing platforms for DNA (cytosine base editor (CBE) and adenine base editor (ABE)) and an RNA base-editing platform. CBE and ABE were developed by fusing either cytosine deaminase or adenine deaminase with an appropriate Cas protein (dCas9 or nCas9) (Liang and Huang 2019). The RNA base editor was developed by fusing type VI CRISPR/Cas effector (dCas13b) with hyper-activated *a*denosine *dea*minase 2 that acts on *R*NA (ADAR2) to create a programmable RNA base editor known as REPAIR (RNA editing for programmable A to I (G) replacement). In base editing, sgRNA position depends on the targeted nucleotide's location in the protospacer region. The targeted nucleotide must be present within the active base editing window on the non-targeted strand, thus deciding the position and orientation of the sgRNA. The size of the active base window (usually four to eight nucleotides) depends on which base editor is used (Thomas et al. 2019). Base-editing efficiency can sometimes be very low at certain positions because these are inaccessible due to nucleosomes.

3.8.6 Designing sgRNA for RNA Editing

An alternative CRISPR/Cas system for regulation at transcriptional level uses CRISPR/Cas13, which specifically targets single-stranded RNA (ssRNA). CRISPR/Cas13 uses CRISPR RNA (crRNA) to recognize and cleave ssRNA (Freije et al. 2019). In bacteria, non-specific cleavage of RNA has been observed after initial cleavage with Cas13. Cas13 is used in a very sensitive diagnostic platform known as the *s*pecific *h*igh-sensitivity *e*nzymatic *r*eporter un*lock*ing (SHERLOCK) assay for differentiating Zika virus strains (Kellner et al. 2019), genotyping human beings, and RNA imaging (Yang et al. 2019). SHERLOCK could also be useful for detecting SARS-CoV-2, the RNA virus that causes coronavirus disease 2019 (COVID-19) (Joung et al. 2020).

3.9 Design Tools for sgRNA

Design tools available to the CRISPR/Cas community have been developed by both academic and commercial institutes. Although the basic objective of these tools is to design and select an optimal sgRNA and provide information about the target site, each tool has its own particular features and benefits. Similarly, these tools all aim to provide sgRNA with minimal off-targets in the genome, but they employ various methods to score these off-targets. For example, off-target scoring in CHOPCHOP is based on empirical data from multiple studies, while Cas-Finder and E-CRISP evaluate off-targets using user-defined values for mismatch number and position.

Some design tools are application- or species-specific. For example, CRISPR-ERA and BE-Designer specifically design sgRNA for transcriptional regulation (CRISPRi/CRISPRa) and base editing, respectively. FlyCRISPR and CRISPR-PLANT are specialized for sgRNA design in *Drosophila* and plants, respectively (Liu et al. 2020). Some sgRNA design tools provide users with additional options for selecting alternative PAM sites, as well as Cas effector. Some useful sgRNA design tools are listed in Table 3.5, after which we discuss some of these potential tools.

3.9.1 CHOPCHOP

More than 200 genomes are available on the CHOPCHOP website; users can input gene name or target sequence. This tool supports gRNA design for multiple applications (KO, KI, CRISPRi, and CRISPRa); users can choose application-specific Cas effector endonuclease. CHOPCHOP ranks potential sgRNAs on position, GC contents, mismatch number, and efficiency scores (Liu et al. 2020).

Table 3.5 Selected sgRNA design tools

Tool name	Input	Nuclease	Available PAM sequence	Webserver/ standalone	Link/website	Reference
GPP sgRNA Designer	DNA sequence	SpCas9 and SaCas9	Human, mouse, rat	Web/ online/ graphic interface	https://portals.broadinstitute.org/gpp/public/analysis-tools/sgrna-design	Doench et al. (2016), Doench et al. (2014), Yennamalli et al. (2017)
CRISPOR	Genomic region, DNA sequence	9 PAMs	Animals (vertebrates and non-vertebrates), plants, and microbes	Webserver	http://crispor.tefor.net/	Haeussler et al. (2016), Yennamalli et al. (2017)
E-CRISP	Gene symbol, gene ID, DNA sequence	Cas9, Custom PAM	Animals (vertebrates and non-vertebrates), plants, and microbes	Webserver	http://www.e-crisp.org/E-CRISP/	Heigwer et al. (2014), Yennamalli et al. (2017)
CRISPRscan	Gene ID, DNA sequence, gene symbol	SpCas9, AsCpf1, LbCpf1	Animals (vertebrates and non-vertebrates)	Webserver	http://www.crisprscan.org/	Moreno-Mateos et al. (2015), Yennamalli et al. (2017)
CRISPRseek	Software package	No	Input sequence of own choice	Standalone	http://www.bioconductor.org/packages/release/bioc/html/CRISPRseek.html (Local R Package)	Zhu et al. (2014), Yennamalli et al. (2017)
CRISPRdirect	Genomic region, DNA sequence, accession numbers	Custom PAM	Animals (vertebrates and non-vertebrates), plants, and microbes	Web/ online/ graphic interface	http://crispr.dbcls.jp/	Naito et al. (2015), Yennamalli et al. (2017)
CHOPCHOP	Gene ID, RefSeq, genomic region	7 PAMs, Custom PAM	Animals (vertebrates and non-vertebrates), plants, and microbes	Webserver	https://chopchop.cbu.uib.no/	Montague et al. (2014), Yennamalli et al. (2017)
Cas-OFFinder	crRNA sequences	16 PAMs	Animals (vertebrates and non-vertebrates), plants, and microbes	Webserver	http://www.rgenome.net/cas-offinder/	Bae et al. (2014), Yennamalli et al. (2017)

(continued)

Table 3.5 (continued)

Tool name	Input	Nuclease	Available PAM sequence	Webserver/ standalone	Link/website	Reference
CCTop	DNA sequence	11 PAMs	Animals (vertebrates and non-vertebrates), plants, and microbes	Webserver	https://cctop.cos.uni-heidelberg.de	Stemmer et al. (2017), Yennamalli et al. (2017)
GuideScan	Genomic region, gene symbol	SpCas9, AsCpf1, LbCpf1	Human	Webserver	http://www.guidescan.com/	Perez et al. (2017)
CRISPR MultiTargeter	DNA sequence, gene symbol, gene ID/transcripts ID	SpCas9, Custom PAM	Animals (vertebrates and non-vertebrates), plants	Webserver	http://www.multicrispr.net/	Prykhozhij et al. (2015), Yennamalli et al. (2017)

3.9.2 Base Editing (BE)-Analyzer and BE-Designer

These are publicly available design tools for base editing. Both tools help researchers select sgRNA for desired region and analyze outcomes of base editing from NGS data. BE-Designer also lists all potential sgRNAs for a given DNA sequence and provides off-targets for a given sgRNA against a large number of species (Hwang et al. 2018).

3.9.3 CRISPOR

One of the best tools for designing efficient sgRNA, CRISPOR contains 19 different PAMs and 417 different genomes. It can accept genome coordinates or user-provided sequences. Each sgRNA will be ranked for off-targets, specificity, and efficiency. Outcome predictions, GC contents, and poly T will also be given for each sgRNA (Liu et al. 2020).

3.9.4 CRISFlash

Like CHOPCHOP, CRISFlash can use sequenced genome or genome sequence to design sgRNA. In addition, it accepts user-defined values to optimize sgRNA and off-targets. CRISFlash is considered a faster tool for sgRNA design and scoring off-targets (Jacquin et al. 2019).

3.10 Prospects

CRISPR/Cas technology is a revolutionary tool for functional genomic human therapeutics and agricultural advances. Because sgRNA plays an indispensable role in CRISPR/Cas-mediated genome editing, numerous tools have been developed for designing efficient and specific sgRNA with minimal off-targets. However, off-targets continue to represent a major challenge for CRISPR/Cas-mediated genome manipulation. Systematic studies show that predictive models for efficient sgRNA design are not always effective for all applications and all species. This makes it imperative that scientists know the weaknesses and strengths of each model for sgRNA design. As new knowledge about CRISPR continues to emerge, it is clear that sgRNA and PAM are not the only influences on CRISPR/Cas cleavage, with additional such factors now including GC contents and chromatin accessibility. The ongoing discovery of new and novel features that contribute to CRISPR/Cas specificity and efficiency will also help minimize off-targets. Moreover, it has become clear that CRISPR/Cas outcomes are specific rather than random. Such findings will facilitate more precise editing with CRISPR/Cas.

In summary, recent advances in our understanding of CRISPR mechanisms and factors affecting specificity and efficiency, combined with the further development of bioinformatics tools, will enable more precision in achieving desired on-target modifications without potential off-targets. Directed evolution using EvolvR may also help scientists to engineer new Cas proteins with improved specificity.

Appendix 1: List of Useful Bioinformatics Tools and Databases for Gene Modification Research

Tool	Description	Link
AlleleID	"AlleleID® is a comprehensive desktop tool designed to address the challenges of bacterial identification, pathogen detection or species identification"	http://premierbiosoft.com/bacterial-identification/index.html
Array Designer 2	It is an Oligo and cDNA Microarray Design Software. "It designs probes for SNP detection, microarray gene expression and gene expression profiling. In addition, comprehensive support for tiling arrays and resequencing arrays is available"	https://array-designer.software.informer.com/4.3/
AutoPrime	Autoprime is a very useful software for designing Reverse Transcription Real Time PCR (Q-RT-PCR) primers that are specific to the exon-intron boundaries	http://www.autoprime.de/
Beacon Designer	"Beacon Designer™ automates the design of real time primers and probes"	http://www.premierbiosoft.com/qOligo/Oligo.jsp?PID=1
Biocomputing Tutorials	The site harbors a number of biocomputational online tools (Cleaner, Translator, NetPlasmit, Aligner, PatSearch, etc. for nucleotide and protein sequences) and half a dozen of software	http://datascience.unm.edu/intro-to-biocomputing/
BioEdit	"BioEdit is a biological sequence alignment editor written for Windows 95/98/NT/2000/XP/7." One can download and then work with the molecular sequences for alignment, restriction mapping, RNA analysis, translation,	https://bioedit.software.informer.com/

(continued)

Tool	Description	Link
	graphical viewing of electropherogram, etc.	
BLAST	Basic local alignment search tool, provided by NCBI	https://blast.ncbi.nlm.nih.gov/
Cas-Database	Cas-Database is a genome-wide gRNA library design tool for Cas9 nucleases from *Streptococcus pyogenes* (SpCas9)	http://www.rgenome.net/cas-database/
Cas-Designer	A bulge-allowed quick guide-RNA designer for CRISPR/Cas-derived RGENs	http://www.rgenome.net/cas-designer/
CINEMA 2.1	CINEMA stands for Color INteractive Editor for Multiple Alignments. It is a free software for sequence alignment with color editor	https://cinemahdapkapp.com/download/
Click2Drug	"Click2Drug contains a comprehensive list of computer-aided drug design (CADD) software, databases and web services. These tools are classified according to their application field, trying to cover the whole drug design pipeline"	http://www.click2drug.org/
Clustal Omega	The latest form of Clustal alignment program. It is online and command-line based. The distinguishing feature of Clustal-omega is its scalability, as several thousands of medium to large-sized sequences can be aligned simultaneously. It will also make use of multiple processors, where present. In addition, the quality of alignments is superior to the previous versions. The algorithm uses seeded guide trees and HMM profile-profile progressive alignments	https://www.ebi.ac.uk/Tools/msa/clustalo/
Clustal W	A very popular site for pairwise and multiple sequence alignment. It runs on Windows, Linux/Unix, and Mac operating systems	https://www.genome.jp/tools/clustalw/
CLUSTAL X	Latest version of ClustalX 2.0 is provided by "Plate-Forme Bio-Informatique de Strasbourge," along with detailed instructions (help) for operating ClustalX. Besides, this site also provides online tools (viz. Actin-	http://www.clustal.org/clustal2/

(continued)

Tool	Description	Link
	Related Proteins Annotation server, EMBOSS, Gene Ontology Annotation, SAGE experiment parameters, GPAT, etc,) and database (SRS, BAliBase, InPACT), Documentation (tutorials to elucidate the parameters of Clustal, GCG, EMBOSS, Bioinformatics protocols, etc.)	
CODEHOP	"The COnsensus-DEgenerate Hybrid Oligonucleotide Primers (CODEHOP) program is hosted by the Fred Hutchinson Cancer Research Center in Seattle, Washington and designs PCR (Polymerase Chain Reaction) primers from protein multiple-sequence alignments"	https://4virology.net/virology-ca-tools/j-codehop/
Comparative RNA Website and Project	The Comparative RNA Web (CRW) Site disseminates information about RNA structure and evolution that has been determined using comparative sequence analysis	http://www.rna.icmb.utexas.edu/
Computational Biology at ORNL	The Computational Biology and Bioinformatics Group of the Biosciences Division of Oak Ridge National Laboratory provides data and bioinformatics tools for prokaryotic and some eukaryotic genome and related analysis. The tools are "Gene Channel," "Generation Microbial Gene Prediction System," "Microbial Gene Prediction System Internet Linked," "Genome Analysis Pipeline," etc.	https://www.ornl.gov/group/cbb
Computational Resources for Drug Discovery	"CRDD (Computational Resources for Drug Discovery) is an important module of the in silico module of OSDD. The CRDD web portal provides computer resources related to drug discovery on a single platform. Following are major features of CRDD"	http://crdd.osdd.net/
Compute pl/Mw	The tool "compute pI/Mw is a tool which allows the computation of the theoretical pI (isoelectric point) and Mw (molecular weight)	https://web.expasy.org/compute_pi/

(continued)

Tool	Description	Link
	for a list of UniProt Knowledgebase (Swiss-Prot or TrEMBL) entries or for user entered sequences"	
COSMID	A Web-based tool for identifying and validating CRISPR/Cas Off-target sites	https://crispr.bme.gatech.edu/
CPHModels 3. 2 Server	"CPHmodels 3.2 is a protein homology modeling server. The template recognition is based on profile-profile alignment guided by secondary structure and exposure predictions"	http://www.cbs.dtu.dk/services/CPHmodels/
CRISPR gRNA Design tool	*CRISPR* gRNA Design tool lets you design gRNA(s) to efficiently engineer your target and minimize off-target effects using ATUM Scoring Algorithms	https://www.dna20.com/eCommerce/cas9/input
CRISPR multitargeter	CRISPR MultiTargeter is a web-based tool for automatic searches of CRISPR guide RNA targets	http://www.multicrispr.net/
CRISPRdb	It enables the easy detection of CRISPR in locally produced data and consultation of CRISPRs present in the data base	http://crispr.u-psud.fr/crispr
CrisprGE	*CrisprGE* is a central hub of CRISPR-based genome editing	http://crdd.osdd.net/servers/crisprge/
CSIR Informatics Portal	This page is maintained by CSIR and harbors the software/tools developed for bioinformatics analysis	http://crdd.osdd.net/info/
DAVID v. 6.7	The Database for Annotation, Visualization and Integrated Discovery (DAVID) v6.7 "provides a comprehensive set of functional annotation tools for investigators to understand biological meaning behind large list of genes"	https://david.ncifcrf.gov/
DeepView: SWISS PDBViewer v. 4.1	"Swiss-PdbViewer (aka DeepView) is an application that provides a user friendly interface allowing to analyze several proteins at the same time. The proteins can be superimposed in order to deduce structural alignments and compare their active sites or any other relevant parts. Amino acid mutations,	https://spdbv.vital-it.ch/download_prerelease.html

(continued)

Tool	Description	Link
	H-bonds, angles and distances between atoms are easy to obtain thanks to the intuitive graphic and menu interface"	
DNA/RNA GC Content Calculator	One can calculate the GC content of a nucleotide sequence	http://www.endmemo.com/bio/gc.php
Dotlet	Dotlet is a free online software used as a tool for diagonal plotting of sequences	https://myhits.sib.swiss/cgi-bin/dotlet
Dotplot(+)	Dot-plot(+) software is used to identify the overlapping portions of two sequences and to identify the repeats and inverted repeats of a particular sequence	http://bip.weizmann.ac.il/education/materials/gcg/dotplot.html
Dotter	Dotter is a graphical dotplot program for detailed comparison of two sequences. It runs on MAC, Linux, Sun solaris, and Windows OS	https://sonnhammer.sbc.su.se/Dotter.html
DRUG DESIGN APPS FOR SMART PHONE	A wonderful site that harbors a number of drug designing applications for smart mobiles	http://click2drug.org/directory_Mobile.php
Drug Designing	This webpage maintains several entries to drug designing. One can learn and make use of these software/links	https://www.hsls.pitt.edu/obrc/index.php?page=drugs_medical
Emboss Align	The European Molecular Biology Open Software Suite (EMBOSS) "is a free Open Source software analysis package specially developed for the needs of the molecular biology (e.g. EMBnet) user community." Some of the applications are prophet (Gapped alignment for profiles), infoseq (Displays some simple information about sequences), water (Smith-Waterman local alignment), pepstats (Protein statistics), etc.	https://www.ebi.ac.uk/Tools/psa/emboss_needle/
Ensembl Genome Browser	"The Ensembl project produces genome databases for vertebrates and other eukaryotic species, and makes this information freely available online"	https://www.ensembl.org/
Ensembl Variant Effect Predictor	"This tool takes a list of variant positions and alleles, and predicts the effects of each of these on overlapping transcripts and regulatory regions annotated in	https://www.ensembl.org/vep

(continued)

Tool	Description	Link
	Ensembl. The tool accepts substitutions, insertions and deletions as input"	
E-RNAi	RNAi construct designer	http://e-rnai.org/
EsyPred3D	"ESyPred3D is an automated homology modeling program. The method gets the benefit of the increased alignment performances of an alignment strategy that uses neural networks"	https://www.unamur.be/sciences/biologie/urbm/bioinfo/esypred/
ExPASY Resource Portal	A resource portal supported by Expert Protein Analysis System and Swiss Institute of Bioinformatics for analyzing bioinformatics data	https://www.expasy.org/
Expasy-Translate tool	It is an online tool that "allows the translation of a nucleotide (DNA/RNA) sequence to a protein sequence"	https://web.expasy.org/translate/
Expert Protein Analysis System	"ExPASy is the SIB Bioinformatics Resource Portal which provides access to scientific databases and software tools (i.e., resources) in different areas of life sciences including proteomics, genomics, phylogeny, systems biology, population genetics, transcriptomics etc."	https://www.expasy.org/
FASTA	This server is hosted by the University of Virginia, USA. It harbors a multiple online software for sequence (nucleic acid and amino acid) comparison, local and global alignment, hydropathy plotting, and protein secondary structure prediction	https://www.ebi.ac.uk/Tools/sss/fasta/
FastPCR	"FastPCR is an integrated tool for PCR primers or probe design, in silico PCR, oligonucleotide assembly and analyses, alignment and repeat searching." This program can be downloaded and run on PCs	https://primerdigital.com/fastpcr.html
Galaxy Platform	"Galaxy is an open, web-based platform for data intensive biomedical research. Whether on the free public server or your own instance, you can perform, reproduce, and share complete analyses"	https://usegalaxy.org/

(continued)

Tool	Description	Link
GAS	GAS is UNIX or DOS-based downloadable, command-line oriented "integrated computer program designed to automate and accelerate the acquisition and analysis of genomic data"	https://bioinformaticssoftwareandtools.co.in/bio_tools.php
Gel Compar II (Paid multimodule, stand-alone software)	It is a commercial product. "GelCompar II consists of the Basic Software and five modules: Cluster analysis, Identification & Libraries, Comparative Quantification and Polymorphism Analysis, Dimensioning techniques & Statistics, and Database Sharing Tools"	https://www.applied-maths.com/modules-and-features-gelcompar-ii
Gelcompar II V. 7.1	For analyzing 1D Gel	https://www.applied-maths.com/download/software
Gel-Quant software	The "Gel-Quant" software is used to analyze one-dimensional gel images. The gel image is saved in "bitmap" format, following electrophoresis and scanning the gel	http://biochemlabsolutions.com/GelQuantNET.html
GeneFisher	"GeneFisher is an interactive web-based program for designing degenerate primers." The underlying assumption is "assumption that genes with related function from different organisms show high sequence similarity, degenerate primers can be designed from sequences of homologues genes." This assumption "leads to isolation of genes in a target organism using multiple alignments of related genes from different organisms"	https://bio.tools/genefisher
GeneCopoeia	GeneCopoeia offers comprehensive tools for microRNA (miRNA) functional analysis so researchers can detect, express, validate, or knockdown microRNA of interest confidently. All known human, mouse, and rat microRNA in miRBase covered	https://www.genecopoeia.com/
geneid	"geneid is a program to predict genes in anonymous genomic sequences designed with a hierarchical structure"	https://genome.crg.cat/geneid.html

(continued)

Tool	Description	Link
geneinfinity	This site contains description and links to various sites pertaining to Protein Secondary Structure. It is a hub for getting a quick look at several servers and metaservers that harbor databases and/or tools for prediction of protein secondary structures	http://www.geneinfinity.org/
GeneMark	GeneMark is a "family of gene prediction programs developed at Georgia Institute of Technology, Atlanta, Georgia, USA"	http://exon.gatech.edu/
Genome Bioinformatics Research Lab	The site harbors "geneid" program which is used to "predict genes, exons, splice sites and other signals along a DNA sequence." This site is also hyperlinked with "Gene prediction on whole genome" which is a "precomputed whole genome prediction data sets"	https://corelabs.ku.edu/genomics-and-bioinformatics-core
Genome Tools	"The GenomeTools genome analysis system is a free collection of bioinformatics tools (in the realm of genome informatics) combined into a single binary named gt. It is based on a C library named "libgenometools" which consists of several modules"	http://genometools.org/
GenomePRIDE 1.0	"GenomePRIDE is primer design program that designs PCR primers or long oligos on an annotated sequence"	http://pride.molgen.mpg.de/genomepride.html
GENSCAN	GENSCAN is a freely available software used for "identification of complete gene structures in genomic DNA." Genscan can be used "for predicting the locations and exon-intron structures of genes in genomic sequences from a variety of organisms"	http://hollywood.mit.edu/GENSCAN.html
Glimmer	Glimmer (Gene Locator and Interpolated Markov ModelER) is a system for finding genes in microbial DNA, especially the genomes of bacteria, archaea, and viruses. Glimmer uses interpolated Markov models (IMMs) to identify the coding regions and distinguish them from noncoding DNA	http://www.cbcb.umd.edu/software/glimmer/glimmer2.jun01.shtml

(continued)

Tool	Description	Link
GreenGenes (16srRNA sequence Alignment)	The greengenes web application provides access to the current and comprehensive 16S rRNA gene sequence alignment for browsing, blasting, probing, and downloading. The data and tools presented by greengenes can assist the researcher in choosing phylogenetically specific probes, interpreting microarray results, and aligning/annotating novel sequences	https://www.ccg.unam.mx/~vinuesa/Using_the_GreenGenes_and_RDPII_servers.html
HHpred	Homology detection and structure prediction by HMM-HMM: used for sequence database searching and structure prediction. It is fast enough and more sensitive in finding remote homologs. HHpred performs pairwise comparison of profile hidden Markov models (HMMs). It can produce pairwise query-template sequence alignments, merged query-template multiple alignments and 3D structural models calculated by the MODELLER software from HHpred alignments	https://toolkit.tuebingen.mpg.de/tools/hhpred
HMMgene 1.1 web server	"HMMgene is a program for prediction of genes in anonymous DNA." "The program predicts whole genes, so the predicted exons always splice correctly. It can predict several whole or partial genes in one sequence, so it can be used on whole cosmids or even longer sequences"	http://www.cbs.dtu.dk/services/HMMgene/hmmgene1_1.php
IDT Antisense Design	To synthesize antisense oligos for a specific target sequence of interest	https://www.idtdna.com/pages/products/functional-genomics/antisense-oligos
I-TASSER Online	I-TASSER is an online bioinformatics platform for predicting protein structure vis-à-vis function. It has been developed by Zhang Lab (University of Michigan). It has topped in the CASP ranking of structure prediction during the years 2007–2010	https://zhanglab.ccmb.med.umich.edu/I-TASSER/
JALVIEW	It is a "multiple alignment editor written in Java." It is used in EBI Clustalw, Pfam protein domain	https://www.jalview.org/

(continued)

Tool	Description	Link
	database; however, it is "available as a general purpose alignment editor and analysis workbench"	
LALIGN	Online free tool for finding local alignment between two sequences (provided in stipulated input format, viz. plain text without header line, Swiss-Prot ID, TrEMBL ID, EMBL ID, EST ID, etc.)	https://embnet.vital-it.ch/software/LALIGN_form.html
LAMP Designer	"LAMP Designer designs efficient primers for Loop-Mediated Isothermal Amplification assays, that amplify DNA and RNA sequences at isothermal conditions, eliminating the necessity of a PCR setup"	https://primerexplorer.jp/e/
MACAW	This link enables you to download Multiple Alignment Construction and Analysis Workbench (MACAW) software. This program is used for "locating, analyzing, and editing blocks of localized sequence similarity among multiple sequences and linking them into a multiple alignment"	http://en.bio-soft.net/format/MACAW.html
MAFFT version 6	"MAFFT is a multiple sequence alignment program for unix-like operating systems. It offers a range of multiple alignment methods, L-INS-i (accurate; for alignment of <~200 sequences), FFT-NS-2 (fast; for alignment of <~10,000 sequences)"	https://mafft.cbrc.jp/alignment/software/
Mapper	Java platform-based online software to map the RE sites on a target sequence	http://www.restrictionmapper.org/
Meth Primer	"MethPrimer is a program for designing bisulfite-conversion-based Methylation PCR Primers"	https://www.urogene.org/methprimer/
MethPrimer	It is a very useful site for designing primers for methylation PCR (Denatured, single-stranded DNA (ssDNA) is modified with sodium bisulfite "followed by PCR amplification using two pairs of primers, with one pair specific for methylated DNA; the other unmethylated DNA")	https://www.urogene.org/methprimer/

(continued)

Tool	Description	Link
mgene	"mGene is a computational tool for the genome-wide prediction of protein coding genes from eukaryotic DNA sequences"	http://mgene.org/
miRNa Body map (Human)	The microRNA body map is a repository of RT-qPCR miRNA expression data and functional miRNA annotation in normal and diseased human tissues	https://sites.google.com/site/mirnatools/mirna-databases
miRNA Target Gene Prediction	This website provides access to 2003 and 2005 miRNA-Target predictions for Drosophila miRNAs	http://www.mirbase.org/help/targets.shtml
miRNA Targets and Expression db	Predicted microRNA targets and target downregulation scores. Experimentally observed expression patterns	http://mirdb.org/
miRNAMap	miRNAMap 2.0 is a collection of "experimental verified microRNAs and experimental verified miRNA target genes in human, mouse, rat, and other metazoan genomes"	http://mirnamap.mbc.nctu.edu.tw/
Mobyle 1.5	This site maintains a number of online bioinformatics programs (assembly, database, display, hmm, phylogeny, protein, sequence, structure, etc.), workflows (alignment, db, phylogeny), and tutorial	http://www.mybiosoftware.com/mobyle-1-0-4-integration-bioinformatics-software-databanks.html
Modbase	It is a database for "comparative protein structure models." The pipeline used is ModPipe	https://modbase.compbio.ucsf.edu/
MODELLER	The homology modeling of Protein 3D structures can be done using downloadable software "MODELLER." It can also be used for the following protein structure-based applications: databases search for amino acid sequences, sequence and structural alignments clustering, de novo modeling of structural loops, model-optimization against user-defined objective function, and so on	https://salilab.org/modeller/
Mol. Modelling Database (MMDB)	It harbors "experimentally resolved structures of proteins, RNA, and DNA, derived from the Protein Data Bank (PDB), with	https://www.ncbi.nlm.nih.gov/Structure/MMDB/mmdb.shtml

(continued)

Tool	Description	Link
	value-added features such as explicit chemical graphs, computationally identified 3D domains (compact substructures) that are used to identify similar 3D structures, as well as links to literature, similar sequences, information about chemicals bound to the structures"	
Molecular Evolution Genetics Analysis (v. 5.1 beta)	A handy package for analyzing sequence data for pair-wise and multiple sequence alignment, phylogenetic tree (include neighbor-joining, maximum parsimony, UPGMA, maximum likelihood and minimum evolution based) construction, and estimation of evolutionary parameters	https://www.ncbi.nlm.nih.gov/pmc/articles/PMC3203626/
MS Utils	Maintains links to several platforms, pipelines, libraries, software for visualization as well as software for proteomic data analysis	https://ms-utils.org/
NEB Cutter	This software is RE site mapper, hosted by New England Biolabs	http://nc2.neb.com/NEBcutter2/
NetPrimer	It is an efficient primer analysis software that can be used for determining the features of the secondary structures of the generated primer sequences	http://www.premierbiosoft.com/netprimer/
NRSP-8 Bioinformatics Online Tools	Explores and utilizes several bioinformatics tools	https://www.animalgenome.org/
Oligo Analyzer Version 3.1 (IDT)	The secondary structures produced by the primer(s) can be checked, and the Gibbs free energy required to produce these structures can be calculated using online Oligo Analyzer Version 3.1 (of IDT)	https://www.idtdna.com/pages/tools/oligoanalyzer
Oligo Tm Determination	Calculates the melting temperature of the oligos	https://worldwide.promega.com/resources/tools/biomath/tm-calculator/
Oligo.Net	"OLIGO Primer Analysis Software is the essential tool for designing and analyzing sequencing and PCR primers, synthetic genes, and various kinds of probes including siRNA and	https://www.oligo.net/

(continued)

Tool	Description	Link
	molecular beacons. Based on the most up-to-date nearest neighbor thermodynamic data, Oligo's search algorithms find optimal primers for PCR, including TaqMan, highly multiplexed, consensus or degenerate primers. Multiple file batch processing is possible. It is also an invaluable tool for site directed mutagenesis"	
Oligonucleotide Properties Calculator	Calculates base-count, thermodynamic properties (ΔS and ΔH), Tm, and GC% values of a given oligo	http://biotools.nubic.northwestern.edu/
Oligos 6.2	"The program helps to design primer combinations given one fixed primer"	https://www.oligo.net/
ORF Finder	"The ORF Finder (Open Reading Frame Finder) is a graphical analysis tool which finds all open reading frames of a selectable minimum size in a user's sequence or in a sequence already in the database"	https://www.ncbi.nlm.nih.gov/orffinder/
PCR PRIMER DESIGN AND REACTION OPTIMISATION	It is a very useful site to learn about the pros and cons of factors affecting PCR	http://www.mcb.uct.ac.za/mcb/resources/pcr/primer
PEDANT	"The pedant genome database provides exhaustive automatic analysis of genomic sequences by a large variety of bioinformatics tools"	http://pedant.gsf.de/
Peptide Mass	This online tool of ExPASy "PeptideMass cleaves a protein sequence from the UniProt Knowledgebase (Swiss-Prot and TrEMBL) or a user-entered protein sequence with a chosen enzyme, and computes the masses of the generated peptides. The tool also returns theoretical isoelectric point and mass values for the protein of interest"	https://web.expasy.org/peptide_mass/
Phylogeny Inference Package	"PHYLIP is a free package of programs for inferring phylogenies. It is distributed as source code, documentation files, and a number of different types of executables"	https://evolution.genetics.washington.edu/phylip.html

(continued)

Tool	Description	Link
PHYRE2	Protein Homology/AnalogY Recognition Engine (PHYRE) is a non-commercial, very popular online protein structure prediction (homology modeling) server. The user friendly GUI is very helpful for the novice in the field of protein structure prediction	http://www.sbg.bio.ic.ac.uk/phyre2
Prediction of miRNA Targets (Mammals)	The tool "searches for predicted microRNA targets in mammals"	http://www.targetscan.org/
Primer Premier	Primer Premier is one of the "most comprehensive software to design and analyze PCR primers." Primers can be designed for standard PCR, SNP genotyping assays, multiplexing assays, along with checking the secondary structures of the designed primers	http://www.premierbiosoft.com/primerdesign/
Primer3 (version 0.4.0)	It is a freely available online software for designing primers and probe from a DNA sequence. It is a very popular software due to availability of several parameters to design primers with high specificity and accuracy	http://bioinfo.ut.ee/primer3-0.4.0/
PrimerBLAST	Extensively used for designing primer and checking the specificity of a given primer	https://www.ncbi.nlm.nih.gov/tools/primer-blast/index.cgi
PrimerQuest	Online primer designing tool provided by IDT	https://www.idtdna.com/primerquest/home/index
Primo Degenerate3.4	"Primo Degenerate 3.4 designs PCR primers based on a single peptide sequence or multiple alignments of proteins or nucleotides. For degenerate primers, the probability of binding to the target is proportional to the effective concentration of the specific primer"	http://www.changbioscience.com/primo/primo.html
Primo Pro 3.2	It is another online primer designing software. Its notable feature is that it can reduce background noise by exercising check on mispriming on non-target DNA sequence. It also "introduces a batch mode option for high throughput PCR primer design"	http://www.changbioscience.com/primo/dihowto.html
Primo Pro 3.4	A java-enabled online primer designing tool	http://www.changbioscience.com/primo/primo.html

(continued)

Tool	Description	Link
Promoter 2. 0 Prediction Server	Promoter2.0 predicts transcription start sites of vertebrate PolII promoters in DNA sequences. It has been developed as an evolution of simulated transcription factors that interact with sequences in promoter regions. It builds on principles that are common to neural networks and genetic algorithms	http://www.cbs.dtu.dk/services/Promoter/
PROMOTERS & TERMINATORS	This site maintains links for different software and tools (viz. PromScan, SCOPE, Promoser, Arnold, WebGesTer) for scanning, predicting promoters and transcription terminators in Eukaryotes and Prokaryotes	https://molbiol-tools.ca/Promoters.htm
Protein Data Bank	PDB is an "information portal to biological macromolecular structure." "The PDB archive contains information about experimentally-determined structures of proteins, nucleic acids, and complex assemblies"	https://www.rcsb.org/
Protein Tertiary Structure	This site contains links to several software for "calculating and displaying the 3-D structure of oligosaccharides and proteins. With the two protein analysis sites the query protein is compared with existing protein structures as revealed through homology analysis"	https://molbiol-tools.ca/Protein_tertiary_structure.htm
ProtParam	"ProtParam is a tool which allows the computation of various physical and chemical parameters for a given protein stored in Swiss-Prot or TrEMBL or for a user entered sequence. The computed parameters include the molecular weight, theoretical pI, amino acid composition, atomic composition, extinction coefficient, estimated half-life, instability index, aliphatic index and grand average of hydropathicity (GRAVY)"	https://web.expasy.org/protparam/
Protscale	"ProtScale allows you to compute and represent the profile produced by any amino acid scale on a selected protein"	https://web.expasy.org/protscale/

(continued)

Tool	Description	Link
QUARK Online	It is online software that applies QUARK algorithm for ab initio protein folding vis-à-vis structure prediction. It is another eminent online tool of Zhang lab that has secured esteemed ranking in CASP	https://zhanglab.ccmb.med.umich.edu/QUARK/
RaptorX	Another efficient protein structure prediction server that predicts the secondary and 3D protein structure. Besides, it also predicts solvent accessibility and disordered regions, and assigns the following confidence scores to indicate the quality of a predicted 3D model. It has been developed by Xu Group of Toyota Technological Institute at Chicago. RaptorX-Binding, another tool available in the homepage of RaptorX, is used for model-assisted protein binding site prediction	http://raptorx.uchicago.edu/
RASMOL	RasMol is a molecular visualization tool for protein in 3-dimension	http://www.openrasmol.org/
RASMOL Home page	"This site is provided for the convenience of users of RasMol and developers of open source versions of RasMol"	http://www.openrasmol.org/
RE specific primer designing	"PCR Designer for Restriction Analysis of Sequence Mutations"	
ReadSeq-Sequence Format Conversion Tool	Online tool for conversion of sequence format	https://www.ebi.ac.uk/Tools/sfc/readseq/
RestrictionMapper	Online, freely available tool for mapping restriction endonuclease sites on a DNA sequence	http://www.restrictionmapper.org/
RNAfold	The RNAfold web server will predict secondary structures of single-stranded RNA or DNA sequences. Current limits are 7500 nt for partition function calculations and 10,000 nt for minimum free energy only predictions	http://rna.tbi.univie.ac.at/cgi-bin/RNAWebSuite/RNAfold.cgi
RNAhybrid	RNAhybrid is a tool for finding the minimum free energy hybridization of a long and a short RNA	https://bibiserv.cebitec.uni-bielefeld.de/rnahybrid/

(continued)

Tool	Description	Link
RNAi Atlas	RNAiAtlas provides a siRNA oligonucleotide data from different sources and companies like Dharmacon (ThermoFisher), Qiagen, and Ambion, esiRNA for humans, and visualizes interactions between siRNA oligo and predicted off-target	https://www.hsls.pitt.edu/obrc/index.php?page=rna_interference
RNAi Explorer-GeneLink-siRNA	A designing tool for siRNA	https://www.genelink.com/sirna/RNAicustomorder.asp
Robetta	Robetta (Beta Version) of Baker Lab, Washington, USA, is a full-chain protein structure prediction tool. It can be used both for ab initio and comparative approaches for protein structure prediction	https://robetta.bakerlab.org/
SANBI Tools	An array of online tools (dPORE-miRNA, TcoF, PROMEX, etc.) are available which are maintained by South African National Bioinformatics Institute	https://www.sanbi.org/resources/infobases/some-tools-developed-in-sanbi-for-use-in-biodiversity-research/
SDSC Biology Workbench	"The Biology WorkBench is a web-based tool for biologists. The WorkBench allows biologists to search many popular protein and nucleic acid sequence databases. Database searching is integrated with access to a wide variety of analysis and modeling tools, all within a point and click interface that eliminates file format compatibility problems"	http://workbench.sdsc.edu/
Secondary Structure Prediction Tools	"These are a collection of protein secondary structure analysis and information sites"	http://www.compbio.dundee.ac.uk/jpred/
Sequence Manipulation Suite-2	A suite available for almost all possible manipulation work that can be done with a given DNA or amino acid sequence, viz. Format change, Sequence splitting, CpG island detection, ORF finding, Pair-wise alignment, RE-Digestion, in silico mutation, etc.	https://www.bioinformatics.org/sms2/
sgRNA Designer	This tool ranks and picks candidate CRISPRko sgRNA sequences for the targets provided, while attempting to maximize on-target activity and minimizing off-target activity	http://www.broadinstitute.org/rnai/public/analysis-tools/sgrna-design

(continued)

Tool	Description	Link
sgRNAcas9	The BiooTools (Biological Online Tools) website is devoted to provide services to assist researchers design specific and efficient CRISPR sgRNA, primer pairs for detecting small ncRNA expression, such as miRNA, piRNA, and siRNA	http://www.biootools.com/
SIDDbase 1.0a.ws1	"SIDDbase-WS is a SOAP based Web Service" that "provides interoperable access to the SIDD software, and access to the repository of stored results from calculations previously performed on complete bacterial genomes"	https://bioinformaticssoftwareandtools.co.in/bio_tools.php
siDesign-Thermo Scientific	The siDESIGN Center is an advanced, user-friendly siRNA design tool, which significantly improves the likelihood of identifying functional siRNA. One-of-a-kind options are available to enhance target specificity and adapt siRNA designs for more sophisticated experimental design	http://www.thermofisher.com/order/genome-database/browse/sirna/keyword/siDESIGN+center
SIM4	A stand-alone program designed to run on Unix-based system. It is used for aligning an expressed DNA sequence with a genomic sequence, allowing for introns	http://nebc.nox.ac.uk/bioinformatics/docs/sim4.html
SIMPA96 Secondary Structure Prediction	An online tool to predict secondary structure of protein	https://npsa-prabi.ibcp.fr/NPSA/npsa_simpa96.html
SimVector	It is "an exceptional tool for drawing publication and vector catalog quality plasmid maps, carrying out restriction analysis and designing cloning experiments"	https://simvector.net/
siRNA Design: How to	A short introduction to siRNA Designing Steps	https://www.rnaiweb.com/RNAi/siRNA_Design/
siRNA Designing-BOCK-iT RNAi Designer	Online siRNA designing tool from Invitrogen	https://rnaidesigner.thermofisher.com/
siRNA Wizard v. 3.1	InvivoGen's siRNA Wizard™ is a software designed to help you select siRNA/shRNA sequences targeting your gene(s) of interest. This program selects siRNA/shRNA sequences that match criteria suggested by studies of	https://www.invivogen.com/sirnawizard/

(continued)

Tool	Description	Link
	RNA interference and which will have the best expression rate in psiRNA vectors	
SOPMA	It is an online protein Secondary structure prediction tool	https://npsa-prabi.ibcp.fr/NPSA/npsa_sopma.html
Splice Predictors	A method to identify potential splice sites in (plant) pre-mRNA by sequence inspection using Bayesian statistical models	http://www.phenosystems.com/www/index.php/links-to-various-tools-and-information/splice-prediction-tools
Statistical Analysis of Protein Sequences (SAPS)	It performs several statistical analysis of the physiochemical properties and other features of the protein sequence, viz. compositional analysis, charge distributional analysis, distribution of other amino acid types, repetitive structures, multiplets, periodicity analysis	https://www.ebi.ac.uk/Tools/seqstats/saps/
Structural Bioinformatics Group	This is the structural bioinformatics-related page maintained by Imperial College London. This site can be used for several purposes, viz. "analysis of protein structure and function with the aim of deriving evolutionary insights, modelling and comparison of biology networks to provide insights into Systems Biology, modelling of the activity and toxicity of small molecules as an aid to the design of novel drugs"	http://bioinformatics.charite.de/
Structural Biology Software Database	Harbors links to several software for docking	https://www.ks.uiuc.edu/Development/biosoftdb/
Swiss Institute of Bioinformatics	"The SIB Swiss Institute of Bioinformatics is an academic, non-profit foundation recognised of public utility." SIB "provides high quality bioinformatics services to the national and international research community"	https://www.sib.swiss/
T-coffee	Tree-based Consistency Objective Function For alignment Evaluation (T-Coffee) is another popular multiple sequence alignment program, developed by Cedric Notredame, CRG Centro de Regulacio Genomica (Barcelona). It allows combining	https://www.ebi.ac.uk/Tools/msa/tcoffee/

(continued)

Tool	Description	Link
	results obtained from several alignment methods. The URL is http://www.ebi.ac.uk/Tools/msa/tcoffee/. The default output format is Clustal, while it accepts sequences in PIR and FASTA format	
The PCR Suite	It is an online primer designing software, hosted by UCSC, that allows users to design primers specific to various types of templates, viz. overlapping amplicons on a template, primers around SNP (in a GenBank), primers flanking exons and cDNA	http://pcrsuite.cse.ucsc.edu/
Translate a DNA Sequence	It is a Java-based free online software, to translate a given input DNA sequences and display one (at a time) of the six possible reading frames according to the selection made by the user. It also displays the graphical output for all the six reading frames together	https://web.expasy.org/translate/
UCSC Human Genome Browser	It is an interactive genome browser dedicated to human genome sequence	https://genome.ucsc.edu/
UnaFold	The likelihood of secondary structure formation by the single-stranded target is checked by UnaFold software of IDT (freely available online)	http://unafold.rna.albany.edu/
Uniprime2	It is a website for universal primer designing	https://bio.tools/uniprime2
User:Jarle Pahr/: Bioinformatics	This page harbors several "links and notes regarding bioinformatics." This is a very useful link since a user can get link to almost all aspects of bioinformatics resources	https://openwetware.org/wiki/User:Jarle_Pahr/Bioinformatics
VBI resources	This site of Virginia Bioinformatics Institute maintains several tools for bioinformatics analysis, viz. "Analysis of Dynamic Algebraic Models," "Complex Pathway Simulator," "Genome Reverse Compiler," etc.	https://www.thevillagefamily.org/content/vbi-resources
VLS3D	This page maintains a "list of in silico drug design online services, standalone and related databases. It is maintained by	https://www.vls3d.com/

(continued)

Tool	Description	Link
	Dr. B. Villoutreix, research director at the French National Medical Research Institute (Inserm)"	
Web Primer	A simple tool for primer designing for PCR or sequencing	http://www.candidagenome.org/cgi-bin/compute/web-primer
Webcutter 2.0	Another RE site detection software (online, free) for linear and circular DNA	https://www.hsls.pitt.edu/obrc/index.php?page=URL1043859576
Webgene	This site maintains several online "tools for prediction and analysis of protein-coding gene structure"	https://www.itb.cnr.it/webgene/
WGE	A website that provides tools to aid with genome editing of human and mouse genomes	http://www.sanger.ac.uk/htgt/wge/
WHAT IF	What If "is a versatile molecular modelling package that is specialized on working with proteins and the molecules in their environment like water, ligands, nucleic acids, etc." The web interface provides a number of tools, viz. Structure validation, Residue analysis, Protein analysis, 2-D graphics, 3-D graphics, Hydrogen (bonds), Rotamer related, Docking, Crystal symmetry, mutation prediction, etc.	https://swift.cmbi.umcn.nl/whatif/WIF1_4.html
YASARA	Yet Another Scientific Artificial Reality Application (YASARA) is used for predicting the rotamers (protein side chain conformations) starting with single point mutations to complete homology models of proteins	http://www.yasara.org/

Appendix 2: List of Commercial and Non-profit Sources of CRISPR/Cas Reagents

Resource	Description	Link
Addgene CRISPR plasmids	A collection of CRISPR plasmids and reagents	http://www.addgene.org/CRISPR/
Beam Therapeutics: Upleveling CRISPR's Precision by Targeting Specific Bases	Beam Therapeutics, a company co-founded recently by leading CRISPR researchers Feng Zhang, David Liu, and J. Keith Young, is developing more precise versions of the CRISPR technology which can effectively swap one base for another in the genome without cutting the DNA or RNA	https://beamtx.com/
Caribou Biosciences: Using CRISPR to Impact Several Industries	Caribou Biosciences (@CaribouBio) is one of the companies using CRISPR technology developing tools that provide transformative capabilities to therapeutics, biological research, agricultural biotechnology, and industrial biotechnology	https://cariboubio.com/
CRISPR Kits	Synthego's CRISPR kits offer economical access to fully synthetic RNA for high fidelity editing and increased precision in genome engineering	https://www.synthego.com/products/crispr-kits
CRISPRflydesign (Bullock Lab)	Offers Cas9 transgenic stocks	http://www.crisprflydesign.org/
Editas Medicine: Using CRISPR to Target Point Mutations in Serious Genetic Disorders	Editas Medicine (@editasmed) is targeting mutations that cause serious genetic diseases and hopes to modify and fix these gene mutations using CRISPR	https://editasmedicine.com/
eGenesis: Using CRISPR to Improve Organ Transplants	eGenesis (@eGenesisBio) is pioneering an especially interesting application of CRISPR-Cas9 technology in the field of human therapeutics. This company is reviving the idea of xenotransplantation, i.e., animal-to-human organ transplants	https://www.egenesisbio.com/
FlyCas9 (Ueda Lab)	Provides reagents, protocols, and online tools for genome engineering by the designer nuclease Cas9 in *Drosophila*	http://www.shigen.nig.ac.jp/fly/nigfly/cas9/index.jsp

(continued)

Resource	Description	Link
flyCRISPR (O'Connor-Giles Lab, Wildonger Lab, and Harrison Lab)	Fly CRISPR resources	http://flycrispr.molbio.wisc.edu/
Goldstein Lab CRISPR	A genome engineering resource for the *Caenorhabditis elegans* research community	http://wormcas9hr.weebly.com/
Inari Agriculture: Using CRISPR to Develop "Customized Seeds"	Inari Agriculture is an agro-biotechnology company that is revolutionizing the agricultural industry through transformational plant breeding technology. Inari uses CRISPR technology to develop seeds with traits optimized to grow best in local conditions	https://www.inari.com/
Inscripta: Increasing CRISPR's Reach	Inscripta (@InscriptaInc) is a Colorado-based CRISPR biotech company that is revolutionizing commercially available CRISPR-associated nucleases. Inscripta's next-generation CRISPR nucleases include natural and synthesized versions of "MADzymes," a nomenclature inspired by the biodiversity found on the island of Madagascar	https://www.inscripta.com/
Intellia Therapeutics: Using Genome Editing for Personalized Disease Treatment	Intellia Therapeutics (@intelliatweets) aims to produce a new class of therapeutic products using a simplified manufacturing process. The company develops CRISPR-based solutions for personalized and curative treatments, and its current in vivo studies are focused on the use of Lipid Nanoparticles (LNPs) for delivery of the CRISPR/Cas9 complex to the liver	https://www.intelliatx.com/
Joung Lab CRISPR	A genome engineering resource for zebrafish research community	http://www.crispr-cas.org/
Ligandal: Establishing the CRISPR Delivery System	Ligandal (@ligandal), one of the companies using CRISPR based in San Francisco, has developed new technology which streamlines the in vivo delivery mechanisms for CRISPR, RNA, and other genetic tools. Ligandal has developed next-generation,	https://www.ligandal.com/

(continued)

Resource	Description	Link
	non-viral protein-based biomaterials to effectively deliver gene therapy materials	
Mammoth Biosciences: Using CRISPR to Advance Clinical Diagnostic	Mammoth Biosciences (@mammothbiosci) has capitalized on CRISPR's unique ability to accurately find and bind to specific sequences of DNA. This company has created the first CRISPR-mediated platform for human disease detection. Their innovative point of care test allows for easy and affordable multiplexed detection of RNA/DNA sequences associated with disease	https://mammoth.bio/
NTrans: Helping CRISPR Edit All Cell Types	NTrans Technologies (@NtransTech), a CRISPR technology company based in the Netherlands, is working to ensure genome engineering can be performed in all cell types. NTrans pioneered a cellular uptake mechanism which circumvents the problems with delivery of CRISPR components for therapeutic purposes	https://www.ntranstechnologies.com/
OxfCRISPR (Liu Lab)	Oxford Fly CRISPR Resources	http://www.oxfcrispr.org/
Pairwise Plants: Using CRISPR to Grow New Varieties of Crops	Pairwise Plants (@PairwisePL) intends to create new crops and modify existing ones using gene editing technology such as CRISPR. The goal is to also assist farmers by providing them with new varieties of crops that require less resources to grow	https://pairwise.com/
Plantedit: Increasing the Worldwide Food Supply using CRISPR	Plantedit (@plantedit) is an Ireland-based CRISPR startup company aiming to produce "DNA-free" non-transgenic sustainable plant products in an attempt to introduce genome editing to food supply enhancement in a regulatory-free manner. The company focuses on creating modified plants that do not contain any foreign genetic material with a goal to meet the ever-increasing demand for "designer" crops	http://plantedit.com/

(continued)

Resource	Description	Link
	while circumventing both the general aversion to ingesting non-plant-based DNA or RNA and the regulatory fences around traditional "GMO."	
Synthetic Genomics: Harnessing CRISPR to Create Sustainable Energy	Synthetic Genomics (@SynGenomeInc) manipulates microalgae for sustainable oil production. Partnering with Exxon Mobil, Synthetic Genomics identified 20 transcription factors thought to be negative regulators of lipid production in microalgae. The company then applied CRISPR-Cas9 to insert loss of function mutations in 18 of the 20 genes. They report a 200% increase in oil production from one of the modified microalgae species with little effect on growth, marking a key advancement in renewable energy biofuels	https://syntheticgenomics.com/
transOMIC	*transOMIC* offers reagents for CRISPR Cas9 gene editing, shRNA constructs, and cDNA and ORF clones	https://www.transomic.com/cms/home.aspx/
Zhang Lab Genome Engineering	CRISPR genome engineering resources website	http://www.genome-engineering.org/

References

Abudayyeh OO, Gootenberg JS, Konermann S, Joung J, Slaymaker IM, Cox DB, Severinov K (2016) C2c2 is a single-component programmable RNA-guided RNA-targeting CRISPR effector. Science 353(6299):aaf5573

Alkan F, Wenzel A, Anthon C, Havgaard JH, Gorodkin J (2018) CRISPR-Cas9 off-targeting assessment with nucleic acid duplex energy parameters. Genome Biol 19(1):1–13

Alkhnbashi OS, Meier T, Mitrofanov A, Backofen R, Voß B (2020) CRISPR-Cas bioinformatics. Methods 172:3–11

Bae S, Park J, Kim JS (2014) Cas-OFFinder: a fast and versatile algorithm that searches for potential off-target sites of Cas9 RNA-guided endonucleases. Bioinformatics 30 (10):1473–1475

Banan M (2020) Recent advances in CRISPR/Cas9-mediated knock-ins in mammalian cells. J Biotechnol 308:1–9

Bassett AR et al (2013) Mutagenesis and homologous recombination in Drosophila cell lines using CRISPR/Cas9. Biol Open 3:42. https://doi.org/10.1242/bio.20137120

Brazelton VA Jr, Zarecor S, Wright DA, Wang Y, Liu J, Chen K, Lawrence-Dill CJ (2015) A quick guide to CRISPR sgRNA design tools. GM Crops Food 6(4):266–276

Brocken DJ, Tark-Dame M, Dame RT (2017) dCas9: a versatile tool for epigenome editing. Curr Issues Mol Biol 26:15–32

Carroll D (2013) Staying on target with CRISPR-Cas. Nat Biotechnol 31(9):807–809

Chen B, Zou W, Xu H, Liang Y, Huang B (2018) Efficient labeling and imaging of protein-coding genes in living cells using CRISPR-Tag. Nat Commun 9(1):1–9

Chuai G, Ma H, Yan J, Chen M, Hong N, Xue D, Gu F (2018) DeepCRISPR: optimized CRISPR guide RNA design by deep learning. Genome Biol 19(1):80

Cong L, Ran FA, Cox D, Lin S, Barretto R, Habib N et al (2013) Multiplex genome engineering using CRISPR/Cas systems. Science 339(6121):819–823

Cui Y, Xu J, Cheng M, Liao X, Peng S (2018) Review of CRISPR/Cas9 sgRNA design tools. Interdiscipl Sci Comput Life Sci 10(2):455–465

Davis CA, Hitz BC, Sloan CA, Chan ET, Davidson JM, Gabdank I, Hilton JA, Jain K, Baymuradov UK, Narayanan AK, Onate KC, Graham K, Miyasato SR, Dreszer TR, Strattan JS, Jolanki O, Tanaka FY, Cherry JM (2018) The Encyclopedia of DNA elements (ENCODE): data portal update. Nucleic Acids Res 46(D1):D794–D801. https://doi.org/10.1093/nar/gkx1081

Doench JG, Hartenian E, Graham DB, Tothova Z, Hegde M, Smith I, Root DE (2014) Rational design of highly active sgRNAs for CRISPR-Cas9–mediated gene inactivation. Nat Biotechnol 32(12):1262–1267

Doench JG, Fusi N, Sullender M, Hegde M, Vaimberg EW, Donovan KF, Virgin HW (2016) Optimized sgRNA design to maximize activity and minimize off-target effects of CRISPR-Cas9. Nat Biotechnol 34(2):184–191

Duan J, Lu G, Xie Z, Lou M, Luo J, Guo L, Zhang Y (2014) Genome-wide identification of CRISPR/Cas9 off-targets in human genome. Cell Res 24(8):1009–1012

Freije CA, Myhrvold C, Boehm CK, Lin AE, Welch NL, Carter A, Yozwiak NL (2019) Programmable inhibition and detection of RNA viruses using Cas13. Mol Cell 76(5):826–837

Fu Y, Sander JD, Reyon D, Cascio VM, Joung JK (2014) Improving CRISPR-Cas nuclease specificity using truncated guide RNAs. Nat Biotechnol 32(3):279–284

Garcia-Doval C, Jinek M (2017) Molecular architectures and mechanisms of Class 2 CRISPR-associated nucleases. Curr Opin Struct Biol 47:157–166

Garneau JE, Dupuis M-È, Villion M, Romero DA, Barrangou R, Boyaval P, Moineau S (2010) The CRISPR/Cas bacterial immune system cleaves bacteriophage and plasmid DNA. Nature 468 (7320):67–71

Gilbert LA, Horlbeck MA, Adamson B, Villalta JE, Chen Y, Whitehead EH, Qi LS (2014) Genome-scale CRISPR-mediated control of gene repression and activation. Cell 159 (3):647–661

Graham DB, Root DE (2015) Resources for the design of CRISPR gene editing experiments. Genome Biol 16(1):260

Haeussler M, Schönig K, Eckert H, Eschstruth A, Mianné J, Renaud JB, Joly JS (2016) Evaluation of off-target and on-target scoring algorithms and integration into the guide RNA selection tool CRISPOR. Genome Biol 17(1):148

Harrington LB, Burstein D, Chen JS, Paez-Espino D, Ma E, Witte IP, Doudna JA (2018) Programmed DNA destruction by miniature CRISPR-Cas14 enzymes. Science 362 (6416):839–842

Heigwer F, Kerr G, Boutros M (2014) E-CRISP: fast CRISPR target site identification. Nat Methods 11(2):122–123

Hirano H, Gootenberg JS, Horii T, Abudayyeh OO, Kimura M, Hsu PD, Nishimasu H (2016) Structure and engineering of Francisella novicida Cas9. Cell 164(5):950–961

Ho SM, Hartley BJ, Julia TCW, Beaumont M, Stafford K, Slesinger PA, Brennand KJ (2016) Rapid Ngn2-induction of excitatory neurons from hiPSC-derived neural progenitor cells. Methods 101:113–124

Hsu PD, Scott DA, Weinstein JA, Ran FA, Konermann S, Agarwala V, Cradick TJ (2013) DNA targeting specificity of RNA-guided Cas9 nucleases. Nat Biotechnol 31(9):827–832
Hwang GH, Park J, Lim K, Kim S, Yu J, Yu E, Bae S (2018) Web-based design and analysis tools for CRISPR base editing. BMC Bioinformatics 19(1):542
Jacquin AL, Odom DT, Lukk M (2019) Crisflash: open-source software to generate CRISPR guide RNAs against genomes annotated with individual variation. Bioinformatics 35(17):3146–3147
Jeon Y, Choi YH, Jang Y, Yu J, Goo J, Lee G, Jeong C (2018) Direct observation of DNA target searching and cleavage by CRISPR-Cas12a. Nat Commun 9(1):1–11
Jinek M, Chylinski K, Fonfara I, Hauer M, Doudna JA, Charpentier E (2012) A programmable dual-RNA–guided DNA endonuclease in adaptive bacterial immunity. Science 337 (6096):816–821
Jinek M, East A, Cheng A, Lin S, Ma E, Doudna J (2013) RNA-programmed genome editing in human cells. elife 2:e00471
Joung J, Ladha A, Saito M, Segel M, Bruneau R, Huang MLW, Greninger AL (2020) Point-of-care testing for COVID-19 using SHERLOCK diagnostics. MedRxiv
Kellner MJ, Koob JG, Gootenberg JS, Abudayyeh OO, Zhang F (2019) SHERLOCK: nucleic acid detection with CRISPR nucleases. Nat Protoc 14(10):2986–3012
Kleinstiver BP, Prew MS, Tsai SQ, Topkar VV, Nguyen NT, Zheng Z, Aryee MJ (2015) Engineered CRISPR-Cas9 nucleases with altered PAM specificities. Nature 523(7561):481–485
Labun K, Montague TG, Gagnon JA, Thyme SB, Valen E (2016) CHOPCHOP v2: a web tool for the next generation of CRISPR genome engineering. Nucleic Acids Res 44(W1):W272–W276
Langmead B, Trapnell C, Pop M, Salzberg SL (2009) Ultrafast and memory-efficient alignment of short DNA sequences to the human genome. Genome Biol 10(3):R25
Lee CM, Cradick TJ, Bao G (2016) The Neisseria meningitidis CRISPR-Cas9 system enables specific genome editing in mammalian cells. Mol Ther 24(3):645–654
Li H, Durbin R (2009) Fast and accurate short read alignment with Burrows–Wheeler transform. Bioinformatics 25(14):1754–1760
Liang P, Huang J (2019) Off-target challenge for base editor-mediated genome editing. Cell Biol Toxicol 35:185
Lino CA, Harper JC, Carney JP, Timlin JA (2018) Delivering CRISPR: a review of the challenges and approaches. Drug Deliv 25(1):1234–1257
Liu H, Wei Z, Dominguez A, Li Y, Wang X, Qi LS (2015) CRISPR-ERA: a comprehensive design tool for CRISPR-mediated gene editing, repression and activation. Bioinformatics 31 (22):3676–3678
Liu L, Chen P, Wang M, Li X, Wang J, Yin M, Wang Y (2017) C2c1-sgRNA complex structure reveals RNA-guided DNA cleavage mechanism. Mol Cell 65(2):310–322
Liu G, Zhang Y, Zhang T (2020) Computational approaches for effective CRISPR guide RNA design and evaluation. Comput Struct Biotechnol J 18:35–44
Luo Y (2019) CRISPR gene editing. Springer, New York, NY
Ma J, Köster J, Qin Q, Hu S, Li W, Chen C, Xu H (2016) CRISPR-DO for genome-wide CRISPR design and optimization. Bioinformatics 32(21):3336–3338
Makarova KS, Koonin EV (2015) Annotation and classification of CRISPR-Cas systems CRISPR. Springer, New York, NY, pp 47–75
Mali P, Yang L, Esvelt KM, Aach J, Guell M, DiCarlo JE, Church GM (2013) RNA-guided human genome engineering via Cas9. Science 339(6121):823–826
Mao Y, Zhang H, Xu N, Zhang B, Gou F, Zhu JK (2013) Application of the CRISPR–Cas system for efficient genome engineering in plants. Mol Plant 6(6):2008–2011
Mohr SE, Hu Y, Ewen-Campen B, Housden BE, Viswanatha R, Perrimon N (2016) CRISPR guide RNA design for research applications. FEBS J 283(17):3232–3238
Montague TG, Cruz JM, Gagnon JA, Church GM, Valen E (2014) CHOPCHOP: a CRISPR/Cas9 and TALEN web tool for genome editing. Nucleic Acids Res 42(W1):W401–W407

Moon SB, Lee JM, Kang JG, Lee NE, Ha DI, Kim SH, Kim YS (2018) Highly efficient genome editing by CRISPR-Cpf1 using CRISPR RNA with a uridinylate-rich 3′-overhang. Nat Commun 9(1):1–11

Moreno-Mateos MA, Vejnar CE, Beaudoin JD, Fernandez JP, Mis EK, Khokha MK, Giraldez AJ (2015) CRISPRscan: designing highly efficient sgRNAs for CRISPR-Cas9 targeting in vivo. Nat Methods 12(10):982–988

Morgens DW, Wainberg M, Boyle EA, Ursu O, Araya CL, Tsui CK, Li A (2017) Genome-scale measurement of off-target activity using Cas9 toxicity in high-throughput screens. Nat Commun 8(1):1–8

Naito Y, Hino K, Bono H, Ui-Tei K (2015) CRISPRdirect: software for designing CRISPR/Cas guide RNA with reduced off-target sites. Bioinformatics 31(7):1120–1123

Nishimasu H, Cong L, Yan WX, Ran FA, Zetsche B, Li Y, Nureki O (2015) Crystal structure of Staphylococcus aureus Cas9. Cell 162(5):1113–1126

Noguchi S, Arakawa T, Fukuda S, Furuno M, Hasegawa A, Hori F, Ishikawa-Kato S, Kaida K, Kaiho A, Kanamori-Katayama M, Kawashima T, Kojima M, Kubosaki A, Manabe RI, Murata M, Nagao-Sato S, Nakazato K, Ninomiya N, Nishiyori-Sueki H, Noma S, Saijyo E, Saka A, Sakai M, Simon C, Suzuki N, Tagami M, Watanabe S, Yoshida S, Arner P, Axton RA, Babina M, Baillie JK, Barnett TC, Beckhouse AG, Blumenthal A, Bodega B, Bonetti A, Briggs J, Brombacher F, Carlisle AJ, Clevers HC, Davis CA, Detmar M, Dohi T, Edge ASB, Edinger M, Ehrlund A, Ekwall K, Endoh M, Enomoto H, Eslami A, Fagiolini M, Fairbairn L, Farach-Carson MC, Faulkner GJ, Ferrai C, Fisher ME, Forrester LM, Fujita R, Furusawa JI, Geijtenbeek TB, Gingeras T, Goldowitz D, Guhl S, Guler R, Gustincich S, Ha TJ, Hamaguchi M, Hara M, Hasegawa Y, Herlyn M, Heutink P, Hitchens KJ, Hume DA, Ikawa T, Ishizu Y, Kai C, Kawamoto H, Kawamura YI, Kempfle JS, Kenna TJ, Kere J, Khachigian LM, Kitamura T, Klein S, Klinken SP, Knox AJ, Kojima S, Koseki H, Koyasu S, Lee W, Lennartsson A, Mackay-Sim A, Mejhert N, Mizuno Y, Morikawa H, Morimoto M, Moro K, Morris KJ, Motohashi H, Mummery CL, Nakachi Y, Nakahara F, Nakamura T, Nakamura Y, Nozaki T, Ogishima S, Ohkura N, Ohno H, Ohshima M, Okada-Hatakeyama M, Okazaki Y, Orlando V, Ovchinnikov DA, Passier R, Patrikakis M, Pombo A, Pradhan-Bhatt S, Qin XY, Rehli M, Rizzu P, Roy S, Sajantila A, Sakaguchi S, Sato H, Satoh H, Savvi S, Saxena A, Schmidl C, Schneider C, Schulze-Tanzil GG, Schwegmann A, Sheng G, Shin JW, Sugiyama D, Sugiyama T, Summers KM, Takahashi N, Takai J, Tanaka H, Tatsukawa H, Tomoiu A, Toyoda H, van de Wetering M, van den Berg LM, Verardo R, Vijayan D, Wells CA, Winteringham LN, Wolvetang E, Yamaguchi Y, Yamamoto M, Yanagi-Mizuochi C, Yoneda M, Yonekura Y, Zhang PG, Zucchelli S, Abugessaisa I, Arner E, Harshbarger J, Kondo A, Lassmann T, Lizio M, Sahin S, Sengstag T, Severin J, Shimoji H, Suzuki M, Suzuki H, Kawai J, Kondo N, Itoh M, Daub CO, Kasukawa T, Kawaji H, Carninci P, Forrest ARR, Hayashizaki Y (2017) FANTOM5 CAGE profiles of human and mouse samples. Sci Data 4:170112. https://doi.org/10.1038/sdata.2017.112

Peng R, Lin G, Li J (2016) Potential pitfalls of CRISPR/Cas9-mediated genome editing. FEBS J 283(7):1218–1231

Perez AR, Pritykin Y, Vidigal JA, Chhangawala S, Zamparo L, Leslie CS, Ventura A (2017) GuideScan software for improved single and paired CRISPR guide RNA design. Nat Biotechnol 35(4):347–349

Pliatsika V, Rigoutsos I (2015) Off-Spotter: very fast and exhaustive enumeration of genomic look alikes for designing CRISPR-Cas guide RNAs. Biol Direct 10:4

Prykhozhij SV, Rajan V, Gaston D, Berman JN (2015) CRISPR multitargeter: a web tool to find common and unique CRISPR single guide RNA targets in a set of similar sequences. PLoS One 10(3):e0119372

Qi LS, Larson MH, Gilbert LA, Doudna JA, Weissman JS, Arkin AP, Lim WA (2013) Repurposing CRISPR as an RNA-guided platform for sequence-specific control of gene expression. Cell 152 (5):1173–1183

Ran FA, Hsu PD, Lin CY, Gootenberg JS, Konermann S, Trevino AE, Scott DA, Inoue A, Matoba S, Zhang Y (2013) Double nicking by RNA-guided CRISPR Cas9 for enhanced genome editing specificity. Cell 154:1380–1389. PubMed: 23992846

Ren B, Liu L, Li S, Kuang Y, Wang J, Zhang D, Zhou H (2019) Cas9-NG greatly expands the targeting scope of the genome-editing toolkit by recognizing NG and other atypical PAMs in rice. Mol Plant 12(7):1015–1026

Sanjana NE, Shalem O, Zhang F (2014) Improved vectors and genome-wide libraries for CRISPR screening. Nat Methods 11(8):783

Sanson KR, Hanna RE, Hegde M, Donovan KF, Strand C, Sullender ME, Doench JG (2018) Optimized libraries for CRISPR-Cas9 genetic screens with multiple modalities. Nat Commun 9 (1):1–15

Shalem O, Sanjana NE, Hartenian E, Shi X, Scott DA, Mikkelsen TS, Zhang F (2014) Genome-scale CRISPR-Cas9 knockout screening in human cells. Science 343(6166):84–87

Shi J, Wang E, Milazzo JP, Wang Z, Kinney JB, Vakoc CR (2015) Discovery of cancer drug targets by CRISPR-Cas9 screening of protein domains. Nat Biotechnol 33(6):661–667

Stemmer M, Thumberger T, del Sol KM, Wittbrodt J, Mateo JL (2015) CCTop: an intuitive, flexible and reliable CRISPR/Cas9 target prediction tool. PLoS One 10(4):e0124633

Stemmer M, Thumberger T, del Sol Keyer M, Wittbrodt J, Mateo JL (2017) Correction: CCTop: an intuitive, flexible and reliable CRISPR/Cas9 target prediction tool. PLoS One 12(4):e0176619

Thomas M, Parry-Smith D, Iyer V (2019) Best practice for CRISPR design using current tools and resources. Methods 164–165:3

Tian P, Wang J, Shen X, Rey JF, Yuan Q, Yan Y (2017) Fundamental CRISPR-Cas9 tools and current applications in microbial systems. Synth Syst Biotechnol 2(3):219–225

Torres-Perez R, Garcia-Martin JA, Montoliu L, Oliveros JC, Pazos F (2019) WeReview: CRISPR tools—live repository of computational tools for assisting CRISPR/Cas experiments. Bioengineering 6(3):63

Uniyal AP, Mansotra K, Yadav SK, Kumar V (2019) An overview of designing and selection of sgRNAs for precise genome editing by the CRISPR-Cas9 system in plants. 3 Biotech 9(6):223

Wang T, Wei JJ, Sabatini DM, Lander ES (2014) Genetic screens in human cells using the CRISPR-Cas9 system. Science 343(6166):80–84

Wang T et al (2015) Identification and characterization of essential genes in the human genome. Science 350(6264):1096

Wang T, Guan C, Guo J, Liu B, Wu Y, Xie Z, Xing XH (2018) Pooled CRISPR interference screening enables genome-scale functional genomics study in bacteria with superior performance. Nat Commun 9(1):1–15

Wheeler EC, Vu AQ, Einstein JM, DiSalvo M, Ahmed N, Van Nostrand EL, Yeo GW (2020) Pooled CRISPR screens with imaging on microraft arrays reveals stress granule-regulatory factors. Nat Methods 17:636–642

Wilson LO, O'Brien AR, Bauer DC (2018) The current state and future of CRISPR-Cas9 gRNA design tools. Front Pharmacol 9:749

Wong N, Liu W, Wang X (2015) WU-CRISPR: characteristics of functional guide RNAs for the CRISPR/Cas9 system. Genome Biol 16(1):1–8

Wu X, Scott DA, Kriz AJ, Chiu AC, Hsu PD, Dadon DB, Chen S (2014a) Genome-wide binding of the CRISPR endonuclease Cas9 in mammalian cells. Nat Biotechnol 32(7):670–676

Wu X, Kriz AJ, Sharp PA (2014b) Target specificity of the CRISPR-Cas9 system. Quant Biol 2 (2):59–70

Wu Z, Zhao J, Qiu M, Mi Z, Meng M, Guo Y, Yuan Z (2018) CRISPR/Cas9 mediated GFP knock-in at the MAP1LC3B locus in 293FT cells is better for bona fide monitoring cellular autophagy. Biotechnol J 13(11):1700674

Xiao A, Cheng Z, Kong L, Zhu Z, Lin S, Gao G, Zhang B (2014) CasOT: a genome-wide Cas9/ gRNA off-target searching tool. Bioinformatics 30:1180

Xie S, Shen B, Zhang C, Huang X, Zhang Y (2014) sgRNAcas9: a software package for designing CRISPR sgRNA and evaluating potential off-target cleavage sites. PLoS One 9(6):e100448

Yamano T, Nishimasu H, Zetsche B, Hirano H, Slaymaker IM, Li Y, Ishitani R (2016) Crystal structure of Cpf1 in complex with guide RNA and target DNA. Cell 165(4):949–962
Yang LZ, Wang Y, Li SQ, Yao RW, Luan PF, Wu H, Chen LL (2019) Dynamic imaging of RNA in living cells by CRISPR-Cas13 systems. Mol Cell 76(6):981–997
Yennamalli RM, Kalra S, Srivastava PA, Garlapati VK (2017) Computational tools and resources for CRISPR/Cas 9 genome editing method. MOJ Proteom Bioinform 5(4):116
Zhang D, Hurst T, Duan D, Chen SJ (2019) Unified energetics analysis unravels SpCas9 cleavage activity for optimal gRNA design. Proc Natl Acad Sci 116(18):8693–8698
Zhu LJ, Holmes BR, Aronin N, Brodsky MH (2014) CRISPRseek: a bioconductor package to identify target-specific guide RNAs for CRISPR-Cas9 genome-editing systems. PLoS One 9(9): e108424

Delivery Methods for CRISPR/Cas Reagents 4

Sidra Ashraf, Nayla Munawar, Muhammad Kashif Zahoor, Amer Jamil, Masooma Hammad, Abdul Ghaffar, and Aftab Ahmad

Contents

S. Ashraf · A. Jamil
Department of Biochemistry, University of Agriculture, Faisalabad, Pakistan

N. Munawar
Department of Chemistry, College of Science, United Arab Emirates University, Al-Ain, UAE

M. K. Zahoor
Department of Zoology, Government College University Faisalabad, Faisalabad, Pakistan

M. Hammad
National Institute for Biotechnology and Genetic Engineering (NIBGE), Faisalabad, Pakistan

A. Ghaffar
Department of Biochemistry, Government College University, Faisalabad (GCUF), Faisalabad, Pakistan

A. Ahmad (✉)
Department of Biochemistry/US-Pakistan Center for Advanced Studies in Agriculture and Food Security, University of Agriculture, Faisalabad, Pakistan
e-mail: aftab.ahmad@uaf.edu.pk

A. Ahmad et al. (eds.), *The CRISPR/Cas Tool Kit for Genome Editing*,
https://doi.org/10.1007/978-981-16-6305-5_4

Abstract

The emergence of the clustered regularly interspaced short palindromic repeats (CRISPR)/CRISPR associated protein (CRISPR/Cas) and its reengineering into a potent genome editing system has revolutionized life sciences. It has brought much excitement and hope in medical and agricultural research for unprecedented control over the redesigning of genomes. Based on CRISPR, many genome engineering tools have been developed and extensively used for the identification of new genes and therapeutic targets, functional genomics, gene therapies, and the development of transgenic animals and plants. The successful applications of CRISPR/Cas depend on the safe and efficient transportation of CRISPR/Cas reagents into the cell nucleus. In this chapter we discuss the merits and demerits of different cargo reagents used for genome editing through CRISPR/Cas. In addition, we detail several delivery methods reported for CRISPR/Cas, including physical, viral, and non-viral delivery methods. We also highlight different emerging delivery methods not currently reported for delivery of CRISPR/Cas reagents. Finally, we discuss available delivering methods of CRISPR/Cas components for plant genome editing.

Keywords

CRISPR/Cas · Reagents of CRISPR/Cas · Delivery tools of CRISPR/Cas · Viral delivery methods · Non-viral methods · Physical methods · Emerging delivery tools

Abbreviations

AAV	Adeno-associated virus
AML	Acute myeloid leukemia
AuNPs	Gold nanoparticles
AV	Adenovirus
CARs	Coxsackie virus β-adenovirus receptors
CNTs	Carbon nanotubes
CPPs	Cell-penetrating peptides
DMD	Duchenne muscular dystrophy
EP	Electroporation
FAH	Fumarylacetoacetate hydrolase
GoF	Gain of function
HDI	Hydrodynamic injection
iTOP	Induced transduction by osmocytosis and propanebetaine
ITR	Inverted terminal repeat
IVT	In vitro transcription

LC-MSPs	Lipid-coated mesoporous silica particles
LNPs	Lipid nanoparticles
LVs	Lentiviral vector
MENDs	Multifunctional envelope-type nanodevices
MI	Microinjection
MOFs	Metal organic frameworks
MRSA	Methicillin-resistant *S. aureus*
MSNPs	Mesoporous silica nanoparticles
NLS	Nuclear localization signal
NSCLC	Non-small cell lung cancer
PEG	Polyethylene glycol
RNPs	Ribonucleoproteins
SiNPs	Silica nanoparticles
SLO	Streptolysin O
TALENs	Transcription activator-like effector nucleases
tracrRNA	Trans-activating CRISPR RNA
TRV	Tobacco rattle virus
VIGS	Virus-induced gene silencing
VSV-G	Vesicular stomatitis virus glycoprotein
ZFNs	Zinc finger nucleases

4.1 Introduction

The discovery of CRISPR/Cas as a functional part of the bacterial adaptive immune system (Liu et al. 2017) and its subsequent development into a powerful genome editing tool have significantly impacted biological research with numerous new developments in animal and plant science. CRISPR/Cas is a versatile tool for genome editing which, due to its simplicity and easy design, has been extensively used in the field of genome engineering (Ran et al. 2017). In contrast to zinc finger nucleases (ZFNs) and transcription activator-like effector nucleases (TALENs)—which use protein structures to recognize DNA followed by nuclease assembly—CRISPR/Cas uses a specific RNA-based recognition of DNA without requiring enzyme engineering. ZFNs and TALENs always require for genome editing a new protein to recognize each new DNA sequence; in CRISPR/Cas, the same protein may be used for all target sequences and only the guide RNA (gRNA) needs to be customized (Lowder et al. 2016). CRISPR/Cas system typically consists of two critical components: a Cas9 endonuclease and a gRNA—which is a fusion of crRNA and trans-activating CRISPR RNA (tracrRNA)—for targeting Cas endonuclease to the specific sequence in the genome. The gRNA can be replaced by artificially synthesized single-guide RNA (sgRNA). In addition, a donor template would only be required for the homology-directed repair (HDR) pathway. In the presence of a protospacer adjacent motif (PAM), usually the sequence 5′-NGG-3′ for Cas9, and complementary base pairing between gRNA and DNA, Cas9 can be directed to any sequence in the DNA to cause double-strand breaks (DSBs). DSBs are naturally repaired through cellular repair mechanisms such as non-homology end joining (NHEJ) or HDR. Without a repair template, the DSB is repaired through NHEJ, which may introduce small indels, resulting in gene knockout (KO). In the presence of donor template, the DSB is repaired through a HDR-based pathway, resulting in defined alterations. Moreover, using multiple gRNA, Cas9 may be targeted to multiple positions in the genome (Sedeek et al. 2019).

CRISPR/Cas systems have been classified into two main classes and six types. In the type I CRISPR/Cas system, multiple Cas proteins are required to form complex with crRNA to recognize and cleave the targeted DNA. In the type III CRISPR system, crRNA is incorporated into a multiple interference complex known as Cmr or Csm to recognize and cleave invasive RNA. In contrast, the type II CRISPR/Cas system requires a single protein to recognize and induce DSBs in the targeted DNA. So, due to its simplicity, specificity, and versatility, the type II system has been widely used for genome editing in plants and animals (Makarova and Koonin 2015; Shmakov et al. 2017). Meanwhile, reengineering of the CRISPR/Cas system, using dead Cas (dCas9), has led to its widespread applications beyond genome editing, including transcriptional modulation, genome imaging, epigenetic modifications, base editing, prime editing and rewriting genetic code. Recently, with the discovery of CRISPR/Cas13, RNA editing has become possible with CRISPR/Cas that is comparable with RNA interference (RNAi). CRISPR interference (CRISPRi) and CRISPR/Cas13 are more advantageous in transcriptional regulation (Adli 2018). With these merits and extensive developments, the CRISPR/Cas system has changed the pace of biological research in almost every field of life sciences. For example, CRISPR/Cas has been used to study the mechanisms of genetic diseases (Khan et al. 2016), development of animal models (Ma et al. 2014b), validation of disease targets (Lu et al. 2017), construction of transgenic plants and animals (Xing et al. 2014; Ma et al. 2017), transcriptional modulation (Gilbert et al. 2013), and the rewriting of epigenetic signatures (Black et al. 2016). This broad range of impacts has led to a rise in publications and patents based on CRISPR/Cas since its first report in 2013. Moreover, CRISPR crops have emerged in world markets, bypassing conventional regulations for transgenic crops (Scheben et al. 2017). Furthermore, CRISPR-based clinical trials in humans have been given a green light in China and the USA (Li et al. 2018).

Despite these merits, powerful applications, and rapid developments, there remain several practical and technical challenges to harnessing the full potential of the CRISPR/Cas system, especially in translational research such as therapeutic applications and transgene-free crops. First, CRISPR/Cas may bind and cleave non-specific sequences to induce off-targets. So, precise cleavage, along with efficient access to a target site, is required to reduce off-targets. Second, control over the DSB repair pathways, NHEJ and HDR, is also required to facilitate switching, according to experimental goals. Third, the low efficiency of the HDR pathway presents enormous challenges in several applications, especially donor-template-mediated site-specific insertions. For HDR-mediated applications, precise repair is desirable in treating diseases, while high efficiency in HDR may facilitate biomedical and agricultural research (Li et al. 2018). Although inhibiting NHEJ through gene silencing or chemicals may increase HDR efficiency, even higher efficiency is required for gene repair (Weber et al. 2015; Maruyama et al. 2015). These issues have encouraged scientists to develop more effective CRISPR/Cas systems. Finally, efficient delivery of CRISPR/Cas reagents into cells, tissues, and organs is a major ongoing challenge for precise genome editing and clinical applications of CRISPR/Cas. Development of novel delivery vehicles for effective delivery of CRISPR cargos is necessary for therapeutic applications and future development of CRISPR/Cas. Efficient delivery of CRISPR/Cas reagents to the target cells is crucial for effective genome editing, reduced off-targets, and improved safety at both the extracellular and intracellular levels (Li et al. 2018). For successful genome editing, RNA-guided Cas9 must be delivered into the nucleus, traversing both the plasma and nuclear membrane. In addition, for therapeutic applications of CRISPR/Cas, other hurdles must be overcome, such as clearance by the mononuclear phagocyte system, tissue specificity, immune responses, and protease or nuclear degradation (Eoh and Gu 2019). In light of these challenges, delivery systems that deliver plasmid DNA encoding Cas9 and sgRNA, in vitro synthesized Cas9 messenger RNA (mRNA) and sgRNA, or Cas9 protein/sgRNA have been used in the past. Current delivery methods include physical delivery methods (microinjection and electroporation), viral delivery methods (adenovirus (AV), adeno-associated virus (AAV), and lentivirus) and non-viral delivery methods (liposomes and nanoparticles) (He et al. 2017). Historically, viral delivery methods are the most effective and common for delivery of CRISPR reagents. Nucleic acids coding for CRISPR/Cas components are packed into integration-deficient lentivirus, AAV, or an AV and delivered into the target cells. Viral vectors are very effective due to high transfection, high efficiency, and diverse tropism of AAV serotypes (Lino et al. 2018). However, different challenges are also associated with viral delivery methods, such as limited insertion size, possible immune response, difficulty in large-scale production, and the risk of carcinogenesis. Non-viral vectors may offer an alternative approach to avoid these problems (Liu et al. 2017). However, they need significant improvement to reach the efficacy of viral vectors. Non-viral vectors offer advantages such as low immunogenicity, larger payload, and ease of large production. Challenges associated with non-viral methods include relatively low gene delivery efficiency and expression, protection of cargo from degradation, opsonization, and achieving delivery to specific cell targets and cellular compartments (Rui et al. 2019). Physical methods such as electroporation (EP) and microinjection (MI) cause a temporary perturbation in membrane, providing a short timeframe for improved delivery across membrane (Çiçek et al. 2019). In addition, MI may allow precise genome editing in a single cell. Although physical methods are good for in vitro applications, their use in vivo is limited. In this chapter we elaborate on these delivery methods for the CRISPR/Cas system. In addition, we discuss the merits and demerits of each method.

4.2 CRISPR/Cas reagents

In CRISPR/Cas system, Cas9 nuclease and sgRNA are the two critical components for site-specific cleavage of DNA. sgRNA is required for targeting of Cas9, while a functional Cas9 is necessary for cleavage of DNA at the site complementary to sgRNA. Both Cas9 and sgRNA can be provided as DNA, RNA, and protein (Çiçek et al. 2019). The choice of reagents affects the outcome of genome editing experiment, and permanent integration of plasmid DNA may result in off-targets. On the other hand, the large size of and positive charge on ribonucleoproteins (RNPs) is a potential problem in the efficient delivery of Cas9 protein (Glass et al. 2018). There are the advantages and disadvantages for each cargo system. Here we provide a detailed discussion of these cargoes.

4.2.1 DNA

Plasmid-based delivery of CRISPR components is the most straightforward strategy due to its simplicity, cost-effectiveness, and avoidance of transfection of different components in the same cells (Liu et al. 2017). Plasmids are the most common form of CRISPR reagent. Both Cas9 and sgRNA (multiple gRNA in multiplex) can be packed into a single plasmid. In addition, a donor template can also be provided in the same plasmid for HDR system. The use of plasmids is also advantageous in terms of stability and prolonged expression of Cas9 for continuous genome editing. Typically, plasmid-based delivery of Cas9 also needs a promoter to begin transcription and a nuclear localization signal (NLS). In plasmid based delivery, sgRNA can be cloned in the same plasmid or in a separate plasmid (Eoh and Gu 2019).

Using plasmids is not free of limitations, as there are challenges associated with the use of a plasmid as a cargo reagent for CRISPR/Cas. First, the large size of Cas9 (4.5 kilobase pairs (kb)) significantly increases the plasmid size and thus limits the delivery and expression of the CRISPR/Cas system. The size of the DNA from the two most commonly used Cas9 genes, derived from *Streptococcus pyogenes* (SpCas9) and *Staphylococcus aureus* (SaCas9), are 4.2 and 3.2 kb, respectively. Moreover, promoter, scaffold, sgRNA, and terminator sequences further increase the size of the plasmid. To address this problem, smaller Cas proteins such as Cpf1 or cloning Cas9 and sgRNA on separate vectors may be helpful.

Second, the plasmid must pass through the plasma membrane and the nuclear membrane into the nucleus for DNA transcription (Liu et al. 2017). Naked plasmids cannot cross the plasma membrane due to their negative charge; hence, positively charged nanoparticles are used to facilitate the delivery of plasmids and protect them from enzymatic degradation (Ibraheem et al. 2014). However, nanoparticles also face obstacles in delivering plasmid, such as aggregation, adsorption to serum proteins, clearance by immune system, premature cargo delivery, cellular uptake by the target cells, escape from the degradative endosomal compartments, and moving across the nuclear membrane (Rui et al. 2019). All these factors lower the efficiency of genome editing.

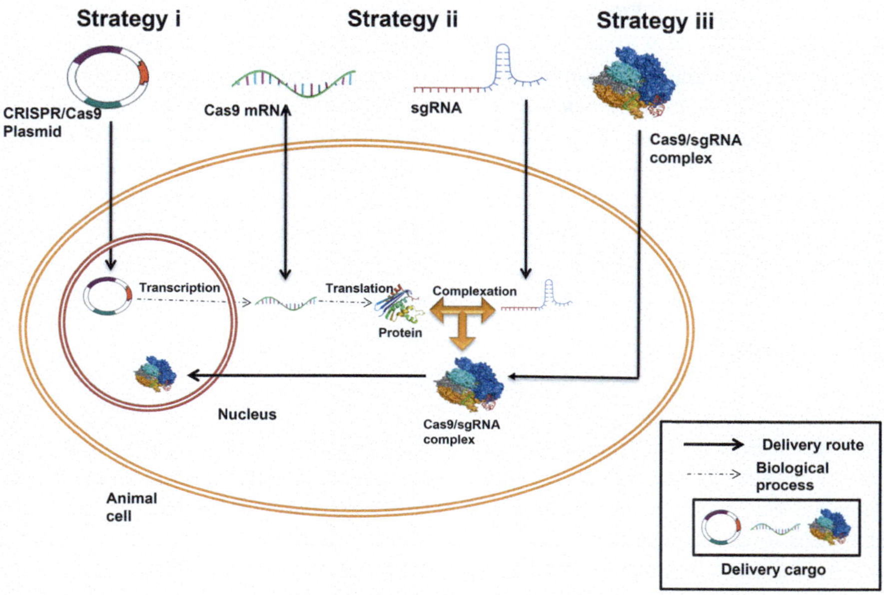

Fig. 4.1 Different forms of CRISPR/Cas reagents. Strategy (i): CRISPR/Cas reagents are delivered as DNA. Plasmids must reach the nucleus for transcription and subsequent translation in cytoplasm to produce Cas9. Cas9 must move back to the nucleus to bind and cleave DNA site specifically. Strategy (ii): Cas9 and sgRNA can be transcribed in vitro and delivered in the form of mRNA in cell cytoplasm. The mRNA will be translated, and protein will move back to the nucleus. Strategy (iii): Cas9 protein and in vitro transcribed RNA for sgRNA can be delivered as ribonucleoproteins (RNPs), subsequently moving to the nucleus for genome editing

In addition, to create a functional CRISPR/Cas9-gRNA complex, plasmid DNA must localize in the nucleus and transcribe as well. This may partially decrease genome editing efficiency and delay therapeutic efficacy. In this case, Cas9 faces an additional trafficking step as proteins are synthesized on cytosolic ribosomes, so Cas9 must move back to the nucleus to bind and cleave the DNA. Cas9 must also bind gRNA in the nucleus to form functional CRISPR/Cas complex (Fig. 4.1). The use of minicircle DNA offers an alternative to some of these challenges, as minicircle DNA is less immunogenic, more efficient on a per mass basis, their small size facilitates cytosolic and nuclear trafficking, and they are less prone to transcriptional silencing associated with plasmid DNA (Vaughan et al. 2006; Chen et al. 2003, 2017; Dad et al. 2014).

Third, delivery of CRISPR/Cas reagents as DNA carries the risk of permanent, unintended genomic integration inducing insertional mutagenesis due to the integration of highly active promoter in genome or the disruption of tumor suppressor genes (Yin et al. 2014a). Although the risk of insertional mutagenesis is quite low with non-viral vectors, this risk must be taken into account for translational therapies. In addition, permanent integration of plasmid DNA into the genome may result in continuous expression of Cas9 which can lead to off-target effects and strong immune response (Rui et al. 2019). Finally, plasmid-based cargo delivery is not feasible for immune cells which may sense the presence of foreign DNA. For example, it has been demonstrated that with plasmid-based delivery 50% editing efficiency was achieved in human embryonic kidney cells, while less than 4% editing efficiency was observed in CD4+ T cells. This was potentially due to the ability of T cells to sense the presence of foreign DNA, consequently leading to an innate immune response. Therefore, for T cells, alternative delivery methods may be helpful (Mandal et al. 2014; Monroe et al. 2014).

4.2.2 mRNA

Direct delivery of in vitro transcribed mRNA of Cas9 and sgRNA is an alternative approach with several advantages over plasmid DNA. For example, mRNA does not need to localize in the nucleus for transcription; consequently, it leads to quick and transient expression of Cas9 (Glass et al. 2018). The mRNA can be directly translated following cytosolic delivery, and Cas protein may be detected as quickly as 4 h post-transfection (Fig. 4.1). In addition, Cas9 delivery as mRNA reduces the risk of integration in the genome, thus limiting the duration of genome editing, compared with DNA (Eoh and Gu 2019). Moreover, mRNA-based CRISPR/Cas cargo also shows less cytotoxicity in primary cells and cell lines (Li et al. 2014). Expression of Cas9 for short durations also reduces the probability of off-targets and insertional mutagenesis. In in vitro applications using Cas9 mRNA as a cargo, protein expression of Cas9 was undetectable 72 h post-transfection. Similarly, the Cas9 expression was undetectable 24 h post-injection in in vivo applications (Jiang et al. 2017; Liang et al. 2015). However, short-term expression of Cas9 also reduces genome editing efficiency. The poor stability of mRNA is an additional challenge in use of this cargo system for CRISPR/Cas, as there are the different lengths and kinetics of expression of Cas9 and mRNA of sgRNA. It has been indicated that in mRNA-based delivery of CRISPR/Cas components, Cas9 mRNA must be delivered 24 h before sgRNA delivery (Miller et al. 2017).

4.2.3 Ribonucleoproteins (RNPs)

Delivering Cas9 along with mRNA of sgRNA offers a straightforward, quick, and direct approach for genome editing. Principally, it is the swiftest way to achieve genome editing without any requirement for transcription and translation of Cas9 (Fig. 4.1). Cas9 protein can efficiently make complex with sgRNA (RNP complex) and move across the nuclear membrane. Genome editing through RNPs avoids issues of poor stability and the risk of permanent integration (Eoh and Gu 2019). Use of RNPs as cargo offers several advantages over mRNA and plasmid DNA, including high efficiency in genome editing, rapid action with transient functionality, no need for promoter and codon optimization, and reduced off-target effects, toxicity, and immune responses (Liu et al. 2017). In addition, RNPs have shown better genome editing efficiency in hard-to-infect cells. Moreover, genome editing using RNPs does not use transcriptional targeting, thus reducing the ability of cargo to enable cell-type specificity. Because of these advantages, the RNP delivery method holds a great potential for translational applications and, consequently, has been extensively investigated for CRISPR/Cas studies involving HDR-mediated insertion applications (Eoh and Gu 2019). However, using RNPs as cargo also has limitations, including the large size of Cas9 (160 kDa) and the positive and negative charges on Cas9 and gRNA, respectively (Glass et al. 2018). Due to the nucleic-acid-binding nature of the Cas9 protein, unmodified SpCas9 holds a net positive charge of +20, while the addition of NLS signal makes it more positively charged. This net positive charge on Cas9 can be neutralized by including a glutamate tag of 20 amino acids through protein engineering. Addition of a glutamate tag has enabled direct cytosolic delivery of Cas9/gRNA complex when assembled with arginine gold nanoparticles. In addition, cost, purity, and bacterial endotoxin contamination also need to be carefully considered when using RNPs. An alternative strategy to increase the plasma membrane permeability of CRISPR/Cas is fusion of multiple viral SV40-NLS domains with Cas9. This method has been reported for genome editing without any addition of vector material both in vivo and in vitro. Moreover, conjugation of cell penetrating peptide (CPP) with Cas9 and gRNA can facilitate uptake and endosomal escape (Mout et al. 2017; Liu et al. 2015; Staahl et al. 2017; Ramakrishna et al. 2014). An additional concern with use of RNPs for CRISPR-mediated genome editing is that Cas9 protein variants from *S. pyogenes* and *S. aureus* have been recognized by antibodies of human patients, which may result in rapid clearance of these proteins upon systemic delivery.

4.3 Delivery Approaches in CRISPR/Cas

Rapid developments in CRISPR/Cas technology during the last decade have changed the landscape of biological sciences, with extensive applications in basic and translational research. However, rapid and efficient delivery of sgRNA and Cas9 in living cells is critical for the continued success of CRISPR/Cas systems, especially in therapeutic and translational applications (Glass et al. 2018). Once introduced in the body, the CRISPR reagent must reach the target organ/cell types, travel through the interstitial space to reach the target cell, and finally move across the plasma and nuclear membranes into the nucleus, avoiding the degradation, clearance, and protective systems of the body (Rui et al. 2019). There are several methods for delivery of CRISPR/Cas reagents, including viral, non-viral, and physical methods (Fig. 4.2). All these methods carry their own merits and demerits; a researcher has to select the most suitable method depending upon the nature of the experiment (Lino et al. 2018). There is no perfect delivery method suitable for all applications, but the following parameters may help in choosing the most appropriate delivery method for CRISPR/Cas reagents.

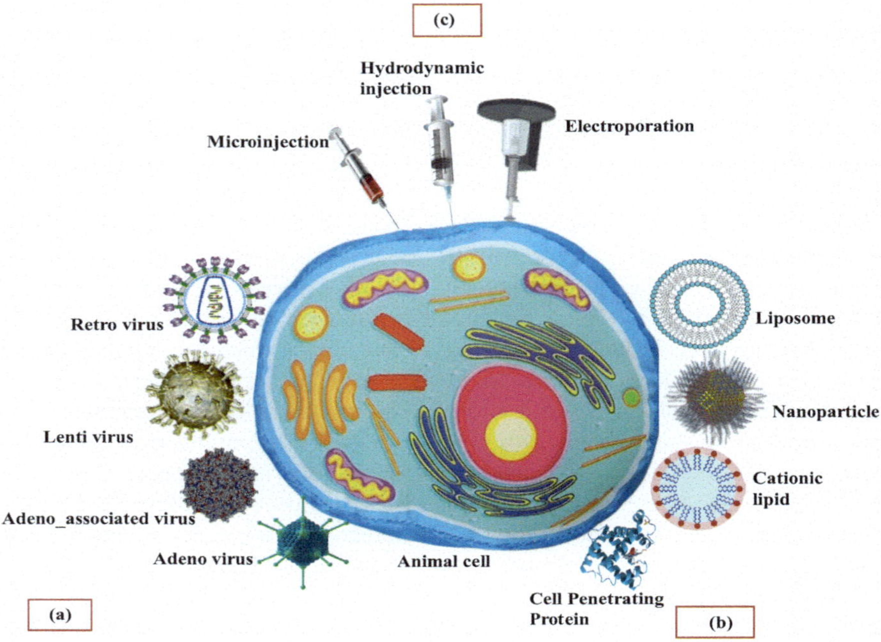

Fig. 4.2 Different delivery methods for CRISPR/Cas-mediated genome editing. (a) Viral delivery methods like adenoviruses, adeno-associated viruses, lentiviruses, retroviruses, and bacteriophages. (b) Non-viral delivery methods such as liposomes, nanoparticles, cationic lipids, and CPPs. (c) Physical delivery methods such as microinjection, hydrodynamic injection, and electroporation

- **In vitro or in vivo applications**. The requirement for in vivo applications such as gene therapy is extensive. For example, the carrier should have no or minimal immune response, and most vectors for in vivo applications should be tissue- and cell-specific (Lino et al. 2018).
- **Size of cargo**. As discussed, viral vectors can only carry a limited payload; for the large Cas9 (4.5 kb), along with sgRNA, viral vectors are not a wise choice. However, non-viral vectors may provide a solution for large cargoes (Glass et al. 2018).
- **Target cell type**. As discussed, RNPs are effective for genome editing in postmitotic cells, although having limited transcription and translational capacity may reduce efficiency of genome editing. The target cell type is also a key factor in choosing an appropriate delivery vector (Rui et al. 2019). For example, neurons, hepatocytes, and myocytes are postmitotic; for these cells, vectors capable of delivering cargoes to non-dividing cells are necessary.
- **Transient or permanent expression**. Transient or permanent expression of Cas9 in the target cell is an important factor in the selection of delivery vectors and the

outcome of an experiment. For example, many viral vectors can integrate the host genome, resulting in permanent expression of Cas9 in the target cell. However, permanent expression may lead to numerous off-targets as well. Vectors for stable and transient expression have their own merits and demerits.

Currently, several methods are available for delivering CRISPR cargo in vivo and in vitro, as shown in Table 4.1. Delivery vectors (viral and non-viral) and physical methods have been widely used for the efficient delivery of CRISPR/Cas. We describe these vectors in detail in the following sections.

4.3.1 Viral Delivery Methods

The ultimate objective in translational applications of CRISPR/Cas is to genetically correct cells in the human body and consequently cure genetic diseases. This requires a delivery system that can target cells, specifically with low cytotoxicity and rapid clearing of CRISPR components after successful genome editing. Over the last three decades, viral vectors have been used for gene delivery, especially in therapeutic and clinical applications. Viral vectors have a natural ability to penetrate the cells and deliver nucleic acid. However, there are safety concerns associated with viral vectors, such as introducing undesired mutations, but to date viral vectors have been used for efficient delivery of CRISPR/Cas components in mammalian cells, both in vivo and in vitro. For safety, the pathogenic parts of many viruses are removed and replaced with the therapeutic transgene of interest to be delivered in the cell (Maggio et al. 2014; Koike-Yusa et al. 2014; Shalem et al. 2014). Viral-mediated delivery is based on two mechanisms: infection and replication. In the infection stage, virus recognizes and enters the specific cells, thereby delivering viral genome into the cell nucleus for replication (In the case of DNA, it is delivered to the nucleus, while for RNA it is delivered to the cytoplasm). Once replication of the viral genome in the cell is completed, they reproduce viroid leaves the cell and infects neighboring cells to start replication again. A number of viral vectors have been developed and used for CRISPR-mediated genome editing (Chandrasekaran et al. 2018).

4.3.1.1 Adenoviruses (AdVs)

AdVs provide an alternative viral-mediated gene delivery method, applicable in both in vivo and in vitro applications (Ramos-Kuri et al. 2015). As an AdV is capable of infecting dividing and non-dividing cells, this indiscriminatory tropism may lead to transduction of non-target cells in living systems. Adenovirus is a non-envelope, double-stranded DNA virus with a genome ranging from 34 to 43 kb in size, flanked by two inverted terminal repeat (ITR) sequences (Xu et al. 2019). Adenoviruses enter the cell through highly expressed cell surface coxsackie virus β-adenovirus receptors (CARs) (Bergelson 1999), which make this virus capable of infecting many types of cells. AdVs can accommodate a DNA cassette up to 8–9 kb in size. In contrast to the other integrating viruses, the AdV genome remains as an

Table 4.1 Summary of delivery vehicles for CRISPR/Cas system and applications

Delivery	Delivery methods	Deliverable cargo	Advantages	Limitations	Applications	References
Viral vector	I. Adenovirus	Plasmid	• High packaging capacity • Highly efficient delivery in vivo	• Difficult to produce in large scale • Inflammatory response	Gene therapy for genetic diseases	Lino et al. (2018), He et al. (2017)
	II. Adeno-associated virus	Plasmid	• Broad cell tropism • Highly efficient delivery in vivo • Minimal Immunogenicity • Safe	• High cost • Low capacity • Difficulty in production • Limited packaging capacity	Gene therapy for genetic diseases	Lino et al. (2018), He et al. (2017), Liu et al. (2017)
	III. Lentivirus	Plasmid	• Highly efficient • Long-term gene expression • High throughput in vivo and in vitro • Large packaging size	• Transgene silencing • Insertional mutagenesis • Potential to rearrangement of cargo gene	Gene function and genomic screen study	Lino et al. (2018), He et al. (2017), Liu et al. (2017)
	IV. Retrovirus	Plasmid	• Highly efficient delivery in vivo	• Oncogene activation • Mutagenesis	Gene therapy for genetic diseases and cancer	He et al. (2017)
Non-viral vector	I. Polymer nanoparticle	RNPs	• Safe • Easy to prepare	• Low efficiency	Gene knockout, repair, and gene-targeted therapy	Liu et al. (2017)
	II. Lipid nanoparticle	Plasmid, mRNA, and protein	• Simple manipulation • Easy to prepare	• Low efficiency • Endosomal degradation of cargo	Gene knockout, repair, and gene-targeted therapy	Lino et al. (2018), Liu et al. (2017)

			• Low cost • Safe			
	III. Cell-penetrating peptides (CPP)	RNPs	• Small in size • Low off-target effects • Safe	• Low efficiency • Difficult for in vivo use • Chemical conjugation is needed • Immunogenicity	Genome editing of cells in vitro	Lino et al. (2018), He et al. (2017), Liu et al. (2017)
	IV. Gold nanoparticle	RNPs	• High delivery efficiency and membrane fusion like delivery • Safe • Inert	• Toxicity in vivo • Non-specific inflammatory response	Gene knockout, repair, and gene-targeted therapy	Lino et al. (2018), Liu et al. (2017)
	V. DNA nanoclew	Plasmid, RNPs	• Safe	• Assembly is complicated	Genome manipulation	Lino et al. (2018)
	VI. iTOP	RNPs	• Highly efficient delivery • Safe	• Not feasible for in vivo use • Non-specific	Gene knockout	Lino et al. (2018), Liu et al. (2017)
Physical method	I. Microinjection (MI)	Plasmid, mRNA, and ribonucleoproteins (RNPs)	• Highly specific • High efficiency in vitro • Reproducible	• Difficult and time-consuming • Low throughput • Induces cell damage	Generation of model animal, genome manipulation for embryo and oocytes	Lino et al. (2018), Liu et al. (2017), He et al. (2017), Chuang et al. (2017)
	II. Electroporation (EP)	Plasmid and mRNA	• Suitable for all cell • High transfection efficiency • Efficient delivery to cell population	• Cytotoxic	Genome editing of cells in vitro	Lino et al. (2018), He et al. (2017), Liu et al. (2017)

(continued)

Table 4.1 (continued)

Delivery	Delivery methods	Deliverable cargo	Advantages	Limitations	Applications	References
	III. Hydrodynamic injection (HDI)	Plasmid, mRNA, and RNPs	• Efficient method for in vivo gene editing • Simple and easy • Low cost	• Non-specific • Low efficiency • Not feasible for large animals • Difficult for clinical use and causes liver expansion, cardiac dysfunction, and tissue trauma	For study gene function in vivo	He et al. (2017), Liu et al. (2017), Lino et al. (2018)

extra-chromosomal episome inside the cell, thus providing transient expression of the gene. Transient expression reduces the risk of off-targets in CRISPR-mediated genome editing. Several efforts have been made to optimize AdVs as a gene delivery tool. In first-generation recombinant AdV vectors, viral E1 gene was removed; however, these vectors cause acute and chronic immune responses induced by the viral capsid and viral gene, respectively. In second-generation AdV vectors, E2 and E4 viral genes were removed to reduce chronic immune response. These vectors can accommodate a payload of around 8 kb (Xu et al. 2019). In the latest-generation AdV vectors, known as helper-dependent or "gutless" AdV vectors, all viral genes were removed. These vectors contain only ITRs and encapsulation, so their packing capacity is increased to 35 kb, making them ideal for packing CRISPR/Cas in one vector. Moreover, these vectors do not cause chronic immune responses. However, viral capsid may still cause an acute immune response. AdV-mediated delivery of CRISPR/Cas components has been used in various applications such as developing disease models, treatment of disease, and drug discovery. For example, Maddalo et al. (2014) established a model of non-small lung cancer with an Eml4 and ALK fusion genes through intratracheal instillation of AdV-mediated CRISPR/Cas9 (Maddalo et al. 2014). In addition, using AdV-mediated delivery of SpCas9, Wang et al. (2015) established a mouse model mimicking non-alcoholic steatohepatitis (NASH), by targeting the PTEN gene (Wang et al. 2015). Voets et al. (2017) used AdV-mediated CRISPR/Cas system to inactivate the SMAD3 gene in normal human lung fibroblast and bronchial epithelial cells to establish the potential of CRISPR/Cas for drug discovery and treatment of existing diseases. Ding et al. (2014) induced loss of function mutation in the PCSK9 gene in mouse lines to reduce the plasma cholesterol levels. Moreover, using AdV-mediated CRISPR/Cas, Li et al. (2015) produced HIV-resistant primary CD4+ T cells by adding cell membrane CCR5 d32 variants.

4.3.1.2 Adeno-Associated Viruses (AAVs)

AAVs are one of the most popular viral vectors used for gene therapy and CRISPR/Cas-mediated genome editing. AAVs of the *Parvoviridae* family (Lino et al. 2018) contain a small single-strand DNA (ssDNA) genome of 4.7 kb in size. There are more than 200 naturally occurring AAVs, as well as genetically engineered variants (Liu et al. 2017). AAVs are able to infect dividing and non-dividing cells and enter the cell through heparin sulfate proteoglycan and integrans (Young et al. 2006). After transduction, the DNA of naturally occurring AAVs can integrate into hotspots in mitochondrial DNA and/or to a site-specific location on chromosome number 19, through action of rep protein (Young et al. 2000). Both integration sites are considered safe and do not cause tumorigenesis. In contrast to the naturally occurring AAVs, genetically engineered AAVs lack rep protein, so they cannot integrate into genome but exist as extra-chromosomal DNA. AAVs are excellent delivery vectors in CRISPR/Cas-mediated genome editing for several reasons. AAVs are not known to cause disease in humans. Hence, due to a favorable safety profile and their therapeutic potential, they have been approved for gene therapy and clinical trials in humans (Lau and Suh 2017). AAVs can efficiently infect cells with little or no innate or adaptive immune response. There is a broad range of known serotypes, capable of infecting dividing and non-dividing cells with different specificities. In addition, permanent integration and long-term existence in cell may be used for prolonged expression of transgene; however, a high rate of off-targets is a disadvantage. The challenge of AAV-mediated delivery for CRISPR/Cas genome editing is the small genome size of the virus, which limits its packing capacity. Cas9 and sgRNA can be packed onto a one plasmid and delivered with a single AAV particle. The size of a SpCas9 and gRNA package is roughly 4.2 kb, while the overall size of AAV (~20 nm in diameter) only permits a package of about 4.5 kb of genomic DNA (Wu et al. 2010), so consistent packing of SpCas9 and gRNA in a single virus particle is challenging. This limitation makes it difficult to include promoter elements, fluorescent tags, multiple gRNA, or a donor template in a single AAV (Lino et al. 2018). Long et al. (2016) used mini-cytomegalovirus promoter/enhancer with derived SpCas9 to correct the Duchenne muscular dystrophy (DMD) mutation in mice, which resulted in enhanced muscle function (Long et al. 2016). The packing limit can be addressed by using truncated SpCas9 or a smaller Cas variant such as SaCas9 or Cas12, which have similar genome editing efficiency as Cas9. However, truncated SpCas9 shows reduced activity. Ran et al. (2015) used SaCas9 and gRNA in a single AAV to target the PCSK9 gene in mice which resulted in remarkable decrease in PCSK9 and total cholesterol level in serum. SaCas9 was also packed with multiple gRNAs into an AAV and showed 60% genome editing efficiency.

An alternative approach for addressing the packing limitation is to deliver gRNA using AAVs in cells already expressing Cas9. Carroll et al. (2016) used AAV to deliver sgRNAs in the cardiomyocytes of mouse embryo, expressing Cas9 to produce a cardiovascular research model. Similarly, Platt et al. (2014) used a similar approach to induce loss-of-function mutation in tumor-suppressing genes and gain-of-function (GoF) mutation in protooncogenes in Cas9-expressing mice. In another approach, a split Cas9 system was used in which the C- and N-terminals of Cas9 were packed in separate AAV vectors. Reconstitution of the halves results in functional Cas9 with similar efficiency to native Cas9. Alternatively, many groups have used a dual AAV system in which Cas9 and sgRNA are packed in separate AAV particles and co-infected them to overcome the size limitation. A dual AAV system was used to disrupt a single gene (MECP2) or multiple genes (Dnmt1, Dnmt3a, and Dnmt3b) in mouse brain (Swiech et al. 2015). Similarly, a dual AAV system was used for therapy of metabolic liver disease in a mouse model. However, such dual systems are more complex than single systems as the delivery of both AAVs into one target cell is challenging.

4.3.1.3 Lentiviral Vectors (LVs)

LVs represent a subclass of retroviruses commonly used as delivery vectors for CRISPR applications because of benefits such as high infectivity, low immunogenicity, and long-term expression. LVs are single-stranded RNA (ssRNA) viruses capable of infecting both proliferating and quiescent cells with a packing capacity of 9 kb. LVs infect non-dividing cells efficiently, which is crucial for genome editing in the liver, brain, and muscles (Liu et al. 2017). Following infection, LV can integrate non-specifically into the host genome, resulting in stable expression of transgene. Stable integration may be helpful in gene augmentation therapies; however, in CRISPR-mediated applications it may increase off-targets. Non-integrating LVs have also been engineered by inducing mutation in the integrase coding region, but most researchers do not have the capacity to generate integrase-deficient LVs (Philippe et al. 2006; Yánez-Munoz et al. 2006; Apolonia et al. 2007). This is one reason why LVs are used less often than AdVs and AAVs. LVs can also be pseudo-typed with different envelope proteins in viral production, thus allowing engineering and alteration of the LVs cellular tropism (Xu et al. 2019). Vesicular stomatitis virus glycoprotein (VSV-G) is the most commonly used envelope protein in recombinant LVs. Use of VSV-G has increased the host cell range as it interacts with the phospholipid component of a number of receptors in cell membrane.

Generally, two plasmids are required to generate LVs, a packing plasmid and the transgene carrier plasmid. Packing plasmid contains gene-encoding structural genes and enzymes required for packing, while transgene-carrier plasmid contains a genome-editing cassette such as Cas9, gRNA, or donor template. However, third-generation LVs split essential gene into three plasmids, thus minimizing the chances of producing a viral particle within the cell. To date, LVs have been successfully used for advancing CRISPR/Cas applications in gene therapy, generating animal models, function-based screening, and eradication of viral infections. For example, Chen et al. (2015) used a CRISPR/Cas-based lentiviral sgRNA library to screen loss of mutation in non-small cell lung cancer (NSCLC) cell lines in mouse and identified genes involved in metastasis and tumor growth. In addition, LV vectors have been commonly used in the delivery of CRISPR/Cas to genetic animal models of cancer. Hecklet al. (2014) used lentiviral-mediated CRISPR/Cas to produce a model of acute myeloid leukemia (AML) by mutating five genes in mouse hematopoietic cells. Similarly, Blasco et al. (2014) used lentiviral-mediated delivery of CRISPR/Cas to generate a cancer model in mouse by inducing chromosomal rearrangement of ALK and EML4. Lentiviral-based delivery has also been used to eradicate viral infection (Blasco et al. 2014). For example, lentiviral-mediated CRISPR/Cas delivery was used to prevent reemergence of HIV by removing HIV DNA from CD4+ T cells in host patients (Kaminski et al. 2016). In addition, lentiviral vectors were used to deliver CRISPR/Cas for removal of latent Epstein-Barr virus in Burkitt's lymphoma cells of patients. Lentiviral vectors have also been used for gene therapy. For example, lentiviral-mediated CRISPR/Cas was used to inhibit hepatitis B virus (HBV) replications in chronic HBV patients.

4.3.1.4 Bacteriophages

Antibiotic resistance is one of the world's major health challenges, as more than 700,000 people die globally each year as a result of antibiotic-resistant infection. Bacteriophages represent a group of viruses that infect bacteria and archaea. There are many families of bacteriophages that vary in structure and target organisms. Bacteriophages hold great potential for combating multidrug-resistant bacteria. However, a major limitation in the development of bacteriophage therapy is specific susceptivity of bacteria. A combination of CRISPR/Cas with bacteriophage has drastically changed the concept of bacteriophage therapy. Bacteriophages have been engineered to deliver a CRISPR/Cas system to target virulence and resistant genes, thus modulating a complex bacterial population (Xu et al. 2019). Bikard et al. (2014) used bacteriophages loaded with a CRISPR/Cas system in a mouse model to specifically target virulent *S. aureus*, while sparing and immunizing non-virulent *S. aureus*. Yosef et al. (2017) used lytic phages in combination with a CRISPR/Cas system to target multidrug-resistant bacteria. In addition, they developed a platform for customizable bacteriophages able to transfer DNA to hosts that were previously difficult to infect (Yosef et al. 2015, 2017).

4.3.2 Non-viral Delivery Methods

Genome editing through CRISPR/Cas using Cas9, sgRNA, and/or donor template requires a vector system that efficiently delivers these components to the target cells. Along with viral vectors, several non-viral vectors have been reported for successful and safe delivery of CRISPR/Cas components to cells. The most widely used non-viral vectors are CPPs, cationic nanocarriers/polymers, and lipid nanoparticles or liposomes. Compared with viral vectors, non-viral vectors are preferred due to easy generation, higher payload capacity, and no immune responses. However non-viral vectors are less efficient than viral vectors in CRISPR/Cas delivery. Therefore, only a limited number of non-viral vectors could act in clinical applications (Rui et al. 2019). In the following section, we summarize the non-viral vectors used for CRISPR/Cas delivery.

4.3.2.1 Polymeric Materials

Polymeric materials such as cationic polymer, CRISPR-Gold, nanoclew coated with poly ethylenimine(PEI), and zinc/imidazole-based metal-organic frameworks (MOFs) have been extensively used for delivery of CRISPR/Cas systems (Chen et al. 2018; Cheng et al. 2018). Polymeric materials have been used to deliver different forms of CRISPR components, including DNA, mRNA, or oligonucleotide. For delivery of a CRISPR/Cas system, polymers are often used in a multicomponent system to enable endosomal escape (Rui et al. 2019). Cationic polymers are suitable for efficient delivery of negatively charged nucleic acid such as gRNA and Cas9. The most commonly used cationic polymer for delivery of CRISPR/Cas components is PEI. In PEI-mediated delivery, molecular weight, number of branched structures, and structural characteristics represent the critical factors that control transfection efficiency and toxicity to the cells (Eoh and Gu 2019). Zuckermann et al. (2015) used PEI-mediated delivery of Cas9 and gRNA encoding plasmid into mouse brain to create a brain tumor model (Zuckermann et al. 2015). PEI was also used to deliver CRISPR/Cas plasmid in HBV-infected mice to inhibit replication of HBV. The cationic polymer bPEI was covalently linked with Cas9 and further coupled with sgRNA to create a CRISPR nanoparticle. This bPEI-conjugated CRISPR/Cas was delivered to methicillin-resistant *S. aureus* (MRSA) with successful editing of the targeted genome. The system showed higher efficiency of genome editing compared with Cas9/gRNA complexed with conventional lipid (Kang et al. 2017; Liu et al. 2017). PEI-mediated delivery of CRISPR/Cas was also successful in vivo, for disruption of tumor suppressive genes such as PTEN, Trp53, and Nf1 in mouse brain. In addition, CRISPR-Gold-mediated delivery, in which the endosomal-disruptive polymer poly(*N*-(*N*-(2-aminoethyl)-2-aminoethyl)aspartamide) was used to coat Cas9-gRNA RNP on gold nanoparticle to deliver CRISPR/Cas components. The CRISPR-Gold was used in vivo for correction of the dystrophin gene in a mouse model for DMD (Lee et al. 2017). CRISPR-Gold-mediated delivery of Cas9 was also used for genome editing in rodent brain using Cas9 and Cpf1 with 50% efficiency (Lee et al. 2018). PEI-coated DNA nanoclews have also been successfully used for in vitro delivery of CRISPR reagents. DNA nanoclew delivery is a unique technology in which a ball-like structure of DNA is synthesized using a rolling circle mode of amplification with palindromic sequences to facilitate self-assembly. Sun et al. (2015) used nanoclews for delivery of Cas9/sgRNA RNPs, demonstrating that the efficiency of genome editing with PEI-coated nanoclews was much higher than for bare Cas9/gRNA with PEI. PEI provides a positive charge to facilitate cellular uptake and help in endosomal escape. Cationic helical polypeptides have also been reported for delivery of CRISPR/Cas plasmids with enhanced genome editing efficiency both in vivo and in vitro (Rui et al. 2019). All these cationic polymers represent a promising approach for local delivery of CRISPR/Cas reagents; however, for systemic delivery, the cationic nature of the particle may present challenges. For systemic delivery of CRISPR reagents, use of the hydrophilic molecule, polyethylene glycol (PEG) (the process is known as "PEGylation") is a common strategy to increase circulation time and reduce fouling of nanoparticle surface (Suk et al. 2016). Zwitterion materials represent an alternative and effective approach for systemic delivery of nucleic acid and have been useful for delivery of silencing RNA (siRNA). For example, the zwitterion materials, cationic quaternary ammonium sulfonamide amino lipids and phosphorylcholine-polycation diblock coplymers, showed improved systemic delivery of siRNA. However, applications of zwitterion in the delivery of RNPs are yet to be explored (Miller et al. 2018; Jackson et al. 2017). For RNP delivery, cationic polymer platforms showed more flexibility for accommodating protein molecules of various charges. Chang et al. (2017) used a system composed of dendrimer end capped with a guanidyl group for the delivery of RNPs. The guanidyl group facilitates protein binding based on hydrogen bonding, salt bridges, and phenyl groups which promote endocytosis and endosomal escape (Chang et al. 2017). The system successfully encapsulated proteins of various sizes, as well as charges. Yan et al. (2010) used an alternative approach in which they synthesized a thin polymer shell around each protein and the polymer shell was covalently linked with protein core. This system enabled efficient cellular uptake of proteins both in vivo and in vitro. In addition, polymer shells protected the protein from protease degradation and retained its activity after cellular uptake (Yan et al. 2010). Hence, this approach could be potentially used for efficient delivery of CRISPR/Cas RNPs. MOFs have been reported for intracellular delivery of proteins. Alsaiari et al. (2018) used a MOF for delivery of CRISPR RNPs; however, the genome editing efficiency was low, necessitating further optimization.

These approaches could be adopted as potential alternative methods for delivery of CRISPR RNPs.

4.3.2.2 Lipid Nanoparticles

Lipid nanoparticles have been extensively used for safe delivery of nucleic acids to the cells. Nucleic acids cannot effectively cross the cell membrane due to their negative charge and hydrophilic in nature. However, cationic lipids can mask the negative charge of nucleic acids through their electrostatic interaction to produce and facilitate delivery into the cells (Liu et al. 2017). Lipid nanoparticles are preferred for several reasons, such as safety, low immune response, and ease of preparation. However, there are certain drawbacks in using LNPs as delivery vectors for CRISPR/Cas. For example, CRISPR/Cas-carrying LNPs passing through the cell are encased in endosome, leading to lysosomal degradation. Further, after endosomal escape, LNPs must cross the nuclear membrane to deliver CRISPR/Cas9 into the cell nucleus—this stage represents a potential failure point which would reduce the efficiency of genome editing (Lino et al. 2018).

LNPs have been used successfully for delivering the CRISPR/Cas system (nucleic acid or RNPs) to different cells (Cong et al. 2013; Mali et al. 2013) for therapeutic purposes or for generating KO animal models (Raghavan et al. 2016; Platt et al. 2014). Lipid nanoparticles are used to deliver RNAi plasmids are also directly applicable for the delivery of CRISPR/Cas plasmids and Cas9: mRNA/gRNA complex to different cell lines for genome editing. Lipofectamine 2000 and 3000 and RNAiMAX are the most common commercially available transfection lipids used as delivery vectors in CRISPR-mediated genome editing. Compared with Lipofectamine, RNAiMAX shows better efficiency to deliver RNPs and low toxicity in the cells (Liang et al. 2015). Horii et al. (2013) used Lipofectamine to deliver CRISPR/Cas plasmid to human pluripotent stem cells to generate model for immunodeficiency-centromeric instability-facial anomalies (ICF) syndrome. In addition, Lipofectamine was used to deliver CRISPR/Cas9 DNA into intestinal stem cells of cystic fibrosis patients for correction of cystic fibrosis transmembrane conductor receptor locus (Schwank and Clevers 2016). Similarly, Liu et al. used Lipofectamine-mediated delivery of CRISPR/Cas9 to inhibits cancer cell growth (Liu et al. 2014). Although commercial lipid can also be directly used to deliver CRISPR/Cas RNPs in vitro, generally they need modification because Cas9 protein is positively charged. To demonstrate the potential of common cationic lipid to deliver CRISPR/RNPs, Zuris et al. (2015) fused negatively charged green flourecent protein (GFP) with Cas9. Delivery of this GFP-fused Cas9 with Lipofectamine showed 80% genome editing efficiency in cultured human cells. In addition, bio-reducible lipids have been used to deliver modified RNPs into mouse brain for genome editing. Bio-reducible lipids facilitate the release of RNPs in cytosol by promoting endosomal release and degradation of nanoparticles (Wang et al. 2016). Therefore, bio-reducible lipids could be used to deliver RNPs in cultured cells in in vivo applications as well.

4.3.2.3 Cell-Penetrating Peptides (CPPs)

CPPs represent a short stretch of amino acids (5–30) with strong translocating capabilities across the cell membrane. They can cross the membrane in an energy-dependent or energy-independent manner. Based on the nature of their constituent amino acids, CCPs are categorized as polycationic, amphipathic, or non-polar. Each class of CPP promotes uptake of different types of proteins in different types of cells. CPPs can be used for both in vivo and in vitro applications; however, extensive optimization is required for each cargo and cell type. Due to their multiple advantages, CPPs are used for delivering therapeutic against multiple diseases (Lino et al. 2018). CCPs can be conjugated with different CRISPR cargos and delivered in a wide range of cells. For example, sgRNA can be complexed with CPP through electrostatic interaction, while Cas9 protein can be conjugated with a non-arginine CPP to facilitate uptake of CRISPR reagents. CPPs have also been used successfully for the delivery of CRISPR/Cas systems in genome editing studies. For example, Ramakrishna et al. (2014) reported separate delivery of CPP-Cas9 and CPP-gRNA in multiple human cell lines. In addition, Axford et al. (2017) showed cellular and subcellular localization of CRISPR RNPs delivered using CPPs (Axford et al. 2017). However, in one study, when a mixture of Cas9/CPP conjugate and sgRNA was delivered, it showed no genome editing effect because gRNA neutralized the positive charge of CPP and consequently reduced its cellular penetration. This suggest that both Cas9 and gRNA should be conjugated with CPP to achieve genome modification (Liu et al. 2017). This indicates that penetration efficiency of CPP across plasma membrane varies with cargo and cell type. Moreover, once inside the cell, the challenge of translocating Cas9:gRNA DNA complex into the nucleus must also be considered (Lino et al. 2018).

4.3.2.4 Gold Nanoparticles (AuNPs)

AuNPs provide a novel delivery platform for RNP-based CRISPR genome editing. Gold nanoparticles can complex with engineered Cas9 (Cas9 with a glutamate peptide tag) and gRNA into nanoparticles, which enters the target cell through a cholesterol-dependent membrane fusion process. Gold nanoparticles show remarkable delivery efficiency (greater than 90%) with 30% genome editing efficiency in a wide range of cells (Mout et al. 2017). The method offers a novel delivery approach for transient genome editing in vitro. However, its potential for genome editing in human primary cells is not yet explored (Liu et al. 2017). Mout et al. (2017) used gold nanoparticles complexed with engineered Cas9: sgRNA RNPs and observed 30% genome editing efficiency in target cells. Lee et al. (2017) used gold nanoparticles to deliver Cas9: sgRNA and ss donor DNA to correct/recover the mutated dystrophin gene in mice suffering from DMD. A single injection of AuNP-Cas9 complex recovered 5.4% of the mutated gene, and treated mice showed reduced fibrosis, as well as potential recovery of muscle function. Although AuNPs are inert and may not trigger immune response, Lee et al. (2017) have shown that AuNPs stimulate immune response (Lee et al. 2017). However, these findings need confirmations. AuNPs represent a promising delivery tool in CRISPR/Cas genome editing for future applications.

4.3.3 Physical Methods

Several physical methods such as microinjection (MI), hydrodynamic injection (HDI), and EP have been used to deliver CRISPR/Cas cargoes to the target cells (Lino et al. 2018). Physical methods use physical force to disrupt cell membrane, consequently enabling CRISPR/Cas reagents to enter cells (Glass et al. 2018). Although, these methods are straightforward and easy to use, special equipment is required. Physical methods are promising tools for delivery of CRISPR/Cas reagents, especially in gene therapy due to advantages such as reproducibility, specificity, and simplicity (Chandrasekaran et al. 2018). Physical methods can be applicable to different cell lines both in vivo and in vitro. Here we discuss commonly used physical methods for CRISPR/Cas-based genome editing.

4.3.3.1 Microinjection (MI)

MI involves direct delivery of CRISPR/Cas reagents into the cell by piercing the cell membrane using a specialized glass micropipette and a microscope. MI is a simple procedure for delivering CRISPR reagents such as nucleic acids (plasmid coding Cas9 and sgRNA or mRNA encoding sgRNA and Cas9) or RNPs (Cas9 protein with gRNA) into single cells with 100% efficiency (Liu et al. 2017). Compared with the viral delivery method, cargo size is not a limitation in delivery through MI. MI can minimize off-targets by delivery of known qualities of the cargo. MI is an excellent method for delivering CRISPR cargoes in vitro and ex vivo; however, use of a microscope to deliver cargoes in target individual cells makes MI applications difficult in a true in vivo setting. In addition, MI does not involve the barriers associated with delivery through cell membrane, the extracellular matrix, and cytoplasmic components (Lino et al. 2018). MI is commonly used for delivery of nucleic acid. CRISPR/Cas DNA or mRNA can be injected with MI in one of the three forms:

- Direct injection of DNA into the nucleus.
- Direct delivery of in vitro transcribed mRNA into the cell nucleus.
- Delivery of in vivo transcribed mRNA into cytoplasm (Lino et al. 2018).

These delivery forms have their merits and demerits. By delivering Cas9 and sgRNA as DNA into the cell nucleus, the cell itself can transcribe and translate both components. This approach may be preferable to avoid lengthy reactions of in vitro transcription (IVT) (Chuang et al. 2017; Nakagawa et al. 2015). However, delivery of circular plasmids or ssDNA can result in permanent and random integration of DNA into the genome. Permanent and random integration may result in constitutive expression with a higher off-targets and also disruption of genes. For delivering CRISPR/Cas mRNA, the ideal case is to deliver Cas9 mRNA into cytoplasm and sgRNA into nucleus. However, making two microinjections in different compartments of a single cell is a technically challenging and laborious process (Yang et al. 2013). Therefore, Cas9 mRNA and sgRNA are delivered into the cytoplasm, whereupon Cas9 mRNA translates into protein, binds with sgRNA, and finally moves into the nucleus for genome modification. Delivery of CRISPR/

Cas9 mRNA through MI often results in transient genome editing, due to the short lifespan of mRNA in eukaryotic cells. Transient action of CRISPR/Cas is desirable to reduced off-targets (Lino et al. 2018). MI-mediated delivery of CRISPR/Cas components has been used in a large number of studies, such as KO of multiple genes with a single injection into rat zygote (Ma et al. 2014a), restoring a cataract-causing mutation in mice (Wu et al. 2013), disruption of genes in cynomolgus monkeys (Niu et al. 2014), and correction of a DMD-causing mutation in mice (Long et al. 2014). MI is commonly used for efficient germline editing by injecting a CRISPR cargo into the zygote (Horii et al. 2014). In addition, injecting Cas9 mRNA and sgRNA into zygote cytoplasm generates normal embryos and mouse pups with desired genome modifications. MI is also useful for generating animal models, as well as transient CRISPRi and CRISPR activation (CRISPRa) regulation of gene (Lino et al. 2018).

4.3.3.2 Electroporation (EP)

EP is a commonly used method for delivering nucleic acids and proteins into mammalian cells. During EP, a high-voltage electrical current transiently increases permeability of the cell membrane, thus allowing nucleic acid and protein flow into the cells. EP is not limited to cell types and can deliver all forms of reagents (DNA, mRNA, or RNPs) for in vitro and ex vivo applications (Liu et al. 2017). However, EP is not useful in vivo because a high voltage is needed to create pores across the cell membrane. In contrast to bacterial cells, mammalian cells are sensitive to the voltage and current time applied in EP. Several researchers have used standard EP equipment, but some have developed technical solutions to increase the efficiency of EP in delivering CRISPR reagents. For example, Hashimoto and Takemoto (2015) achieved a high efficiency of CRISPR delivery with a custom EP chamber. Qin et al. (2015) used standard EP equipment to deliver CRISPR reagents in zygote with high efficiency. EP was used to deliver plasmid DNA encoding Cas9 and sgRNA to generate a colorectal cancer model in human intestinal organoids (Matano et al. 2015) and an Alzheimer's model in human cells (Paquet et al. 2016). In addition, EP was used to deliver plasmid encoding CRISPR components to correct mutations causing DMD (Ousterout et al. 2015). "Nucleofection" is a specialized procedure to deliver cargoes directly into the nuclei of mammalian cells without breaking the nuclear membrane; the procedure has also been used for CRISPR-mediated genome editing. Nucleofection has been used to deliver plasmid encoding CRISPR/Cas for various applications such as correction of a cataract-causing mutation in mouse stem cells (Wu et al. 2015), engineering resistance to HIV infection (Ye et al. 2014), generating a lung cancer model (Choi and Meyerson 2014), and conferring resistance to herpes virus infection (Wang and Quake 2014). EP has also been used to deliver CRISPR/Cas9 cargoes for editing genes in vivo (Zuckermann et al. 2015). In the delivery of CRISPR RNPs to primary human cells (Kim et al. 2014; Schumann et al. 2015), EP showed, compared with plasmid transfection of RNPs, reduced off-targets and less stress on cells.

4.3.3.3 Hydrodynamic Injection (HDI)

HDI is an efficient and simple procedure to deliver nucleic acids, mainly to the liver but also to the kidney, muscle, heart, and lungs. It involves rapid administration of a large volume (8–10% of body weight) of nucleic acid in solution into the bloodstream of mice using the tail vein (Al-Dosari et al. 2005). The rapid injection of a large volume of cargo in solution develops a hydrodynamic pressure which temporarily increases permeability of endothelial cell, consequently forcing the cargo to move across the membrane into the cells. HDI has been utilized to deliver DNA, protein, siRNA, and even cancer cells. HDI-based delivery of the CRISPR/Cas system has been primarily used for genome editing applications (Liu et al. 2017). For example, Yin et al. (2014b) used HDI delivery of CRISPR plasmids for in vivo correction of fumarylacetoacetate hydrolase (FAH) mutation in mouse hepatocytes in a model of hereditary tyrosinemia (Yin et al. 2014b). In addition, Guan et al. (2016) used HDI-mediated delivery of CRISPR/Cas plasmid to restore homeostasis in edited mice. HDI of plasmids was also used for CRISPR-mediated genome editing such as mutation in tumor suppression genes to induce liver tumor (Xue et al. 2014) and editing the HBV genome in HBV-infected mice (Dong et al. 2015) to demonstrate the therapeutic potential of CRISPR/Cas. Despite its simplicity and success, HDI is not considered safe for therapeutic and clinical applications. HDI-mediated delivery may cause trauma, cardiac dysfunction, increased blood pressure, potential physiological complications, and even death (Suda et al. 2007; Bonamassa et al. 2011).

4.3.3.4 Gesicles

Micro-vesicles are the heterogeneous group of extracellular membranous organelles involved in cellular function such as cell communication, proliferation, and immunity. Gesicles are produced from human cells by spiking them with glycoprotein of vesicular stomatitis virus (VSV-G). Gesicles represent a new delivery method for therapeutic agents that do not provoke an immune response in the cell. For cargo delivery by gesicle, an expression cassette coding CRISPR/Cas components is transfected into producer cells ex vivo, which are then induced by glycoprotein to produce gesicles loaded with delivery cargoes. These gesicles are isolated and used for delivery. Quinn et al. (2016) used a ligand-dependent dimerization method to deliver CRISPR RNPs into human-induced pluripotent stem cells, with controlled release and no off-targets (Quinn et al. 2016).

4.3.3.5 iTOP

Induced transduction by osmocytosis and propanebetaine (iTOP) is a novel method that allows cellular uptake of cargoes through macro-pinocytosis triggered by manipulation of osmatic potential. iTOP has been successfully used for delivery of RNPs. Cas9 protein and sgRNA were delivered through iTOP in human embryonic stem cells with 26% genome editing efficiency (D'Astolfo et al. 2015). In contrast to other delivery methods such as CPP, EP, and cationic lipid, iTop showed a lower efficiency of genome editing. Moreover, iTOP is not suitable for in vivo applications (Liu et al. 2017).

4.3.3.6 Mechanical Cell Deformation

Mechanical cell deformation or micro-constriction results in transient disruption of the cell membrane, consequently allowing uptake of the cargo through passive diffusion. Mechanical cell deformation has been used by researchers to deliver various cargoes into cells because of lipid efficiency and low rate of cell death (Worthen et al. 2017; Sharei et al. 2013). While using a microfluidic device, Han et al. (2015) delivered CRISPR plasmid into different cell lines through micro-constriction. With this delivery method, EGF gene was knocked out in breast cancer cells (MDA-MB-231) and lymphoma cells (Sv-DHL-1) with more than 90% and 70% efficiency, respectively (Han et. 2015). However, use of a microfluidic device has not been reported for in vivo applications.

4.3.4 Potential Future Delivery Tools

Along with viral, non-viral, and physical delivery methods, some emerging delivery methods have great potential and may be adopted for CRISPR/Cas-mediated genome editing. Here we discuss these emerging methods.

4.3.4.1 Inorganic Nanoparticles

Inorganic nanoparticles have potential to deliver CRISPR components into cells—indeed, some have already been reported for CRISPR/Cas system such as AuNPs. Gold nanoparticles are a novel delivery tool for transient genome editing applications in a variety of cell types; however, other inorganic particles such as carbon nanotubes (CNTs), bare mesoporous silica nanoparticles (MSNPs), and dense silica nanoparticles (SiNPs) are not reported for CRISPR applications. Compared with viral and lipid/polymer-based delivery tools, inorganic nanoparticles offer several advantages such as ease of generation and characterization, reproducible composition, and long-term stability. Therefore, we may expect increased use of inorganic nanoparticles for delivering CRISPR cargoes in both in vivo and in vitro applications in the future (Lino et al. 2018).

4.3.4.2 Lipid-Coated Mesoporous Silica Particles (LC-MSPs)

LC-MSPs are composed of an internal silica nanoparticle core coated with lipid bilayer (Liu et al. 2009). It is a biological delivery system with several attractive features. For example, the internal silica core provides a large surface area which is useful for high cargo capacity. Moreover, customizable pore size, pore chemistry, and overall size make them useful for loading different types of cargoes (Du et al. 2014; Durfee et al. 2016), and the outer lipid coat can be customized to increase circulation time, precise targeting, and controlled release of cargo. Although LC-MSPs are not used for CRISPR/Cas, their properties make them attractive for delivery of CRISPR reagents. However, several challenges remain, especially in the packing of large cargoes, which needs to be optimized as CRISPR reagents are larger than reagents delivered with LC-MSPs. So far, LC-MSPs have been used to deliver imaging agents, photography agents, and chemotherapeutic agents, both in vivo and in vitro (Lino et al. 2018).

4.3.4.3 Streptolysin O (SLO)

The SLO toxin produced by *Streptococcus* bacteria creates pores in cell membrane, allowing the entry of cargo into the cell. Although SLO is toxic, a system has been developed for reversing the toxin's effect (Lino et al. 2018). SLO has been used to deliver siRNA (Brito et al. 2008) and as an imaging agent for live cell therapy (Teng et al. 2017). Although use of SLO in vivo could be challenging, it has great potential for in vitro delivery of CRISPR/Cas.

4.3.4.4 Multifunctional Envelope-Type NanoDevices (MENDs)

A MEND consists of condensed plasmid and PLL core, coated with a lipid film (Kogure et al. 2004). Presence of lipid film over the plasmid/PLL core increases the transfection rate tenfold. In addition, the transfection rate was increased by a factor of 1000 folds by adding a CPP, stearyl octa-arginine, to the lipid shell. The lipid shell can be tailored with other functional components, such as PEG, to increase circulation time, ligands for tissue specific targeting, and lipid for endosomal escape (Lino et al. 2018). Tetra-lamellar MENDs (T-MENDs) have been developed to target cell nucleus and mitochondria, while PEG-peptide-dioleoylphosphatidylethanolamine (DOPE)-conjugated MENDs were developed to target bladder cancer cells (Nakamura et al. 2012). MENDs have been used for delivery of siRNA and hold great potential for delivering CRISPR/Cas reagents; however, for in vivo applications, more work is required.

4.4 Delivery Methods in Plants

DNA is the most common form of CRISPR reagent used for plant genome editing. DNA cassettes coding for Cas9 and gRNA could be degraded or stably integrated into the genome at random sites. Stable integration may result in higher off-targets, consequently limiting the commercial applications of genome-edited plants. To avoid this problem, genetic segregation may be used to select transgene-free plants. However, genetic segregation does not work for asexually propagating crops such as potato, cassava, and banana. Another option is to use the suicide genes CMS2 and BARNASE to kill transgene-containing pollen and embryos produced by the T0 plant (Chen et al. 2019). Alternatively, transient gene expression of CRISPR/Cas reagents could be used to produce transgene-free, genome-edited plants. This method has been used for wheat, as well as base editing in other plants. Although this method has shown success, degraded DNA fragments may still integrate into the genome. To avoid the issues of off-targets, segregation, and integration of degraded fragments, in vitro transcribed mRNA and sgRNA or RNPs have been used to produce transgene-free plants (Chen et al. 2019). Delivery methods in plants can be divided into direct (physical and chemical) or indirect (agrobacterium or viral) methods.

4.4.1 Direct Methods

Direct delivery methods such as biolistic, protoplast, whiskers, pollens, liposomes, and electroporation involve physical or chemical methods to deliver CRISPR reagents in plants. Direct methods are not genotype dependent, as long as a regeneration method is available. Moreover, direct methods can deliver multiple constructs simultaneously and editing reagents in various forms such as DNA, RNA, and protein. Biolistic is the most commonly used direct delivery method for CRISPR/Cas genome editing (Ran et al. 2017). It can deliver multiple constructs with reasonable efficiency; however, integration of multiple copies of the transgene in the transformants is one of the limiting factors of this method. Biolistic has been used to deliver CRISPR/Cas reagents in soybean (Li et al. 2015), maize (Svitashev et al. 2016), barley, rice, or wheat (Shan et al. 2013a, b). Protoplast is another method which involves direct delivery of CRISPR reagents as DNA, mRNA, and RNPs into the protoplast by EP or PEG treatments. It is a suitable method for transfection of a large number of cells with multiple components. Moreover, it is useful for modifying genes with repair templates. However, regeneration is one of the main limitations in protoplast-mediated transfection (Ran et al. 2017). Several studies have demonstrated the potential of CRISPR/Cas with protoplast such as potato (Andersson et al. 2017) and lettuce (Woo et al. 2015). Other methods such as nanoparticles, CPP, and whiskers have been used as carriers of DNA, protein, or mRNA, but not specifically CRISPR/Cas reagents.

4.4.2 Indirect Methods

Indirect methods involve delivery of plasmids expressing CRISPR components into the target cells through agrobacterium or plant virus systems (Ran et al. 2017).

4.4.2.1 Agrobacterium

Agrobacterium is the most commonly used delivery system reported for model systems, crop species, vegetables, fruits, and tree species. Agrobacterium has been used to deliver CRISPR reagents for modifying single genes or targeting multiple genes simultaneously (Ran et al. 2017). For example, Nelles et al. (2015) and Xie et al. (2015) targeted multiple genes in rice. Although agrobacterium is the most convenient, cheap, and easy method for delivering DNA constructs in plants, it has drawbacks. For example, it cannot deliver small DNA fragments, RNA, or proteins. In addition, in using agrobacterium, transgene will always integrate into the genome, and agrobacterium transformation is still dependent on recipient genotype, especially in monocots (Ran et al. 2017).

4.4.2.2 Virus-Mediated Delivery of CRISPR Reagents

Viral delivery methods in plants are not as common as in mammalian cells. Viral-mediated delivery was first employed in 1995 using tobacco mosaic virus for virus-induced gene silencing (VIGS) of an endogenous gene in *Nicotiana benthamiana*

(Kumagai et al. 1995). Most plant viruses contain ssRNA genome, so in vitro transcribed RNA can be used for VIGS. Inoculation with DNA viruses such as geminiviruses is simple as it requires only viral DNA. In an alternative method, cDNA of viral genome is inserted into binary vector and subsequently introduced into the plant cell through agro-infection. Tobacco rattle virus (TRV) is an efficient system used for VIGS, as well as genome editing (Ali et al. 2015; Marton et al. 2010). It is an ssRNA virus, having a bipartite genome consisting of two positive-sense single-stranded RNAs, designed as RNA1 and RNA2. It is an efficient system for producing transgene-free genome-edited plants because viral RNA does not integrate into plant genome. Geminiviruses consist of single-stranded circular DNA genome; thus, they enable direct infection with plasmid DNA. Geminiviruses have also been used to deliver genome editing reagents into plant cells especially for gene modifications using the HDR pathway (Baltes et al. 2014; Čermák et al. 2015; Gil-Humanes et al. 2017). Virus-mediated delivery represents an ideal system for future because viruses infect a wide range of plant species and could be used for producing transgene-free CRISPR-edited plants.

4.5 Challenges and Future Prospect

CRISPR/Cas is a powerful and simple genome editing tool that has been extensively applied for curing genetic diseases, disease modeling, therapeutics, and translational research. With recent developments in this technology such as rewriting genetic code, base editing, prime editing and RNA editing, its applications will certainly be more extensive in the future. However, researchers face several challenges to realize the full potential of CRISPR/Cas, especially in clinical and therapeutic applications. Off-targeting, ethical issues, and lack of safe and efficient delivery tools are the major challenges the scientific community faces in CRISPR/Cas applications. Extensive efforts have been made to predict and reduce off-targets; however, our understanding of sgRNA, binding, and mismatch tolerance leading to off-targets remains poor. Similarly, ethical and regulatory issues are a challenge in plant as well as animal sciences. In 2018, a Chinese scientist claimed CRISPR editing of human embryos with germline modification to fight HIV. This claim highlighted ethical and regulatory issues for CRISPR/Cas all over the world. Cargo delivery tools in CRISPR/Cas genome editing remain one of the biggest challenges in CRISPR applications, and an all-purpose delivery tool with few problems is yet to emerge. Methods ranging from physical to viral delivery tools have been reported for various applications. Each method has its merits and demerits, and some are specifically for in vitro applications only. In contrast to DNA and mRNA, minimal off-target effects have resulted from delivery of RNPs. There are a number of options for smaller cargo sizes; however, the relatively large size of the Cas9 gene, as well as the protein-nucleic acid complex, makes their delivery problematic. The development of new and effective delivery methods for RNPs will have a significant impact in clinical and translational applications of CRISPR/Cas. Safety and specificity are additional concerns in delivery tools for CRISPR/Cas applications both in vivo and in vitro. In living organisms, safety is the major concern, and the specific and controlled delivery of CRISPR reagents to the target cells will certainly reduce off-targets and improve safety.

Long-term studies on the safety of nanoparticles are especially needed to assess any toxicity associated with their use. Despite these challenges, progress in CRISPR applications in various fields such as animal models, medicines, and therapeutics has been impressive. As CRISPR/Cas technology evolves, delivery tools will become more precise and safer, consequently increasing the therapeutic potential of this marvelous genome editing technology. In conclusion, despite the challenges, rapid advancement in CRISPR/Cas with improved delivery tools is expected to pave the way for clinical applications of CRISPR/Cas in the near future.

References

Adli M (2018) The CRISPR tool kit for genome editing and beyond. Nat Commun 9(1):1–13

Al-Dosari MS, Knapp JE, Liu D (2005) Hydrodynamic delivery. Adv Genet 54:65–82

Ali Z, Abul-Faraj A, Li L, Ghosh N, Piatek M, Mahjoub A, Aouida M, Piatek A, Baltes NJ, Voytas DF (2015) Efficient virus-mediated genome editing in plants using the CRISPR/Cas9 system. Mol Plant 8(8):1288–1291

Alsaiari SK, Patil S, Alyami M, Alamoudi KO, Aleisa FA, Merzaban JS, Khashab NM (2018) Endosomal escape and delivery of CRISPR/Cas9 genome editing machinery enabled by nanoscale zeolitic imidazolate framework. JACS 140(1):143–146

Andersson M, Turesson H, Nicolia A, Fält A-S, Samuelsson M, Hofvander P (2017) Efficient targeted multiallelic mutagenesis in tetraploid potato (Solanum tuberosum) by transient CRISPR-Cas9 expression in protoplasts. Plant Cell Rep 36(1):117–128

Apolonia L, Waddington SN, Fernandes C, Ward NJ, Bouma G, Blundell MP, Thrasher AJ, Collins MK, Philpott NJ (2007) Stable gene transfer to muscle using non-integrating lentiviral vectors. Mol Ther 15(11):1947–1954

Axford DS, Morris DP, McMurry JL (2017) Cell penetrating peptide-mediated nuclear delivery of Cas9 to enhance the utility of CRISPR/Cas genome editing. FASEB J 31(1_Suppl):909.904

Baltes NJ, Gil-Humanes J, Cermak T, Atkins PA, Voytas DF (2014) DNA replicons for plant genome engineering. Plant Cell 26(1):151–163

Bergelson JM (1999) Receptors mediating adenovirus attachment and internalization. Biochem Pharmacol 57(9):975–979

Bikard D, Euler CW, Jiang W, Nussenzweig PM, Goldberg GW, Duportet X, Marraffini LA (2014) Exploiting CRISPR-Cas nucleases to produce sequence-specific antimicrobials. Nat. Biotechnol 32(11):1146-1150

Black JB, Adler AF, Wang H-G, D'Ippolito AM, Hutchinson HA, Reddy TE, Pitt GS, Leong KW, Gersbach CA (2016) Targeted epigenetic remodeling of endogenous loci by CRISPR/Cas9-based transcriptional activators directly converts fibroblasts to neuronal cells. Cell Stem Cell 19(3):406–414

Blasco RB, Karaca E, Ambrogio C, Cheong T-C, Karayol E, Minero VG, Voena C, Chiarle R (2014) Simple and rapid in vivo generation of chromosomal rearrangements using CRISPR/Cas9 technology. Cell Rep 9(4):1219–1227

Bonamassa B, Hai L, Liu D (2011) Hydrodynamic gene delivery and its applications in pharmaceutical research. Pharm Res 28(4):694–701

Brito JL, Davies FE, Gonzalez D, Morgan GJ (2008) Streptolysin-O reversible permeabilisation is an effective method to transfect siRNAs into myeloma cells. J Immunol Methods 333(1–2):147–155

Carroll KJ, Makarewich CA, McAnally J, Anderson DM, Zentilin L, Liu N, Olson EN (2016) A mouse model for adult cardiac-specific gene deletion with CRISPR/Cas9. PNAS 113(2):338–343

Čermák T, Baltes NJ, Čegan R, Zhang Y, Voytas DF (2015) High-frequency, precise modification of the tomato genome. Genome Biol 16(1):232

Chandrasekaran AP, Song M, Kim K-S, Ramakrishna S (2018) Different methods of delivering CRISPR/Cas9 into cells. In: Progress in molecular biology and translational science, vol 159. Elsevier, Amsterdam, pp 157–176

Chang H, Lv J, Gao X, Wang X, Wang H, Chen H, He X, Li L, Cheng Y (2017) Rational design of a polymer with robust efficacy for intracellular protein and peptide delivery. Nano Lett 17(3):1678–1684

Chen Z-Y, He C-Y, Ehrhardt A, Kay MA (2003) Minicircle DNA vectors devoid of bacterial DNA result in persistent and high-level transgene expression in vivo. Mol Ther 8(3):495–500

Chen S, Sanjana NE, Zheng K, Shalem O, Lee K, Shi X, Sharp PA (2015) Genome-wide CRISPR screen in a mouse model of tumor growth and metastasis. Cell 160(6):1246–1260

Chen Z, Liu F, Chen Y, Liu J, Wang X, Chen AT, Deng G, Zhang H, Liu J, Hong Z (2017) Targeted delivery of CRISPR/Cas9-mediated cancer gene therapy via liposome-templated hydrogel nanoparticles. Adv Funct Mater 27(46):1703036

Chen T-T, Yi J-T, Zhao Y-Y, Chu X (2018) Biomineralized metal–organic framework nanoparticles enable intracellular delivery and endo-lysosomal release of native active proteins. J Am Chem Soc 140(31):9912–9920

Chen K, Wang Y, Zhang R, Zhang H, Gao C (2019) CRISPR/Cas genome editing and precision plant breeding in agriculture. Annu Rev Plant Biol 70:667–697

Cheng G, Li W, Ha L, Han X, Hao S, Wan Y, Wang Z, Dong F, Zou X, Mao Y (2018) Self-assembly of extracellular vesicle-like metal–organic framework nanoparticles for protection and intracellular delivery of biofunctional proteins. J Am Chem Soc 140(23):7282–7291

Choi PS, Meyerson M (2014) Targeted genomic rearrangements using CRISPR/Cas technology. Nat Commun 5(1):1–6

Chuang C-K, Chen C-H, Huang C-L, Su Y-H, Peng S-H, Lin T-Y, Tai H-C, Yang T-S, Tu C-F (2017) Generation of GGTA1 mutant pigs by direct pronuclear microinjection of CRISPR/Cas9 plasmid vectors. Anim Biotechnol 28(3):174–181

Çiçek YA, Luther DC, Kretzmann JA, Rotello VM (2019) Advances in CRISPR/Cas9 technology for in vivo translation. Biol Pharm Bull 42(3):304–311

Cong L, Ran FA, Cox D, Lin S, Barretto R, Habib N, Hsu PD, Wu X, Jiang W, Marraffini LA (2013) Multiplex genome engineering using CRISPR/Cas systems. Science 339(6121):819–823

D'Astolfo DS, Pagliero RJ, Pras A, Karthaus WR, Clevers H, Prasad V, Lebbink RJ, Rehmann H, Geijsen N (2015) Efficient intracellular delivery of native proteins. Cell 161(3):674–690

Dad A-B, Ramakrishna S, Song M, Kim H (2014) Enhanced gene disruption by programmable nucleases delivered by a minicircle vector. Gene Ther 21(11):921–930

Ding Q, Strong A, Patel KM, Ng SL, Gosis BS, Regan SN, Musunuru K (2014) Permanent alteration of PCSK9 with in vivo CRISPR-Cas9 genome editing. Circ Res 115(5):488–492

Dong C, Qu L, Wang H, Wei L, Dong Y, Xiong S (2015) Targeting hepatitis B virus cccDNA by CRISPR/Cas9 nuclease efficiently inhibits viral replication. Antivir Res 118:110–117

Du X, Shi B, Tang Y, Dai S, Qiao SZ (2014) Label-free dendrimer-like silica nanohybrids for traceable and controlled gene delivery. Biomaterials 35(21):5580–5590

Durfee PN, Lin Y-S, Dunphy DR, Muñiz AJ, Butler KS, Humphrey KR, Lokke AJ, Agola JO, Chou SS, Chen I-M (2016) Mesoporous silica nanoparticle-supported lipid bilayers (protocells) for active targeting and delivery to individual leukemia cells. ACS Nano 10(9):8325–8345

Eoh J, Gu L (2019) Biomaterials as vectors for the delivery of CRISPR–Cas9. Biomater Sci 7(4):1240–1261

Gilbert LA, Larson MH, Morsut L, Liu Z, Brar GA, Torres SE, Stern-Ginossar N, Brandman O, Whitehead EH, Doudna JA (2013) CRISPR-mediated modular RNA-guided regulation of transcription in eukaryotes. Cell 154(2):442–451

Gil-Humanes J, Wang Y, Liang Z, Shan Q, Ozuna CV, Sánchez-León S, Baltes NJ, Starker C, Barro F, Gao C (2017) High-efficiency gene targeting in hexaploid wheat using DNA replicons and CRISPR/Cas9. Plant J 89(6):1251–1262

Glass Z, Lee M, Li Y, Xu Q (2018) Engineering the delivery system for CRISPR-based genome editing. Trends Biotechnol 36(2):173–185

Guan Y, Ma Y, Li Q, Sun Z, Ma L, Wu L, Wang L, Zeng L, Shao Y, Chen Y (2016) CRISPR/Cas9-mediated somatic correction of a novel coagulator factor IX gene mutation ameliorates hemophilia in mouse. EMBO Mol Med 8(5):477–488

Han X, Liu Z, Chan JM, Zhang K, Li Y, Zeng Z, Qin L (2015) CRISPR-Cas9 delivery to hard-totransfect cells via membrane deformation. Sci Adv 1(7):1500454

Hashimoto M, Takemoto T (2015) Electroporation enables the efficient mRNA delivery into the mouse zygotes and facilitates CRISPR/Cas9-based genome editing. Sci Rep 5:11315

He Z-Y, Men K, Qin Z, Yang Y, Xu T, Wei Y-Q (2017) Non-viral and viral delivery systems for CRISPR-Cas9 technology in the biomedical field. Sci China Life Sci 60(5):458–467

Horii T, Morita S, Kimura M, Kobayashi R, Tamura D, Takahashi RU, Hatada I (2013) Genome engineering of mammalian haploid embryonic stem cells using the Cas9/RNA system. PeerJ 1: 230

Horii T, Arai Y, Yamazaki M, Morita S, Kimura M, Itoh M, Abe Y, Hatada I (2014) Validation of microinjection methods for generating knockout mice by CRISPR/Cas-mediated genome engineering. Sci Rep 4:4513

Ibraheem D, Elaissari A, Fessi H (2014) Gene therapy and DNA delivery systems. Int J Pharm 459(1–2):70–83

Jackson MA, Werfel TA, Curvino EJ, Yu F, Kavanaugh TE, Sarett SM, Dockery MD, Kilchrist KV, Jackson AN, Giorgio TD (2017) Zwitterionic nanocarrier surface chemistry improves siRNA tumor delivery and silencing activity relative to polyethylene glycol. ACS Nano 11(6):5680–5696

Jiang C, Mei M, Li B, Zhu X, Zu W, Tian Y, Wang Q, Guo Y, Dong Y, Tan X (2017) A non-viral CRISPR/Cas9 delivery system for therapeutically targeting HBV DNA and pcsk9 in vivo. Cell Res 27(3):440–443

Kaminski R, Chen Y, Fischer T, Tedaldi E, Napoli A, Zhang Y, Karn J, Hu W, Khalili K (2016) Elimination of HIV-1 genomes from human T-lymphoid cells by CRISPR/Cas9 gene editing. Sci Rep 6(1):1–15

Kang YK, Kwon K, Ryu JS, Lee HN, Park C, Chung HJ (2017) Nonviral genome editing based on a polymer-derivatized CRISPR nanocomplex for targeting bacterial pathogens and antibiotic resistance. Bioconjug Chem 28(4):957–967

Khan FA, Pandupuspitasari NS, Chun-Jie H, Ao Z, Jamal M, Zohaib A, Khan FA, Hakim MR, ShuJun Z (2016) CRISPR/Cas9 therapeutics: a cure for cancer and other genetic diseases. Oncotarget 7(32):52541

Kim S, Kim D, Cho SW, Kim J, Kim J-S (2014) Highly efficient RNA-guided genome editing in human cells via delivery of purified Cas9 ribonucleoproteins. Genome Res 24(6):1012–1019

Kogure K, Moriguchi R, Sasaki K, Ueno M, Futaki S, Harashima H (2004) Development of a non-viral multifunctional envelope-type nano device by a novel lipid film hydration method. J Control Release 98(2):317–323

Koike-Yusa H, Li Y, Tan E-P, Velasco-Herrera MDC, Yusa K (2014) Genome-wide recessive genetic screening in mammalian cells with a lentiviral CRISPR-guide RNA library. Nat Biotechnol 32(3):267–273

Kumagai MH, Donson J, Della-Cioppa G, Harvey D, Hanley K, Grill L (1995) Cytoplasmic inhibition of carotenoid biosynthesis with virus-derived RNA. Proc Natl Acad Sci 92(5):1679–1683

Lau C-H, Suh Y (2017) In vivo genome editing in animals using AAV-CRISPR system: applications to translational research of human disease. F1000Research 6:2153

Lee K, Conboy M, Park HM, Jiang F, Kim HJ, Dewitt MA, Mackley VA, Chang K, Rao A, Skinner C (2017) Nanoparticle delivery of Cas9 ribonucleoprotein and donor DNA in vivo induces homology-directed DNA repair. Nat Biomed Eng 1(11):889–901

Lee B, Lee K, Panda S, Gonzales-Rojas R, Chong A, Bugay V, Park HM, Brenner R, Murthy N, Lee HY (2018) Nanoparticle delivery of CRISPR into the brain rescues a mouse model of fragile X syndrome from exaggerated repetitive behaviours. Nat Biomed Eng 2(7):497–507

Li L, Natarajan P, Allen C, Peshwa M (2014) CGMP-compliant, clinical scale, non-viral platform for efficient gene editing using CRISPR/Cas9. Cytotherapy 16(4):S37

Li Z, Liu Z-B, Xing A, Moon BP, Koellhoffer JP, Huang L, Ward RT, Clifton E, Falco SC, Cigan AM (2015) Cas9-guide RNA directed genome editing in soybean. Plant Physiol 169(2):960–970

Li L, Hu S, Chen X (2018) Non-viral delivery systems for CRISPR/Cas9-based genome editing: challenges and opportunities. Biomaterials 171:207–218

Liang X, Potter J, Kumar S, Zou Y, Quintanilla R, Sridharan M, Carte J, Chen W, Roark N, Ranganathan S (2015) Rapid and highly efficient mammalian cell engineering via Cas9 protein transfection. J Biotechnol 208:44–53

Lino CA, Harper JC, Carney JP, Timlin JA (2018) Delivering CRISPR: a review of the challenges and approaches. Drug Deliv 25(1):1234–1257

Liu J, Stace-Naughton A, Jiang X, Brinker CJ (2009) Porous nanoparticle supported lipid bilayers (protocells) as delivery vehicles. J Am Chem Soc 131(4):1354–1355

Liu Y, Zeng Y, Liu L, Zhuang C, Fu X, Huang W, Cai Z (2014) Synthesizing AND gate genetic circuits based on CRISPR-Cas9 for identification of bladder cancer cells. Nat Commun 5(1):5393. https://doi.org/10.1038/ncomms6393

Liu J, Gaj T, Wallen MC, Barbas CF III (2015) Improved cell-penetrating zinc-finger nuclease proteins for precision genome engineering. Mol Ther Nucleic Acids 4:e232

Liu C, Zhang L, Liu H, Cheng K (2017) Delivery strategies of the CRISPR-Cas9 gene-editing system for therapeutic applications. J Control Release 266:17–26

Long C, McAnally JR, Shelton JM, Mireault AA, Bassel-Duby R, Olson EN (2014) Prevention of muscular dystrophy in mice by CRISPR/Cas9–mediated editing of germline DNA. Science 345(6201):1184–1188

Long C, Amoasii L, Mireault AA, McAnally JR, Li H, Sanchez-Ortiz E, Bhattacharyya S, Shelton JM, Bassel-Duby R, Olson EN (2016) Postnatal genome editing partially restores dystrophin expression in a mouse model of muscular dystrophy. Science 351(6271):400–403

Lowder L, Malzahn A, Qi Y (2016) Rapid evolution of manifold CRISPR systems for plant genome editing. Front Plant Sci 7:1683

Lu Q, Livi GP, Modha S, Yusa K, Macarrón R, Dow DJ (2017) Applications of CRISPR genome editing technology in drug target identification and validation. Expert Opin Drug Discovery 12(6):541–552

Ma Y, Shen B, Zhang X, Lu Y, Chen W, Ma J, Huang X, Zhang L (2014a) Heritable multiplex genetic engineering in rats using CRISPR/Cas9. PLoS One 9(3):e89413

Ma Y, Zhang X, Shen B, Lu Y, Chen W, Ma J, Bai L, Huang X, Zhang L (2014b) Generating rats with conditional alleles using CRISPR/Cas9. Cell Res 24(1):122–125

Ma T, Tao J, Yang M, He C, Tian X, Zhang X, Zhang J, Deng S, Feng J, Zhang Z (2017) An AANAT/ASMT transgenic animal model constructed with CRISPR/Cas9 system serving as the mammary gland bioreactor to produce melatonin-enriched milk in sheep. J Pineal Res 63(1): e12406

Maddalo D, Manchado E, Concepcion CP, Bonetti C, Vidigal JA, Han YC, Ventura A (2014) In vivo engineering of oncogenic chromosomal rearrangements with the CRISPR/Cas9 system. Nature 516(7531):423–427

Maggio I, Holkers M, Liu J, Janssen JM, Chen X, Gonçalves MA (2014) Adenoviral vector delivery of RNA-guided CRISPR/Cas9 nuclease complexes induces targeted mutagenesis in a diverse array of human cells. Sci Rep 4:5105

Makarova KS, Koonin EV (2015) Annotation and classification of CRISPR-Cas systems. In: CRISPR. Springer, Berlin, pp 47–75

Mali P, Aach J, Stranges PB, Esvelt KM, Moosburner M, Kosuri S, Yang L, Church GM (2013) CAS9 transcriptional activators for target specificity screening and paired nickases for cooperative genome engineering. Nat Biotechnol 31(9):833–838

Mandal PK, Ferreira LM, Collins R, Meissner TB, Boutwell CL, Friesen M, Vrbanac V, Garrison BS, Stortchevoi A, Bryder D (2014) Efficient ablation of genes in human hematopoietic stem and effector cells using CRISPR/Cas9. Cell Stem Cell 15(5):643–652

Marton I, Zuker A, Shklarman E, Zeevi V, Tovkach A, Roffe S, Ovadis M, Tzfira T, Vainstein A (2010) Nontransgenic genome modification in plant cells. Plant Physiol 154(3):1079–1087

Maruyama T, Dougan SK, Truttmann MC, Bilate AM, Ingram JR, Ploegh HL (2015) Increasing the efficiency of precise genome editing with CRISPR-Cas9 by inhibition of nonhomologous end joining. Nat Biotechnol 33(5):538–542

Matano M, Date S, Shimokawa M, Takano A, Fujii M, Ohta Y, Watanabe T, Kanai T, Sato T (2015) Modeling colorectal cancer using CRISPR-Cas9–mediated engineering of human intestinal organoids. Nat Med 21(3):256–262

Miller JB, Zhang S, Kos P, Xiong H, Zhou K, Perelman SS, Zhu H, Siegwart DJ (2017) Non-viral CRISPR/Cas gene editing in vitro and in vivo enabled by synthetic nanoparticle co-delivery of Cas9 mRNA and sgRNA. Angew Chem Int Ed 56(4):1059–1063

Miller JB, Kos P, Tieu V, Zhou K, Siegwart DJ (2018) Development of cationic quaternary ammonium sulfonamide amino lipids for nucleic acid delivery. ACS Appl Mater Interfaces 10(3):2302–2311

Monroe KM, Yang Z, Johnson JR, Geng X, Doitsh G, Krogan NJ, Greene WC (2014) IFI16 DNA sensor is required for death of lymphoid CD4 T cells abortively infected with HIV. Science 343(6169):428–432

Mout R, Ray M, Yesilbag Tonga G, Lee Y-W, Tay T, Sasaki K, Rotello VM (2017) Direct cytosolic delivery of CRISPR/Cas9-ribonucleoprotein for efficient gene editing. ACS Nano 11(3):2452–2458

Nakagawa Y, Sakuma T, Sakamoto T, Ohmuraya M, Nakagata N, Yamamoto T (2015) Production of knockout mice by DNA microinjection of various CRISPR/Cas9 vectors into freeze-thawed fertilized oocytes. BMC Biotechnol 15(1):33

Nakamura T, Akita H, Yamada Y, Hatakeyama H, Harashima H (2012) A multifunctional envelope-type nanodevice for use in nanomedicine: concept and applications. Acc Chem Res 45(7):1113–1121

Nelles DA, Fang MY, Aigner S, Yeo GW (2015) Applications of Cas9 as an RNA-programmed RNA-binding protein. BioEssays 37(7):732–739

Niu Y, Shen B, Cui Y, Chen Y, Wang J, Wang L, Kang Y, Zhao X, Si W, Li W (2014) Generation of gene-modified cynomolgus monkey via Cas9/RNA-mediated gene targeting in one-cell embryos. Cell 156(4):836–843

Ousterout DG, Kabadi AM, Thakore PI, Majoros WH, Reddy TE, Gersbach CA (2015) Multiplex CRISPR/Cas9-based genome editing for correction of dystrophin mutations that cause Duchenne muscular dystrophy. Nat Commun 6(1):1–13

Paquet D, Kwart D, Chen A, Sproul A, Jacob S, Teo S, Olsen KM, Gregg A, Noggle S, Tessier-Lavigne M (2016) Efficient introduction of specific homozygous and heterozygous mutations using CRISPR/Cas9. Nature 533(7601):125–129

Philippe S, Sarkis C, Barkats M, Mammeri H, Ladroue C, Petit C, Mallet J, Serguera C (2006) Lentiviral vectors with a defective integrase allow efficient and sustained transgene expression in vitro and in vivo. Proc Natl Acad Sci 103(47):17684–17689

Platt RJ, Chen S, Zhou Y, Yim MJ, Swiech L, Kempton HR, Dahlman JE, Parnas O, Eisenhaure TM, Jovanovic M (2014) CRISPR-Cas9 knockin mice for genome editing and cancer modeling. Cell 159(2):440–455

Qin W, Dion SL, Kutny PM, Zhang Y, Cheng AW, Jillette NL, Wang H (2015) Efficient CRISPR/Cas9-mediated genome editing in mice by zygote electroporation of nuclease. Genetics 200(2):423-430

Quinn T, Morell M, Garachtchenko T, Lee L, Fong M, Haugwitz M, Farmer A (2016) 313. Gesicle mediated delivery of a Cas9-sgRNA protein complex. Mol Ther 24:S126

Raghavan A, Wang X, Rogov P, Wang L, Zhang X, Mikkelsen TS, Musunuru K (2016) High-throughput screening and CRISPR-Cas9 modeling of causal lipid-associated expression quantitative trait locus variants. bioRxiv:056820

Ramakrishna S, Dad A-BK, Beloor J, Gopalappa R, Lee S-K, Kim H (2014) Gene disruption by cell-penetrating peptide-mediated delivery of Cas9 protein and guide RNA. Genome Res 24(6):1020–1027

Ramos-Kuri M, Rapti K, Mehel H, Zhang S, Dhandapany PS, Liang L, García-Carrancá A, Bobe R, Fischmeister R, Adnot S (2015) Dominant negative Ras attenuates pathological ventricular remodeling in pressure overload cardiac hypertrophy. Biochim Biophys Acta Mol Cell Res 1853(11):2870–2884

Ran FA, Cong L, Yan WX, Scott DA, Gootenberg JS, Kriz AJ, Zhang F (2015) In vivo genome editing using Staphylococcus aureus Cas9. Nature 520(7546):186–191

Ran Y, Liang Z, Gao C (2017) Current and future editing reagent delivery systems for plant genome editing. Sci China Life Sci 60(5):490–505

Rui Y, Wilson DR, Green JJ (2019) Non-viral delivery to enable genome editing. Trends Biotechnol 37(3):281–293

Scheben A, Wolter F, Batley J, Puchta H, Edwards D (2017) Towards CRISPR/Cas crops—bringing together genomics and genome editing. New Phytol 216(3):682–698

Schumann K, Lin S, Boyer E, Simeonov DR, Subramaniam M, Gate RE, Haliburton GE, Chun JY, Bluestone JA, Doudna JA (2015) Generation of knock-in primary human T cells using Cas9 ribonucleoproteins. Proc Natl Acad Sci 112(33):10437–10442

Schwank G, Clevers H (2016) CRISPR/Cas9-mediated genome editing of mouse small intestinal organoids. Springer, pp 3–11

Sedeek KE, Mahas A, Mahfouz M (2019) Plant genome engineering for targeted improvement of crop traits. Front Plant Sci 10:114

Shalem O, Sanjana NE, Hartenian E, Shi X, Scott DA, Mikkelsen TS, Heckl D, Ebert BL, Root DE, Doench JG (2014) Genome-scale CRISPR-Cas9 knockout screening in human cells. Science 343(6166):84–87

Shan Q, Wang Y, Chen K, Liang Z, Li J, Zhang Y, Zhang K, Liu J, Voytas DF, Zheng X (2013a) Rapid and efficient gene modification in rice and Brachypodium using TALENs. Mol Plant 6(4):1365–1368

Shan Q, Wang Y, Li J, Zhang Y, Chen K, Liang Z, Zhang K, Liu J, Xi JJ, Qiu J-L (2013b) Targeted genome modification of crop plants using a CRISPR-Cas system. Nat Biotechnol 31(8):686–688

Sharei A, Zoldan J, Adamo A, Sim WY, Cho N, Jackson E, Mao S, Schneider S, Han M-J, Lytton-Jean A (2013) A vector-free microfluidic platform for intracellular delivery. Proc Natl Acad Sci 110(6):2082–2087

Shmakov S, Smargon A, Scott D, Cox D, Pyzocha N, Yan W, Abudayyeh OO, Gootenberg JS, Makarova KS, Wolf YI (2017) Diversity and evolution of class 2 CRISPR–Cas systems. Nat Rev Microbiol 15(3):169–182

Staahl BT, Benekareddy M, Coulon-Bainier C, Banfal AA, Floor SN, Sabo JK, Urnes C, Munares GA, Ghosh A, Doudna JA (2017) Efficient genome editing in the mouse brain by local delivery of engineered Cas9 ribonucleoprotein complexes. Nat Biotechnol 35(5):431–434

Suda T, Gao X, Stolz D, Liu D (2007) Structural impact of hydrodynamic injection on mouse liver. Gene Ther 14(2):129–137

Suk JS, Xu Q, Kim N, Hanes J, Ensign LM (2016) PEGylation as a strategy for improving nanoparticle-based drug and gene delivery. Adv Drug Deliv Rev 99:28–51

Sun W, Ji W, Hall JM, Hu Q, Wang C, Beisel CL, Gu Z (2015) Self-assembled DNA nanoclews for the efficient delivery of CRISPR–Cas9 for genome editing. Angew Chem Int 127(41):12197–12201

Svitashev S, Schwartz C, Lenderts B, Young JK, Cigan AM (2016) Genome editing in maize directed by CRISPR–Cas9 ribonucleoprotein complexes. Nat Commun 7(1):1–7

Swiech L, Heidenreich M, Banerjee A, Habib N, Li Y, Trombetta J, Sur M, Zhang F (2015) In vivo interrogation of gene function in the mammalian brain using CRISPR-Cas9. Nat Biotechnol 33(1):102–106

Teng KW, Ishitsuka Y, Ren P, Youn Y, Deng X, Ge P, Lee SH, Belmont AS, Selvin PR (2017) Labeling proteins inside living cells using external fluorophores for fluorescence microscopy. elife 6:e25460

Vaughan EE, DeGiulio JV, Dean DA (2006) Intracellular trafficking of plasmids for gene therapy: mechanisms of cytoplasmic movement and nuclear import. Curr Gene Ther 6(6):671–681

Voets O, Tielen F, Elstak E, Benschop J, Grimbergen M, Stallen J, Essrich C (2017) Highly efficient gene inactivation by adenoviral CRISPR/Cas9 in human primary cells. PLoS One 12(8):0182974

Wang J, Quake SR (2014) RNA-guided endonuclease provides a therapeutic strategy to cure latent herpesviridae infection. Proc Natl Acad Sci 111(36):13157–13162

Wang D, Mou H, Li S, Li Y, Hough S, Tran K, Li J, Yin H, Anderson DG, Sontheimer EJ (2015) Adenovirus-mediated somatic genome editing of Pten by CRISPR/Cas9 in mouse liver in spite of Cas9-specific immune responses. Hum Gene Ther 26(7):432–442

Wang M, Zuris JA, Meng F, Rees H, Sun S, Deng P, Han Y, Gao X, Pouli D, Wu Q (2016) Efficient delivery of genome-editing proteins using bioreducible lipid nanoparticles. Proc Natl Acad Sci 113(11):2868–2873

Weber T, Wefers B, Wurst W, Sander S, Rajewsky K, Kühn R (2015) Increasing the efficiency of homology-directed repair for CRISPR-Cas9-induced precise gene editing in mammalian cells. Nat Biotechnol 33(5):543–548

Woo JW, Kim J, Kwon SI, Corvalán C, Cho SW, Kim H, Kim S-G, Kim S-T, Choe S, Kim J-S (2015) DNA-free genome editing in plants with preassembled CRISPR-Cas9 ribonucleoproteins. Nat Biotechnol 33(11):1162–1164

Worthen CA, Rittié L, Fisher GJ (2017) Mechanical deformation of cultured cells with hydrogels. In: Fibrosis. Springer, Berlin, pp 245–251

Wu Z, Yang H, Colosi P (2010) Effect of genome size on AAV vector packaging. Mol Ther 18(1):80–86

Wu Y, Liang D, Wang Y, Bai M, Tang W, Bao S, Yan Z, Li D, Li J (2013) Correction of a genetic disease in mouse via use of CRISPR-Cas9. Cell Stem Cell 13(6):659–662

Wu Y, Zhou H, Fan X, Zhang Y, Zhang M, Wang Y, Xie Z, Bai M, Yin Q, Liang D (2015) Correction of a genetic disease by CRISPR-Cas9-mediated gene editing in mouse spermatogonial stem cells. Cell Res 25(1):67–79

Xie K, Minkenberg B, Yang Y (2015) Boosting CRISPR/Cas9 multiplex editing capability with the endogenous tRNA-processing system. Proc Natl Acad Sci 112(11):3570–3575

Xing H-L, Dong L, Wang Z-P, Zhang H-Y, Han C-Y, Liu B, Wang X-C, Chen Q-J (2014) A CRISPR/Cas9 toolkit for multiplex genome editing in plants. BMC Plant Biol 14(1):327

Xu CL, Ruan MZ, Mahajan VB, Tsang SH (2019) Viral delivery systems for CRISPR. Viruses 11(1):28

Xue W, Chen S, Yin H, Tammela T, Papagiannakopoulos T, Joshi NS, Cai W, Yang G, Bronson R, Crowley DG (2014) CRISPR-mediated direct mutation of cancer genes in the mouse liver. Nature 514(7522):380–384

Yan M, Du J, Gu Z, Liang M, Hu Y, Zhang W, Priceman S, Wu L, Zhou ZH, Liu Z (2010) A novel intracellular protein delivery platform based on single-protein nanocapsules. Nat Nanotechnol 5(1):48–53

Yánez-Munoz RJ, Balaggan KS, MacNeil A, Howe SJ, Schmidt M, Smith AJ, Buch P, MacLaren RE, Anderson PN, Barker SE (2006) Effective gene therapy with nonintegrating lentiviral vectors. Nat Med 12(3):348–353
Yang H, Wang H, Shivalila CS, Cheng AW, Shi L, Jaenisch R (2013) One-step generation of mice carrying reporter and conditional alleles by CRISPR/Cas-mediated genome engineering. Cell 154(6):1370–1379
Ye L, Wang J, Beyer AI, Teque F, Cradick TJ, Qi Z, Chang JC, Bao G, Muench MO, Yu J (2014) Seamless modification of wild-type induced pluripotent stem cells to the natural CCR5Δ32 mutation confers resistance to HIV infection. Proc Natl Acad Sci 111(26):9591–9596
Yin H, Kanasty RL, Eltoukhy AA, Vegas AJ, Dorkin JR, Anderson DG (2014a) Non-viral vectors for gene-based therapy. Nat Rev Genet 15(8):541–555
Yin H, Xue W, Chen S, Bogorad RL, Benedetti E, Grompe M, Koteliansky V, Sharp PA, Jacks T, Anderson DG (2014b) Genome editing with Cas9 in adult mice corrects a disease mutation and phenotype. Nat Biotechnol 32(6):551–553
Yosef I, Manor M, Kiro R, Qimron U (2015) Temperate and lytic bacteriophages programmed to sensitize and kill antibiotic-resistant bacteria. Proc Natl Acad Sci 112(23):7267–7272
Yosef I, Goren MG, Globus R, Molshanski-Mor S, Qimron U (2017) Extending the host range of bacteriophage particles for DNA transduction. Mol Cell 66(5):721–728. e723
Young SM, McCarty DM, Degtyareva N, Samulski RJ (2000) Roles of adeno-associated virus Rep protein and human chromosome 19 in site-specific recombination. J Virol 74(9):3953–3966
Young LS, Searle PF, Onion D, Mautner V (2006) Viral gene therapy strategies: from basic science to clinical application. J Pathol 208(2):299–318
Zuckermann M, Hovestadt V, Knobbe-Thomsen CB, Zapatka M, Northcott PA, Schramm K, Belic J, Jones DT, Tschida B, Moriarity B (2015) Somatic CRISPR/Cas9-mediated tumour suppressor disruption enables versatile brain tumour modelling. Nat Commun 6(1):1–9
Zuris JA, Thompson DB, Shu Y, Guilinger JP, Bessen JL, Hu JH, Liu DR (2015) Cationic lipidmediated delivery of proteins enables efficient protein-based genome editing in vitro and in vivo. Nat Biotechnol 33(1):73–80

Reengineering of the CRISPR/Cas System **5**

Zulqurnain Khan, Tahmina Sattar, Maria Siddique, Zulfiqar Ali, Asif Ali Khan, Syed Ovais Aftab, Muhammad Zubair Ghouri, Qaisar Sultan, Nauman Gulzar, and Farooq Ahmad

Contents

Z. Khan (✉) · T. Sattar · M. Siddique · Z. Ali · A. A. Khan
Institute of Plant Breeding and Biotechnology, Muhammad Nawaz Sharif University of Agriculture, Multan, Pakistan
e-mail: zulqurnain.khan@mnsuam.edu.pk

S. O. Aftab
Center for Agricultural Biochemistry and Biotechnology (CABB), University of Agriculture, Faisalabad, Pakistan

M. Z. Ghouri
Center for Agricultural Biochemistry and Biotechnology (CABB), University of Agriculture, Faisalabad, Pakistan

Center for Advanced Studies in Agriculture and Food Security (CAS-AFS), University of Agriculture, Faisalabad, Pakistan

Q. Sultan
Department of Botany, Government College University, Faisalabad, Punjab, Pakistan

N. Gulzar
Department of Biochemistry, University of Agriculture, Faisalabad, Punjab, Pakistan

F. Ahmad
Department of Zoology, The Islamia University of Bahawalpur, Bahawalpur, Punjab, Pakistan

A. Ahmad et al. (eds.), *The CRISPR/Cas Tool Kit for Genome Editing*,
https://doi.org/10.1007/978-981-16-6305-5_5

Abstract

Engineering genomes through programmable nucleases (PNs) is a versatile approach used to create double-stranded breaks (DSBs) which can be repaired subsequently by non-homologous end joining (NHEJ)-mediated knockout or homology-directed repair (HDR)-mediated knock-in pathways. This system is well established in many organisms, including humans and animals. In this book chapter, we have collated information about reengineering of versatile CRISPR/Cas9 system to increase specificity and reduce the off-target effects. Specificity of CRISPR/Cas9 system can be increased by minimization of undesired mutations, generating Cas9 orthologs with expanded PAM requirements and have reduced off-target effects. In addition, we describe approaches to be used to minimize off-target mutations such as modification of sgRNAs, chemical modification of guide RNA, and increasing PAM specificity. Finally, we discussed multiplex genome editing through CRISPR/Cas9 system and its potential applications and challenges.

Keywords

CRISPR/Cas9 · NHEJ · HDR · PAM · Knock-in · Knock-out

Abbreviations

BAC	Bacterial artificial chromosome
BLESS	Breaks labelling, enrichment on streptavidin and next-generation sequencing
ChIP	Cross-linking chromatin immunoprecipitation
CRISPR	Clusters of regularly inter-paced small palindromic repeats
dCas9	Dead Cas9
Digenome-seq	Digested genome sequencing
DSBs	Double-stranded breaks
FKBP	FK506 binding protein 12
FRB	FKBP rapamycin binding
GFP	Fluorescent protein
GMOs	Genetically modified organisms

gRNA	Guided ribose-nucleic acid
HDR	Homology-directed repair
HR	Homologous recombination
HTGTS	High-throughput genome-wide translocation sequencing
IDLV	Integrase-defective lentiviral vectors
KO	Knockout
NHEJ	Non-homologous end joining
PAM	Protospacer adjacent motif
PNs	Programmable nucleases
RAG1	Recombination activating gene 1
RFN	RNA-guided FokI-dCas9 nuclease
RGEN	RNA-guided engineered nucleases
RNA	Ribose-nucleic acid
RNP	Ribonucleoprotein
sgRNAs	Single-guide RNAs
SpCas9	*Streptococcus pyogenes* Cas9
TALENs	Transcription activator-like effector nucleases
TAT	Tyrosine aminotransferase
tru-gRNAs	Truncated gRNAs
WAS	Wiskott-Aldrich syndrome
ZFNs	Zinc finger nucleases

5.1 Generating Knockout and Knock-In with CRISPR/Cas9

Genetically modified organisms (GMOs) have been playing a massive role in molecular biology and biotechnology by producing innovative and disease-free organisms. Genome editing tools like zinc finger nucleases (ZFNs), transcription activator-like effector nucleases (TALENs), and clustered regularly interspaced short palindromic repeats (CRISPR)/CRISPR-associated (Cas) nucleases are used to generate GM animals through microinjection into fertilized eggs (Yoshimi et al. 2016). CRISPR/Cas9 technology has proven to be an efficient system for the generation of targeted modifications in animals as well as plant species. PNs recognize the long DNA sequences and cleave the target by creating a double-stranded break (DSB), which then be repaired by NHEJ via the small insertion or deletion at the site of cleavage. NHEJ pathway is generally used for the generation of organisms with efficient knockout (KO) alleles at the targeted regions in the genome (Hu et al. 2013). PNs are usually used to create targeted knock-ins (KI) using the pathway of homology recombination (HR). In the HR pathway, the DNA fragment of donor plasmid is utilised, and for the selection, green fluorescent protein (GFP) is usually used (Ma et al. 2014). In mammalian or embryo cells, DSB repairs through the pathway of HR which is less successful than that of NHEJ pathway (Sonoda et al. 2006; Peng et al. 2014). During the stages of cell cycle, the NHEJ is the only preferred pathway that takes place, but the other repairing pathways are confined to cell phases. HR-mediated repair pathway occurs during S/G2 phase of the cell cycle, which is precise and involved in the creation of gene KIs. The successful reports of HR-mediated KIs were observed in many cells and organisms like human pluripotent stem cells CpFTSY and ZEP gene knock-in in *Chlamydomonas reinhardtii* (Baek et al. 2016), Pax6-IRES-EGFP knock-in in mouse (Inoue et al. 2018), primary human T cells (Schumann et al. 2015), bovine-induced pluripotent cells (Heo et al. 2015), and eGFP and Cre knock-in in rats (Ma et al. 2014). We are using different platforms and reagents provided by Addgene, VectorBuilder, GeneCopoeia, Home – Nootropics Frontline, etc. Some of the CRISPR-based knock-in/knockouts reported so far are summarized in Table 5.1.

5.2 Increasing Specificity of Cas Protein

The RNA-guided Cas9 endonuclease emerges as an effective, precise, and powerful tool for genome editing of living organisms with relative ease and equal success. CRISPR/Cas9 technology has been adopted widely by many of researchers working in life sciences. The major outlook of CRISPR/Cas9 system is to increase its specificity to the target sites for reducing the off-targeting (Ryan et al. 2018). Genetic disorders have been treated by directly targeting the mutations sites with CRISPR/Cas9 that hold the greatest potential for therapeutic studies (Tycko et al. 2016). The main disadvantage of this technology is the DNA cleavage by Cas9 protein at off-target sites. Researchers are working to enhance the efficiency of Cas9 protein, and it is still in progress which included the significant improvements in guide RNA selection, use of novel enzymes, assays used to detect off-target effects, gRNA and protein engineering, Cas9 variants and orthologs, coupling Cas9 to artificial inhibitory domains, inhibition of non-homologous end joining, increased Cas9 specificity, and chemical modifications in sgRNAs (Vartak and Raghavan 2015; Tycko et al. 2016; Cebrian-Serrano and Davies 2017; Ryan et al. 2018; Aschenbrenner et al. 2020). Additionally, the efforts have also been made by researchers to understand and enhance the specificity of Cas9 protein (Makarova and Koonin 2015). Presently, in wild-type Cas9, various modifications have been done, which is widely being used for genome editing and derived from *Streptococcus pyogenes*.

5.2.1 Guide RNA (gRNA) Selection Through Computational Approaches and Predictive Models

For in silico gRNA designing, several online tools and computational models have been developed. Cas-OFFinder is an online tool for the designing of guide RNA, which identifies the off-targets and ranks RNAs via relative orthogonality, while other tools rank the gRNAs using specificity score calculation (Hsu et al. 2014). These in silico tools rank gRNAs and provide valuable information about their optimization, increasing the efficacy of Cas9 to cleave the specific sites in genomic

Table 5.1 CRISPR-based gene knock-in/knockouts in animal cells and crop plants

Crops/ organisms	Target gene	Type of edit	Results	References
Chinese hamster ovary (CHO) cells	IGF-IR gene	Gene knockout and/or knock-in	Tenfold increase in IGF-1 productivity	Romand et al. (2016)
Primary human T cells	CXCR4, PD-1 (PDCD1)	Knock-in	Tumor immunotherapy	Schumann et al. (2015)
Human pluripotent stem cells	SOX2, PAX6, OTX2, AGO2	Knockout	Temporal expression cells for study of gene functions in human	Chen et al. (2015)
Bovine-induced pluripotent cells	NANOG locus	Knock-in	Improve transgenic livestock production	Heo et al. (2015)
Sheep	MSTN	Knockout	Inhibition of myostatin production	Crispo et al. (2015)
Mouse	Pax6-IRES-EGFP	Knock-in	Neurodevelopmental dynamics	Inoue et al. (2018)
Pig	H11 locus	Knock-in	Stable and robust transgene expression	Ruan et al. (2015)
	Igfbp4 and AqpI genes	Gene knockout	Increase in adaptation of CHO-K1 cells to suspension culture	Lee et al. (2016a, b)
Crop plants				
Banana	Phytoene desaturase	Gene knockout	Decreased chlorophyll and total carotenoid contents	Kaur et al. (2018)
Cotton	CLCuD IR and rep regions	Gene knockout	Targeted cleavage activity of multiple viruses for multiple infection and associated DNA satellites, such as CLCuD-complex	Iqbal et al. (2016)
Theobroma cacao	TcNPR3, a suppressor of the defense response	Gene knockout	Increased resistance to *Phytophthora tropicalis* (cocoa pathogen)	Fister et al. (2018)
Rice	*OsSWEET11*, *OsSWEET14* (rice bacterial blight susceptibility genes)	Knockout of promoter region	The promoter of the blight susceptibility gene was disrupted	Xu et al. (2019)
	CAO1	Knockout	Reduction of pale-green leaf	Jung et al. (2021)
	OsAnn3 gene	Knockout	Increase cold tolerance in rice	Shen et al. (2017)
Wheat		Gene knockout	The number of mildew microcolonies formed	Wang et al. (2014)

(continued)

Table 5.1 (continued)

Crops/ organisms	Target gene	Type of edit	Results	References
	TaMLO-A1, *TaMLO*-B1, and *TaMLO*-D1		on the leaves was significantly reduced against the control, and no apparent fungal growth was observed on the leaves of edited plants	
Cassava	ALS	Knock-in	Developing tolerance to sulfonylurea herbicides	Endo et al. (2016), Gomez et al. (2017)
A. thaliana	NPTII	Knock-in	Resistant to kanamycin	Schiml et al. (2014), Mao et al. (2013)
	TFL1	Knock-in	Expression using eGFP	Zhao et al. (2016)
N. benthamiana	NtPds	Knockout	Phenotypic expression is albino	Gao et al. (2015)
	NtPDR6	Knockout	Having more than two number of branches	Gao et al. (2015)
S. lycopersicum	SlAGO7	Knockout	Tinny pointer-like and muscular leaves	Brooks et al. (2014)
	ANT1	Knock-in	Ectopic accumulation of pigment	Čermák et al. (2015)
H. vulgare	HvPM19	Knockout	Grain dormancy	Lawrenson et al. (2015)
B. oleracea	BolC.GA4	Knockout	Phenotypic expression is dwarf	Lawrenson et al. (2015)
Z. mays	ALS2	Knock-in	Resistant to herbicides sulfonylurea and replacement of gene	Svitashev et al. (2015)
	ARGOS8	Gene knock-in	Drought stress tolerance	Shi et al. (2017)
Populus trichocarpa	potPDS	Knockout	Phenotypic expression is albino	Ma et al. (2016)
	4CL1	Knockout	Accumulation of lignin reduced	Tsai et al. (2020)
	4CL2	Knockout	Tannins content reduced	Tsai et al. (2020)
G. max	DD43 region	Knock-in	Hygromycin resistance	Kleinstiver et al. (2015)
	ALS1	Knock-in	Resistance to sulfonylurea herbicides, gene replacement	Kleinstiver et al. (2015)

(continued)

Table 5.1 (continued)

Crops/ organisms	Target gene	Type of edit	Results	References
Camelina sativa	FAD2	Gene knockout	Decreased fatty acids, i.e., polyunsaturated in nature	Jiang et al. (2017)
Grapefruit	CsLOB1 promoter	Gene knockout	Alleviated citrus canker	Jia et al. (2016)
Orange	CsLOB1 promoter	Gene knockout	Resistant to citrus canker disease	Peng et al. (2017)
Cucumber	eIF4E	Gene knockout	Virus resistance	Chandrasekaran et al. (2016)
Potato	Wx1	Gene knockout	High contents of amylopectin	Andersson et al. (2017)
Mushroom	PPO	Gene knockout	Anti-browning phenotype	Waltz (2016)
Flax	EPSPS	Gene knock-in	Resistant to herbicide	Sauer et al. (2016)

DNA. The Cas9 specificity also depends on other variables including cell type, species, dosage amount, and delivery methods. In general, well-designed gRNAs avoid off-targets by one or two mismatches at target sites. In addition, improved models predict off-target by three or four mismatches at target sites. In addition, online tools may provide ease in practical usage for most of the users, but offline tool such as CRISPRseek and many others provide the ease and flexibility to bioinformaticians. Some online/offline tools for gRNA designing are summarized in Table 5.2.

5.2.2 Cas9 Orthologs and Variants

At present, various orthologs of Cas9 proteins have been engineered for genome editing such as *Staphylococcus aureus* Cas9 (SaCas9), smaller in size, about 1 kb than the previously used SpCas9, and another is packed AVV which is paired with gRNA cassettes for genome editing (Ran et al. 2015). SaCas9 exhibits promising nuclease activity with target sequence of about 21–24 nt using a 5′-NNGRRT PAM as compared to SpCas9 with 20 nt using 5′-NGG PAM recognition sites. Variants of Cas9s derived from *Neisseria meningitidis* (NmCas9) and *Streptococcus thermophilus* (StCas9) in mammalian cells are through the system of CRISPR1 and CRISPR3 (Müller et al. 2016). These StCas9 and NmCas9 are more efficient than SpCas9 as these require more complex PAM recognition sites 5′-NNNNGAT and long crRNA spacer (21–24 nt) (Lee et al. 2016a, b). So, by using longer PAM recognition sequence, off-targets can be minimized by limiting the target range: for example, small PAM sequence NGG motifs occur every 8 bp, but the longer PAM sequence NNNNGATT was found every 128 bp which reduces the off-target effects in the given target sites. Interestingly, with improved specificity or on-targets, Cas9

Table 5.2 Online/offline tools for designing gRNA

Tools	Specificity score	Off-target analysis	Species support	Websites	References
Benchling	CFD	Yes	22	http://benchiling.com	Hsu et al. (2013)
Optimized CRISPR-design	Off-target scoring	Yes	15	https://horizondiscovery.com/en/products/tools/CRISPR-Design-Tool	Prykhozhij et al. (2015)
ZiFit	Orthogonality of off-targets	No	9	http://zifit.partners.org/ZiFiT/	Hwang et al. (2013)
E-CRISP	Orthogonality is based on the relative scoring	Yes	34	http://www.e-crisp.org/E-CRISP/	Heigwer et al. (2014)
CHOPCHOP	Off-target scoring	Yes	19	https://chopchop.cbu.uib.no/	Montague et al. (2014)
SgRNACas9	Off-target scoring	Yes	N.A.	https://www.takarabio.com/learning-centers/gene-function/gene-editing/gene-editing-tools-and-information/sgrna-design-tools	Xie et al. (2014)
SgRNA designer	Activity score-type II	No	N.A.	https://portals.broadinstitute.org/gpp/public/analysis-tools/sgrna-design	Doench et al. (2014)
CRISPRseek	Off-target scoring	Yes	N.A.	http://bioconductor.org/packages/release/bioc/html/CRISPRseek.html	Zhu et al. (2014)
CRISPR multitargeter	Activity score-type II	Yes	N.A.	https://multicrispr.net/	Prykhozhij et al. (2015)
CRISPR direct	Off-target scoring	Yes	18	https://crispr.dbcls.jp/	Naito et al. (2015)
EuPaGDT	Comparative ranking on the basis of orthogonality	Yes	28	http://grna.ctegd.uga.edu	Tycko et al. (2016)
Broad GPP portal	CFD	No	2	https://portals.broadinstitute.org/gpp/public/analysis-tools/sgrna-design	Doench et al. (2016)

(continued)

Table 5.2 (continued)

Tools	Specificity score	Off-target analysis	Species support	Websites	References
CROP IT	No	No	2	http://cheetah.bioch.virginia.edu/AdliLab/CROP-IT/	Singh et al. (2015)
Cas-OFFinder	No	Yes	34	http://www.rgenome.net/cas-offinder/	Bae et al. (2014)
CCTop	CFD	Yes	N.A.	https://crispr.cos.uni-heidelberg.de/	Stemmer et al. (2017)

orthologs have been discovered and engineered for future use (Friedland et al. 2015; Ran et al. 2015).

Recently, new and improved protein of CRISPR/Cas9 has been discovered, i.e., Cpf1, which is efficiently used in genome editing of human cell lines. Cpf1 targets the single-guide RNA and recognize PAM sequence of 5′-TTN at the site of target and create the overhang with staggered DSB proximal to the sequence of PAM (Yamano et al. 2016). Cpf1 genome-wide specificity can be visualized by using Digenome-seq and GUIDE-seq; eight orthologs of Cpf1 were screened in human cells, out of which two are involved for cutting the specific target site (Kleinstiver et al. 2015; Kim et al. 2016). Other putative proteins such C2C1, C2C2, and C2C3 are known as CRISPR effector proteins involved in the metagenomic datasets. In bacterial cells, C2C2 protein acts as RNA-guide RNA-targeting nuclease. When C2C2 protein binds to complementary sequence of target RNA, it acts as RNase and hampers the growth of cells. It remains ambiguous to identify in which manners these new CRISPR proteins carry diverse properties. The targeting specificity is affected by the PAM sequences, but the engineering and discovery open up new avenues for genome editing in the future beyond SpCas9.

5.2.3 Protein Engineering Manifest PAMs

SpCas9 has emerged as rapidly optimizing tool with increased precision in their structure and molecular mechanism by the efforts of biological engineers. Cas9 has the two different types of nuclease domains, i.e., RuvC and HNH, according to the homology in their sequence (Sapranauskas et al. 2011). SpCas9 can be converted into nickase after creating point mutation in the RuvC domain (Cong et al. 2013). Single DNA nicks repaired with greater efficiency than DSBs; however, paired gRNAs create offset nicks which resulted in indel formation, so specificity is 1500-folds less, compared to single DNA nicks (Friedland et al. 2015). Mutation in the second nuclease domain of Cas9, HNH, results in production of fully inactive nuclease “dCas9” (Nishimasu et al. 2014). Specificity can also be improved by expressing paired gRNA which creates dimerization-dependent system by combining the dCas9 with fok1 nuclease domain (Tsai et al. 2014; Guilinger et al. 2014; Bolukbasi et al. 2015).

However, additional components and longer transgene are the requirements for the abovementioned techniques to work properly. These limitations can be overwhelmed by using other strategies such as a crystal structures and strategy of rational engineering which generate SaCas9 and SpCas9 nucleases with "enhanced specificity" (eSaCas9 and eSpCas9 respectively). These Cas9's differs at non-target DNA strand groove with substitution of just three or four codons (Slaymaker et al. 2016).

The specificity of Cas9 usually depends on the PAM sites, so alternative PAM sequences increase the target sites in the genome. Similarly, it is a powerful strategy of directed evolution of PAM sequences as previously described for SpCas9 and SaCas9 (Kleinstiver et al. 2015). Directed evolution forms single-point mutations for SpCas9 with improved specificity for 5′-NGG PAMs over the 5′-NAG PAMs (Hsu et al. 2013; Jiang et al. 2013a, b). In SaCas9, directed evolution is used to modify the PAM 5′-NNGRRT into 5′-NNNRRT which enhances the target specificity from ~2- to ~4-folds (Kleinstiver et al. 2015). Directed evolution approach holds significant importance because it does not require any structural information. This approach also has significant importance in the therapeutic target and just requires specific editing and lacking the canonical PAMs. So, altered PAM sequences increase the specificity and fidelity, and other approaches like protein engineering and directed evolution are beneficial approaches for the generation of modified Cas9s that recognize AT-rich PAMs which is usually used to target AT-rich genomic sites.

5.2.4 Improvement in Cas9 Specificity Via dCas9

The catalytically inactivated dCas9 (dCas9) fuses with other effector domains and controls the factors related to epigenetic and transcriptional rate (Hsu et al. 2014). The mechanism of active Cas9 shows that initially Cas9 scan the genomic DNA for PAM site, afterwards, 5 bp seed sequence of gRNA stabilizes the transient binding state (Wu et al. 2014; Jiang et al. 2013a, b, 2015; Kuscu et al. 2014). The dCas9 off-targeting is obtained by using the mixture of biased or unbiased approaches. The unbiased approach like whole-transcriptome RNA-seq also used dCas9-effector for the assessment of off-target mutations (Chen et al. 2013; Briner et al. 2014). These assays confirm that dCas9 activator upregulates the genes at the transcription site (Dang et al. 2015). These results showed the importance of considering the off-target effects associated with Cas9.

5.2.5 Other Modifications to Increase Specificity of Cas Protein

The strategies to engineer Cas9 and to increase its specificity by modifying the gRNA-based specificity of Cas protein for improvement are described below. There are six functional modules which are characterized by the single or dual gRNA system for type II CRISPR system (Briner et al. 2014). On-targeting efficacy can be achieved via mutation, extension, and/or deletion of these modules, e.g., transcription rate is enhanced by disturbing the downstream site of the spacer DNA, whereas extension up to 5 nt in sgRNA duplex region can improve RNA structure (Chen et al. 2013; Dang et al. 2015). Further studies revealed that gRNA is an important component for improving specificity in CRISPR technology. The Cas9 recognition and site-specific cleavage is enhanced by using 2–3 bp shortened gRNA which is called tru-gRNAs, which determines RNA-DNA complementary region away from the PAM sequence. In human cells, tru-gRNA has the ability to reduce the possibility of off-targets to the extent of about 5000-folds or more when compared to the efficiency of gRNA (Fu et al. 2014). The GUIDE-seq analysis for the specificity of Cas9 with various tru-RNAs reveals that the off-targeting effects in a genome were significantly reduced by 2–5-folds in contrast to Cas9 (Tsai et al. 2015). GUIDE-seq assays revealed that all the off-target effects are not reduced. Few new off-target sites are also created after the shortening of gRNA.

Another way to elevate Cas9 site specificity to reduce off-targeting is to use modified gRNAs having two extra "G" nucleotides at 5′end (Cho et al. 2014). Due to increase in the length of gRNA, on-target activity is compromised (Kim et al. 2015). Limiting the duration of Cas9 activity can increase its specificity along with reduced off-target effects in a genome. Among several methods used for Cas9/gRNAs delivery in the cell, the most frequently used method is to transfect the plasmid DNA vector into the cell for initiating the Cas9/gRNA expression. Cas9/gRNA existence in the cell for an extended period has greater chances of causing off-target effects in a particular genome. Therefore, other delivery methods such as RNPs and electroporation provide limited time for Cas9/gRNA to retain in cells that limits the off-target effects. The Western blot assay revealed that Cas9 protein delivered by RNPs degrades within 24 h, but continues to express for several days if delivered via plasmid; thus, on-targeting efficiency of Cas9 protein with RNPs is 13-fold greater than that delivered with plasmid in human cancer cell line (Kim et al. 2014). Inducible Cas9 is active only in the supplied period of stimulus that reduced the activity time of Cas9.

Split Cas9 consists of two domains, i.e., FK506 binding protein 12 (FKBP) and FKBP rapamycin binding (FRB). Split Cas9 in combination with lentiviral vector limits the off-target activity (Zetsche et al. 2015; Wright et al. 2015). Using inducible Cas9 the on-target specificity is enhanced in human cells up to 25 by intein-based inducible system (Davis et al. 2015). A recent way to enhance Cas9 specificity is to use photoactivatable Cas9, a system called optogenetic that usually controls the activity of Cas9 (Nihongaki et al. 2015).

Finally, some alternative ways are present to modify the 20-nt gRNA sequence which could enhance the targeting specificity of Cas9. Still there are lots of methods that are developed for robust and on-target gRNAs designing (Dellinger et al. 2011). The chemical modification in the three first and last nucleotides of sgRNAs could enhance the CRISPR/Cas9 indels rate and HDR pathway (Hendel et al. 2015).

The terminal modification of nucleotides in sgRNAs enhances the resistance in cells against exonucleases. Furthermore, crRNA modification enhances the rate of indel formation in the human cells (Rahdar et al. 2015; Lee et al. 2017). The chemical modification can also enhance the targeting specificity of tru-crRNAs using ribo-phosphate backbone with 2′-*O*-methine-4′ bridge phosphonothioates and 2′-fluoro-substituents in a special combination (Rahdar et al. 2015). CRISPR/Cas9 with increased specificity and their modifications are discussed in Table 5.3.

5.3 Reducing off-Target

Flexibility of CRISPR/Cas9 system is very beneficial for targeted genetic modification, but it has unintended side effects in the genome that are not related to the target regions. Off-targeting effects are linked to Cas9 and gRNAs in RNA-guided engineered nucleases (RGENs). The optimum PAM sequence recognized by Cas9 protein for cleavage in *S. pyogenes* is 5′-NGG-3′. Cas9, on the other hand, has different PAM recognition sites, i.e., 5′-NGA-3′ or 5′-NAG-3′. Due to these PAM sequences, the specificity of Cas9 is reduced resulting in unwanted mutations in the targeted genome (Hsu et al. 2013). RGENs accept 20 nt mismatches in the gRNA and targeted DNA sequence. RGENs produce DNA or RNA bulge after creating a cleavage site with some additional or missed nucleotides (Lin et al. 2014). The inaccurate DNA repairing in off-target site response to cleavage the chromosomal rearrangements causing deletion (Lee et al. 2010), translocations (Brunet et al. 2009; Cho et al. 2014), and inversions (Park et al. 2014). Chromosomal rearrangements are linked with cancer as associated with the activation of oncogenes in the genome, so the off-target effects must be considered and eliminated as much as possible (Koo et al. 2015).

A number of techniques have been used to reduce these mutations such as a high-fidelity variant of SpCas9 (SpCas9-HF1) and a design to reduce the non-specific contacts to the DNA strand via the alteration in the amino acid sequences. Improving or optimizing the genome-wide specificities of other Cas9 orthologs and engineered variants has been suggested as a strategy for decreasing off-targeting. The efficacy of SpCas9-HF1 was analyzed with 37 single-guide RNAs in human cells, indicating that SpCas9-HF provides on-target hits >85% compared to the wild-type SpCas9 (Kleinstiver et al. 2016). When eight different sgRNAs were targeted to the non-repetitive sequences in human cells, most off-target effects created by SpCas9-HF1 were not detected. In research and therapeutic applications, SpCas9-HF1 used as an alternative method due to minimized off-targeting and enhanced precision.

5.3.1 CRISPR/Cas9 and Methods to Reduce off-Target Mutations

The choice of a distinct target site is one of the important points to be kept in mind for minimal off-targets. In addition, the dissimilarity from other sequences in the

Table 5.3 Improvements towards CRISPR/Cas9 specificity

Improvement strategies	Description	Advantages	Disadvantages	References
Computational guide selection				
Specificity score and CFD	Scoring and off-target site detection	Differentiate guide RNAs designing and ranks off-target sites	SpCas9 data provides scoring of 20-mer guides	Doench et al. (2016)
WGS with respect to reference genome	WGS of related animals, patient, and various cell lines	Due to genetic changes, new off-target sites are identified that are absent in the reference genome (i.e., hg38)	Costly	Yang et al. (2014)
Protein engineering				
Single or paired nickases	Nickase activity is generated by mutating either one domain of the Cas9	The repairing pathway for nicks are HDR and efficiently repaired than the DSBs	On-target editing is less effective	Cong et al. (2013)
SpCas9 PAM variant D1135E	A single-point mutation generates 5′-NGG PAM and thus increases the specificity	Important reduction in editing at 5′-NAG and 5′-NGA PAMs	On target efficiency may be compromised	Kleinstiver et al. (2015)
eSpCas9	Cas9 activity is weakened by creating 3 mutations in the nt groove	Off-target effects can be detected by using BLESS and deep sequencing	On-target efficiency may be compromised	Slaymaker et al. (2016)
Neisseria meningitidis Cas9	This protein belongs to class II type II Size of protein is about 1109 amino acid PAM recognition site "NNNNGATT," target length 23–24 nt	Like other class II type II proteins	Like other class II type II proteins	Hou et al. (2013)
Acidaminococcus Cas12a (Cpf1)	Class II type V amino acids, i.e., 1308–1310 and PAM recognition	Staggered DSB Short PAM	PAM requirement T-rich	Zetsche et al. (2015)

(continued)

Table 5.3 (continued)

Improvement strategies	Description	Advantages	Disadvantages	References
	site is "TTTV" target length 23, 24 nt			
Leptotrichia buccalis Cas13a (C2c2)	This protein belongs to class II type VI and has 1159 amino acids	Target RNA	Long target site	East-Seletsky et al. (2016)
Dimeric dCas9-FokI	This engineered protein has size of about 1817 amino acids, and PAM recognition site of this protein is "NGG," and targeted length is about 20 nt	Big size Need two domains of FokI	Less off-targeting	Sakuma et al. (2016)
High-fidelity *Streptococcus pyogenes* Cas9 SpCas9-HF	This engineered protein has size of about 1368 amino acids, and PAM recognition site of this protein is "NGG," and targeted length is about 20 nt	–	Great potential for genome editing in plants especially base editing and epigenetic modifications/ gene regulation	Kleinstiver et al. (2015)
Streptococcus pyogenes Cas9 nickase	This engineered protein has size of about 1368–1424 amino acids, and PAM recognition site of this protein is "NGG," and targeted length is about 20 nt	Produces single cut needs nickase	Less off-targeting	Eggenschwiler et al. (2016)
Streptococcus pyogenes Cas9: eSpCas9 with enhanced specificity	This engineered protein has size of about 1424 amino acids, and PAM recognition site of this protein is "NGG," and	High specificity and precision	Off-targeting	Slaymaker et al. (2016)

(continued)

Table 5.3 (continued)

Improvement strategies	Description	Advantages	Disadvantages	References
	targeted length is about 20 nt			
RNA modifications				
Tru-guides	Tru-guides have 17–18 nt rather than 20 nt that allows less mismatches	Implementation is so simple and accurate	Efficiency is low, more off-targeting, and high rate of sequence homology	Fu et al. (2014), Cho et al. (2014)
Measurement				
Targeted deep sequencing	NGS of off-target sites using computational method and quantify the reads that are close to the PAM site	Quantifiable, complex, and accessible	Biased to off-target sites	Ran et al. (2013)
GUIDE-seq	DSBs identified by using dsODN that further provide site for sequencing	Wet-lab or web-based method, process the data using online tools	Competent delivery method requires for dsODN, toxic for some cell types not used for in vivo modeling	Tsai et al. (2015)
Digenome-seq	In vitro gDNA samples are digested using Cas9-RNP complex	Employed for any type of cell as digestion of extracted gDNA sample is difficult in nature	WGS may be costly	Kim et al. (2015)
Delivery				
SaCas9 (plasmid-based)	Cas9 ortholog, i.e., *Staphylococcus aureus* with size of about 3.2 kb, gRNA 20–23 nt with PAM 5′-NNGRRT	Smaller size as AVV Packing in one vector with 1 or 2 gRNAs BLESS, GUIDE-seq used to analyze specificity as well as targeted sequencing	Ortholog, less efficient and characterized	Ran et al. (2015)

(continued)

Table 5.3 (continued)

Improvement strategies	Description	Advantages	Disadvantages	References
RNP	Delivered using Cas9-RNP complex into the cells through the methods of lipofection or electroporation	On-target editing due to delivery of packed construct	Good for ex vivo experiments	Lin et al. (2014)

genome with 2–3 nt change is promising to avoid off-targeting mutation (Cho et al. 2014). Computer-based software have been used such as Cas-OFFinder (http://www.rgenome.net/cas-offinder/) that can search the potential off-target sites and unique target sites in the genome of 21 different organisms including humans. Another valuable tool, CRISPR-Design (https://horizondiscovery.com/en/products/tools/CRISPR-Design-Tool), is also available which predicts the sgRNAs with limited off-target mutations on the basis of site score. The user-friendly sgRNA designer program associated with the web tool such as CHOPCHOP (https://chopchop.cbu.uib.no/) can also be used to design gRNAs with least off-target effects.

For selecting the targets from the input sequences, the CRISPR-direct (https://crispr.dbcls.jp/) is an efficient and helpful functional tool to reduce the off-targeting mutation. CRISPR is an advanced editing technique that is used for the site-specific cleavage of the genomic DNA. The gRNA binds with the target sequence (20 nt) and develops the Cas9 endonuclease complex. The target sequence has two different types of requirements: Firstly, the target sequence must be present adjacent to the 5′NGG3′ called as protospacer adjacent motif (PAM). Secondly, the target sequence must be specific in the whole genome. If the target sequence is not specific in the entire genome, then the chance of developing off-target is increased. CRISPR-direct is an important software that is successfully used for selecting the specific target sequence and also avoids off-targeting. Moreover, this tool has the genomic sequence of multiple organisms and crops such as pig, mice, mouse, frog, chicken, rat, silkworm, rice, budding yeast, sorghums and many more (Naito et al. 2015).

Second, by combining four separate modified sgRNAs, off-target mutations can be reduced, while mutation efficiencies at on-target sites in the genome are maintained. sgRNAs have two additional 5′ guanine nucleotides that precisely cut the targeted DNA in human cells than typical sgRNAs. If the sgRNAs are modified by the accumulation of two guanine nucleotides, the transcriptional activity of U6 and T7 promoter is effectively controlled. The mechanism of these guanine nucleotides to control the transcriptional rate is still unclear. Another modified sgRNA is tru-sgRNA having 17 nt in place of 20 nt that could improve the targeting specificity in the cell (Alipanahi et al. 2015). This tru-sgRNA is more prone to cause off-target mutations due to less energy for binding at the site of sgRNA-DNA complementarity. So, it is recommended to use sgRNAs with 17 nt to enhance on-target cleavage.

The third type of modified sgRNA is paired nickase, which is used to create nicks or single-stranded breaks on either strand of DNA (Mali et al. 2013). Cas9 and ZFN nickases can be created by inactivating one of the domains of Cas9 as each domain cleaves either strand of DNA. D10A Cas9 nickase forms are more precise and effective than the H840A Cas9. The requirement for creating nicks is to use two sgRNAs and two PAM sequences for recognition.

Fourthly, recombinant Cas9 protein which is more on-target than the plasmid-encoded Cas9. The off-targeting associated with the RNP complex is minimized by using endogenous proteases immediately after the delivery in the cells. Furthermore, delivery using plasmid DNA is more prone to cause off-targeting as Cas9-sgRNA complex is expressed in the cell for many days after the delivery. The more efficient way to deliver Cas9-SgRNA complex is through RGEN in combination with RNP, as this delivery involves electroporation resulting in zero off-targeting (Kim et al. 2015). Another method of delivery is protein transduction or lipofection that gives transient expression and reduces off-targeting (Ramakrishna et al. 2014). Additionally, dimerization-dependent Cas9 is another alternate method to minimize off-targeting (Tsai et al. 2014; Guilinger et al. 2014). By combining tru-sgRNAs with dimerization-dependent RFNs, the undesirable cleavage activity can be reduced (Tsai and Joung 2016).

The inactive Cas9 with FokI domain (fCas9) further improves the on-target DNA cleavage activity. This fCas9 improves on-target DNA cleavage up to >140-fold than WTCas9 and fourfold than paired nickase (Guilinger et al. 2014). So, there is a need to further characterize and improve the Cas9 specificity and cleavage activity. Further, tru-RFN having 19 bp missing in their complementary site enhances the on-target activity up to 40% relative to the wild RFNs. The abovementioned approach helps in the editing of human cells efficiently and precisely (Wyvekens et al. 2015).

The wise use of Cas9 and double nicking approach reduced the off-target effects as well as opened up a new avenue for efficiently using these approaches in therapeutic applications (Mali et al. 2013). Furthermore, CRISPR/Cas9 activity is improved by the inhibition of NHEJ pathway that uses Scr7 inhibiting domain that directly targets the DNA ligase IV (Bischoff et al. 2020). Additionally, the effective way to deliver modified gRNA and Cas9 protein into the cells is through ribonucleoprotein-based method. This method reduced the cytotoxicity compared to plasmid-based method (Hendel et al. 2015).

5.3.2 Methods for Off-Target Detection

5.3.2.1 Computational Method

TagScan (https://ccg.epfl.ch//tagger/tagscan.html) is the software that is efficiently used for recognizing the off-target sequences in the mammalian genome, specifically in humans (Iseli et al. 2007). With the help of PAM sequence, the guide sequence has a single base insertion and deletion at the different sites of the genome. Moreover, COSMID is a web tool that is used for searching the off-site in a genome.

Due to indel mutation in the target DNA, COSMID is used efficiently to recognize the off-target activity in a genome. Different types of primers are used for this purpose, and these primers are designed for amplifying the required sequence in a genome. Numerous methods are being established to detect off-target effects in the DNA including cell-based and cell-free (in vitro) method along with their pros and cons.

5.3.2.2 Cell-Based Assays

This assay includes whole-genome sequencing (WGS) that is considered as unbiased method to recognize nuclease specificity. This method is usually used to identify single-cell clones and non-mosaic F1 animals whose genomes are modified by genome engineering (Smith et al. 2014). Some limitations are associated with this method as its lacks the sensitivity that extensively describes off-target effects or those effects that take place at lower frequencies in the cells (Tsai et al. 2014). It is difficult to perform WGS on millions of reads of cellular genome. In case of WGS of a large heterogeneous genome, the detection of off-targeting frequency becomes low. So, it is impractical to perform WGS for off-targeting detection for therapeutic application.

Several methods are developed to screen off-target mutations in silico and in vitro, but for in vivo only a few methods are available such as linear double-stranded integrase-defective lentiviral vectors (IDLV) (Fu et al. 2013; Mali et al. 2013). IDLV, fused with DNA to tag nuclease, is used to examine activity of DSBs in vivo, this method is usually used to identify off-target sites of zinc finger nucleases (Zhang et al. 2015). The modified system of IDLV detects off-target sites with frequency of about 1% when CRISPR/Cas9 and TALEN systems are used to target the genes in tyrosine aminotransferase (TAT) and Wiskott-Aldrich syndrome (WAS). So, this approach is considered as effective to enhance target efficiency of CRISPR/Cas9 in therapeutic and medicinal applications (Wang et al. 2015).

Cross-linking chromatin immunoprecipitation (ChIP) is a method to identify the protein-DNA interaction. When ChIP is combined with second-generation sequencing (ChIP-seq), it makes possible to identify the protein-DNA interaction at the genomic level (Park 2009). This method uses the dCas9 for stable binding of dCas9-sgRNA complex to the target site. ChIP-seq reported that when Trp53 locus is targeted using Cas9-gRNA complex, on-target cleavage occurs with only one off-target (Cencic et al. 2014; Kuscu et al. 2014). This method gives the information about the potential binding sites of versatile tool, i.e., CRISPR/Cas9. So, the major aim of these methods is to detect DSBs generated by Cas9-gRNAs during off-targeting mutations. Some other approaches to detect off-target sites in the genome such as IDLV, GUIDE-seq, and LAM-HTGTS are discussed later on (Teytelman et al. 2013).

The detection of mutation that occurs at off-target sites are still a challenge to the scientific community (Gabriel et al. 2015). Previously, T7 endonuclease I assay was used to detect off-target sites, but this assay has some limitations, such as not detecting the off-target sites that occurs <1% in the genome of an individual, high cost, and the difficulty in large-scale screening (Kim et al. 2009). Recently, many other methods have been developed for the detection of off-target sites including deep sequencing (detect frequency ranges from 0.01–0.1 to 11%), ChIP-seq, and web-based prediction tools (Heigwer et al. 2014). Web-based method is less practical as it missed the off-target sites that have less similarity of sequence. Many review papers suggest that active Cas9 nuclease produced very fewer off-target sites (Duan et al. 2014).

The off-target detection method GUIDE-seq consists of two step processes: (Adiego-Pérez et al. 2019) RNA-guided endonuclease (RGEN) create DSBs in human cells which are then tagged by 34 bp double-stranded oligodeoxynucleotide (dsODN) that leaves blunt ends that are further repaired by NHEJ pathway (Tsai et al. 2015; Alipanahi et al. 2015). In genomic DNA dsODN sites are identified at the nucleotide level using next-generation sequencing (Zhang et al. 2015).

High-throughput genome-wide translocation sequencing (HTGTS) is another approach to detect off-target sites in the DNA (Frock et al. 2015). This approach is based on the detection of translocations between the nuclease-induced "bait" DSB and off-target "prey" DSBs. HTGTS is useful because it can be applied in vivo to detect off-target effects. It just requires the active complex of Cas9-gRNA. There are few drawbacks that are associated with HTGTS: translocations induced by nucleases are very rare events in nature, and it needs a very large genome as an input.

In fixed cells the method for detecting DSBs is called BLESS (breaks labelling, enrichment on streptavidin and next-generation sequencing) (Yan et al. 2017). In this method snapshot of transient DSBs occurs in a cell through the direct in situ ligation of adaptors called biotinylated hairpin. BLESS is used to detect in vivo off-target sites in a tissues without being dependent on the internal cellular DNA repairing machinery (Ran et al. 2015). The disadvantage of using this method is that it only detects off-target mutation that occurs at a specific time, not the DSBs that are already cleaved in the permeabilized cells. This method requires approximately ten million cells, and it is challenging due to technical expertise requirements.

5.3.2.3 In Vitro Assay

Another in vitro method called digested genome sequencing (Digenome-seq) is used to detect DSBs created by nucleases (Kim et al. 2015). This method is done in vitro by digesting genomic DNA (after extraction from desired cells) as well as Cas9-gRNA ribonucleoprotein (RNP complexes). These genomic fragments of DNA are then subjected to sequencing with about 500 million reads. The cleaved sites of Cas9 are considered as those sites that have the same starting and end mapping positions. In vitro condition assay is made using purified DNA, though cell-based factors are not compromised, i.e., epigenetic factors, subcellular localization, and chromatin remodeling. Additionally, if the concentration of RNP complex increases, the detection of very small cleavage sites is possible that usually not detected by the cell-based methods.

Digenome-seq has some limitation as sequencing is done without the improvement of cleavage sequences which results in incomplete detection of nuclease-induced DSBs. Digenome-seq is performed using HiSeq X Ten system which uses

Table 5.4 Comparison of off-target detection assays

Assays	Types	Advantages	Disadvantages
T7E1 assay	Cell-based	Simple assay	Poor sensitivity, not cost-effective
GUIDE-seq (genome-wide unbiased identification of DSBs enabled by sequencing)	Cell-based or live cells	Simple, easy, and precise tool, and for data analysis different types of open sources are available	Cytotoxic, limited to cell type, less sensitive, and not for in vivo
ChIP-seq	Cell-based	Used for identifying the Cas9 sequence in the genome	All off-target sites detected by dCas9 are not cleaved by Cas9
HTGTS (high-throughput genome-wide translocation sequencing)	Cell-based or lived cells	Used as in vivo, delivery through edited complex	Comparatively insensitive
BLESS (breaks labelling, enrichment on streptavidin and next-generation sequencing)	Cell-based or fixed cells	Used for those tissues where Cas9 delivered in the in vitro condition	Challenging, just identify those DSBs that occur in a specific time
IDLV	Cell-based	Sensitive to 1% and easily programmable	Detection of all off-target sites is not possible
Digenome-seq	Non-cell-based	Sensitive 0.1% or less, unbiased, and effective in cost	Not used broadly

multiple samples in one run. Features and comparison of off-target detection assays are given in Table 5.4.

5.4 Expanding PAM Requirements

CRISPR/Cas9 system possesses many advantages such as cost-effective, robust, easy to handle, and versatile applicability in diverse organisms (Makarova et al. 2011). There are many types and classes of CRISPR/Cas9 system, among them, SpCas9 is the most effective as it creates targeted DSBs in the genome (Dickinson et al. 2013). The effectivity of SpCas9 is due to distinct PAM requirement, 5′-NGG-3′ is the mostly used PAM sequence for SpCas9. Additionally, modified and alternative PAM sequences increase the targeting specificity as well as provide flexibility for targeting knock-in mutations. For genome editing in human cell lines, different variants of Cas9 orthologs with modified PAM recognition sites have been developed such as *Neisseria meningitidis* (NmCas9), *Streptococcus thermophilus* (St1Cas9), and *Staphylococcus aureus* (SaCas9) (Karvelis et al. 2015). SaCas9 is smaller in size and has a large PAM recognition sequence, i.e., 5′-NNGRRT-3.

Additionally, SaCas9 variant KKH SaCas9 with partially relaxed PAM sequence 5′-NNNRRT-3′ has been developed which increases the targeting range of SaCas9.

Engineered variants of *Streptococcus pyogenes* Cas9 (SpCas9) have been in practice to eliminate PAM requirement. Researchers have developed a variant named SpG, capable of targeting an expanded set of NGN PAMs, and further optimized this enzyme to develop a near-PAMless SpCas9 variant named SpRY (NRN > NYN PAMs) Q (Bell et al. 2016).

In *C. elegans*, the functional editing is enhanced by using Cas9 variants reported by Anders et al. (2014). Recently, two variants of Cas9 have also been reported: one is VQR that is mutated in nature with amino acid substitution D1135V, R1335Q, and T1337R with NGA PAM requirement, and the second variant is VRER which also contains mutated residues of D1135V, G1218R, R1335E, and T1337R with NGCG PAM recognition sequence (Anders et al. 2014).

The main drawback of CRISPR/Cas9 system is the requirement of PAM sequence at the target region. Although, the engineered Cas9 or variants of Cas9 could not work effectively in the mammalian cells. The xCas9 works effectively with NGG or non-NGG PAM sequences compared to the engineered SpCas9. These findings increase the targeting range of CRISPR/Cas9 without compromising Cas9 editing efficiency, PAM compatibility, and DNA specificity.

The targeted site of Cas9 contains PAM sequence which is used for the recognition of Cas9 nucleases. The SpCas9 requires NGG PAM that occurs in the genome in every 16 bp. This limits the targeting specificity for application in base editing where PAM requirements are about 13–17 bp from the target site. Due to these limitations, there is a need to explore other genome editing tools. To overcome these limitations, scientists engineered the CRISPR/Cas9 systems accepting different variants of PAM. In mammalian cells, the natural CRISPR nucleases effectively work such as *Staphylococcus aureus* Cas9 (SaCas9), *Acidaminococcus* sp. Cpf1 II, *Lachnospiraceae* bacterium Cpf1 II, *Campylobacter jejuni* Cas9, *Streptococcus thermophilus* Cas9, and *Neisseria meningitidis* Cas9.

Usually PAM is located immediately after the protospacer sequence (Horvath et al. 2008). Genome editing method is derived from *Streptococcus pyogenes* and has been engineered for on-target cleavage and nicking. Furthermore, distinct PAM sites help in the targeting specificity of Cas9 protein (Cong et al. 2013). Different variants of Cas9 and their PAM requirements are presented in Table 5.5.

5.5 CRISPR/Cas12 for Editing Genome

Genome engineering research has been revolutionized by CRISPR/Cas system. CRISPR is better than the traditional genome engineering methods with its prospective benefits of being marker-free, time-saving, flexible, efficient, and precise editing tool in genomics research. It requires less screening and is a simple and easy method. It is more specific as it does not manipulate the genomic sequence based on multiple-protein effector complex, and one single cad protein CRISPR/Cas system is classified as Class 1 and Class 2. Due to the complexity and signature protein, it is

Table 5.5 Cas9 nuclease variants or orthologs and their PAM specificities

Cas9 nuclease variants or orthologs	Origin	PAM sites (5′-3′)	Specification	Functions	References
Native Cas9 (SpCas9)	*Streptococcus pyogenes*	NGG	100-nt-long gRNA	Introduce double-stranded break; create blunt ends	Mojica et al. (2005)
SpCas9 VQR	Engineered	NGAG	Altered PAM specificity	Unrevealed in plants	Kleinstiver et al. (2015)
SpCas9 VRER	Engineered	NGCG	Altered PAM specificity	Unrevealed in plants	Kleinstiver et al. (2015)
eSpCas9	Engineered	NGG	Altered PAM specificity	Unrevealed in plants	Slaymaker et al. (2016)
NmCas9	*Neisseria meningitidis*	NNNNGATT	Longer crRNA component (24 nt)	Reduced off-target effects	Hou et al. (2013)
StCas9	*Streptococcus thermophilus*	NNAGAAW	On-target cleavage activity	Reduced off-target effects	Horvath et al. (2008)
SaCas9	*Staphylococcus aureus*	NNGRRT or NNGRR(N)	On-target cleavage activity	Reduced off-target effects	Esvelt et al. (2013)
Cas9-DD (destabilized Cas9)	Engineered from *S. pyogenes*	NGG	Conjugation of destabilized domain to Cas9	Temporal, spatial, and locus-specific control of gene expression increased NHEJ-mediated gene insertion efficiency	Geisinger et al. (2016)
Cpf1	*Prevotella Francisella 1*	NTT	Contain a RuvC-like endonuclease domain, lack HNH endonuclease domain 42-nt-long gRNA	Require one RNA (crRNA); generate staggered cut; easier to deliver low-capacity vector, e.g., AVV	Zetsche et al. (2015)

AsCpf1	*Acidaminococcus* sp. BV3L6	TTTN	Altered PAM specificity; evaluated in plants	Reduced off-target effects	Zetsche et al. (2015)
Cas9 nickase (Cas9n)	Engineered from *S. pyogenes*	NGG	Mutations in native Cas9	Generate single-stranded break; efficient HDR mechanism	Cong et al. (2013)
BICas9	*Brevibacillus laterosporus*	NNNNCNAA	Altered PAM specificity; evaluated in plants	Reduced off-target effects	Karvelis et al. (2015)

further divided into six types (Types I–VI). The Cpf1 consists of Cas9, and Cas12a is the most developed and widely studied in bacteria (Yao et al. 2018). Cas12 is a compact system that can create real cuts in dsDNA. Cas X has higher multiplexing ability as it can create its own gRNA. It is best used for epigenome editing. After activation by a target DNA molecule matching its sequence, the Cas12 can cut the ssDNA randomly. Thus, it has the ability to detect tiny amount of target DNA in a mixture. CRISPR/Cas12 does not require tracrRNA; it only requires the expression of crRNA (about 43 nucleotides). It can target multiple loci at the same time by multiplexing. For CRISPR/Cas9 genome editing, each sgRNA expression is achieved by separate vectors or multi-cassette vectors as a single transcript. Cas12 has the ability to recognize the PAM by yielding multiple variants. This ability of Cas12 genome editing permits the targeting of different DNA sequences. Cas12 is very sensitive to undesirable side effects; it has high on-target specificity and shows very low tolerance for non-seed mismatches. If any mismatches occur between crRNA and target DNA, Cas12 activity ceases. Cas12 can delete the old target site by producing DSBs which can lead to larger mutations. Thus, Cas12 is more specific and efficient than the Cas9 because it causes longer on-target cuts in the DNA (Yao et al. 2018). Applicability of Cas12/Cpf1 may be beneficial where the chances of repair of DSB are high as in the case of DNA viruses.

5.5.1 Application

CRISPR/Cas system is implied in food and farming industry for the engineering of probiotics and cultures. In agriculture industry, it is used for the development of desired traits like increase in crop yield, development of resistance against drought, and improvement in the nutritious value of crops. It can also be used in the inactivation of the genes during the culturing of human cells. Because of the immense applications of CRISPR/Cas9, it has been used for editing the genome of crops, bacteria, and fungi; in short it can be used for every organism, like modification of mosquitoes through gene drive to make them malaria- and dengue-free (Wedell et al. 2019), genetically modified strains and production of biofuel, treatment of sickle cell anemia by editing the faulty genes that are even transmitted through heredity (Park et al. 2017), etc. Similarly, another variant, Cas12, has also been used in the scientific experiments for modification of a broad range of foreign species as it can induce DNA recombination in other bacteria by heterologous expression; furthermore, it has been reported that Cas12 can also be used to alter the cardiomyocytes of mice and human (Safari et al. 2019). It is assumed that CRISPR/Cas and its variants even have the potential to produce new species or those which have been extinct (Safari et al. 2019). Some of the important applications of CRISPR/Cas12 in yeast, plants, insects, vertebrates, invertebrates, and bacteria are summarized in Table 5.6.

Table 5.6 Application of CRISPR/Cas12 in different organisms

Organism	Strategy for genome editing	Citation
Yeast (*Saccharomyces cerevisiae*)	Multiple crRNAs in a single transcript, simple transformation and expression of DNA	Zhang et al. (2019)
Plants (*Chlamydomonas reinhardtii*)	Transfection of LbCas12a-crRNA complexes and ssDNA delivery through electroporation	Ferenczi et al. (2017)
Insects (*Drosophila melanogaster*)	Cas12a and crRNA integration in the genome, plasmid expression	Port and Bullock (2016)
Nonmammalian invertebrates *Danio rerio* (zebrafish) and *Xenopus tropicalis* (western clawed frog	LbCas12a-mediated genome editing, transformation, and plasmid expressions	Moreno-Mateos et al. (2017)
Bacteria *E. coli*, *Yersinia pestis*, *Mycobacterium smegmatis*, and *Corynebacterium glutamicum*	Cas12a-stimulated DNA recombination by heterologous expression of FnCas12a, crRNAs, and proteins	Zetsche et al. (2015)
Mammals Human HEK293 cells	FnCas12a-mediated genome editing	Tu et al. (2017)

5.6 Genome Editing Through CasX

CasX is the third tool in the CRISPR/Cas toolkit which is one of the attractive and advanced editing tools. CasX has been isolated from a non-pathogenic bacterium through metagenomic analysis which is found in groundwater. CasX is a dual RNA-guided DNA nuclease, very small in size, and has remarkable DNA cleavage characteristics. It recognizes a 5′-TTCN PAM and is therefore different from the other Cas9 endonucleases. It has RuvC domain located at C terminus like other Cas9 endonucleases, and it is smaller than the Cas12 about 320 aa. It can also produce double-stranded break like other CRISPR systems. It can generate site-specific repression in bacterial genome. It has the ability of editing the genome of human cells (Khanzadi and Khan 2020).

CasX is also known as Cas12e5. CasX is a hybrid enzyme of Cas9 and Cas12a having RNA folds and protein domains. CasX has the ability of plasmid interference in bacteria and ceases the bacterial transformation along with its associated crRNAs. The RuvC domain has similarity with the domain of Cas9 and Cas12a enzymes. The phylogenetic analysis shows that in ancestral CRISPR loci, an event-independent insertion happened which leads to the appearance of CasX from a TnpB-type transposase. Evolutionary inconsistency suggests that the structure and molecular mechanism of CasX is different from other CasX systems. CasX has a quaternary structure. The CasX has unique overlapping exon in every gene of 6–45 bp. This makes the CasX system a common genome editor. The CasX-driven CRISPR system is derived from *Deltaproteobacteria* (DpbCasX) and *Planctomycetes* (PlmCasX) (Roberson 2019).

The RuvC nuclease domain of CasX causes sequential cleavage like Cas12e. It can produce 5′ overhang of almost 1 bp nucleotides by staggered cuts which can cleave the NTS within protospacer. It can digest the single-stranded DNA non-specifically. CasX works as eukaryotic genome editing effector efficiently. The expression of CasX in cultured human cells induces targeted mutagenesis. The biophysical analyses predict that the CasX would be useful for mechanistic understanding and future genome engineering (Sontheimer 2019).

5.6.1 Applications of CasX

5.6.1.1 Human and *E. coli* Genome Editing

Adaptive immunity is provided against bacteriophage by RNA-guided CRISPR/Cas9 and Cas12a proteins. Double-stranded cleavage of DNA is achieved through CRISPR/CasX whose structure is unique and programmed for this action. In vivo and biochemical data illustrate that CasX protein is designed for the human genome and *E. coli* modification. Eight cryo-EM structures of CasX in different states of assembly with its guide RNA and double-stranded DNA substrates reveal an extensive RNA scaffold and a domain required for DNA unwinding. The functional components of CasX are different from Cas9 and Cas12 (Liu et al. 2019).

5.6.1.2 Other Gene Editing Proteins

Through the metagenomic study of bacteria, native groundwater RNA-guided DNA nuclease was revealed. This RNA-guided DNA nuclease recognizes a 5′-TTCN PAM and interferes in the plasmid of *E. coli* when introducing sgRNA (covalently linked crRNA-tracrRNA). No correspondence was reported with the rest of the Cas endonucleases except a domain present at the C-terminus called RuvC domain. These features of CasX have similarity with the type 5 Cas12, but there is a difference in size of CasX and Cas12 which is ~980 aa and ~ 1200 aa, respectively (Yang and Patel 2019).

5.7 Multiplex Genome Editing Through CRISPR/Cas

Today, scientists are looking for new ways of genome editing. Genome editing is a type of genetic engineering in which DNA fragment is inserted, deleted, or replaced in the genome of a living organism by using engineered nucleases or molecular markers. Currently, the simplest, most versatile, and precise method of gene editing, which is revolutionizing the genetic engineering and biotechnology industry, is the CRISPR/Cas9 system.

CRISPR/Cas system is based upon the archaeal and bacterial adaptive immune system which is used for genome engineering in a variety of living cells. The short RNA molecules lead the way of effector endonucleases which are encoded with

CRISPR arrays. Sequence of spacers that emerge from organisms that are invaders by direct repeat is contained by indigenous CRISPR arrays. The indigenous CRISPR arrays contain sequence of spacers that emerge from the invader organisms by direct repeat (Adiego-Pérez et al. 2019). The most common conditions in industrial microorganisms are aneuploidy and polyploidy. There is a decrease in CRISPR editing efficiencies due to non-haploid strain as its requirement is to stimulate targeting of multiple alleles (Mertens et al. 2019).

5.7.1 Multiplex Genome Editing

The multiplex genome editing includes multiple gene editing at a time after activating the bacterial metabolic pathway. In the era when CRISPR is not yet introduced, frequentative gene editing is done by the selection of markers which is necessary to make strains that express multiple genes by using a pathway called multiple-gene expression pathway. This technique saves time and resources that are used in initiative strain construction. The most recent techniques used today under CRISPR are Cas9 and Cas12a. Table 5.7 explains multiplexing and its applications in important microorganisms.

5.7.2 Applications

In microbes: The first microbe used to check the capabilities of Cas9 multiplexing was indigenous Cas9 carrying bacteria, *Streptococcus pneumonia.* Two spacers

Table 5.7 Multiplex genome editing techniques used in organisms

Organism	Strategy for multiplexed gRNA expression	Citation
Escherichia coli	Plasmid expression and several sgRNA expression constitutive cassettes	Zhang et al. (2017)
Streptococcus pneumoniae	Genome-integrated and native-like CRISPR array	Adiego-Pérez et al. (2019)
Streptomyces lividans	Plasmid expression (pSG5rep) and several sgRNA expression constitutive cassettes	Cobb et al. (2015)
Saccharomyces cerevisiae	Multicopy/plasmid expression or several sgRNA expression cassettes	Mans et al. (2015)
Ogataea polymorpha CBS 11895 (n)	Centrometric plasmid expression and synthetic array of ribozyme-flanked sgRNA	Juergens et al. (2018)
Penicillium chrysogenum (DS68530)	Transient expression and protoplast-mediated transformation, in vitro synthetized sgRNA in RNP	Pohl et al. (2016)
Aspergillus nidulans (IBT27263 (n))	Centromeric plasmid expression or synthetic array of tRNA-flanked sgRNAs	Nødvig et al. (2018)
Scheffersomyces stipitis (UC7, (n))	RNA pol III promoter as a viral sgRNA expression cassettes or centromeric/plasmid expression	Cao et al. (2018)

were expressed by the integration of synthetic array into the genome (Adiego-Pérez et al. 2019).

Higher plants: More than 100 contemporary targeting events were reported in 2019 which consist of multiplex genome editing in plants that begin with focusing on input characters such as herbicide resistance, biosynthesis of hormones and their precipitation, plant development, metabolic engineering, and molecular farming (Raza et al. 2019).

Genome, strain, and metabolic engineering: Multiplexing by CRISPR made it easier to edit genomes which was difficult to achieve by traditional techniques of genetic engineering. A study was recently performed in which large-scale programmable genome assembly and rearrangement were made by the expression of single gRNAs. For the chemical synthesis of 61-codon *E. coli* genome, DNA mega-chunks were positioned on the BACs and incorporated into the genome through the cleavage of double-stranded DNA and lambda red recombination. This was all done by using CRISPR multiplexing technology. Occurrance of genomic truncations due to DSBs may reduce the use of Cas9 for the implementation of specific mutations (McCarty et al. 2020).

Genotype to phenotype; combinatorial mapping: Deletion of genes and change of the phenotype of these genes are studied in a wide range of organisms by scientists for decades. This helps in our understanding of how can we regulate the behavior by the genome-encoded information. In the past, getting this information is difficult, but today by the use of CRISPR tools we are able check all this type of information in less time with efficiency (McCarty et al. 2020).

In diagnostics and biosensing: For the detection and cleavage of specified DNA/RNA sequences, Cas enzymes are programmed; it makes them a good choice to be used in the development of biosensors. Multiplex technologies are also used in diagnostics as the detection of several pathogens using guide RNAs in a single run is cost-effective and time-saving technology with high accuracy (McCarty et al. 2020).

Challenges in multiplexing approach: Recent data indicate that the field of multiplex genome editing has achieved a remarkable progress. In Fig. 5.1, the major challenges are highlighted for the development of well-organized multiplex CRISPR/Cas system in different microorganisms.

5.8 Future Perspectives

CRISPR/Cas9 system is an advanced tool in the field of plant breeding and plays an important role in therapeutics and medicinal applications. Cas9 nuclease-mediated mutations are highly specific and efficient than that of the random mutagenesis. Knockout (KO) and knock-in (KI) alleles have been generated by using CRISPR/

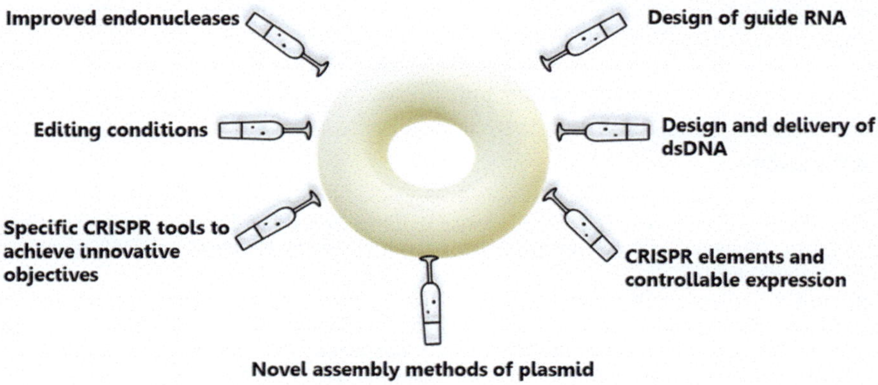

Fig. 5.1 Challenges faced by multiplexing (Adiego-Pérez et al. 2019)

Cas9 system. There are many successful studies of HR-mediated knock-in and NHEJ-mediated knockouts reported. Several modifications in CRISPR/Cas9 protein have been made to increase specificity, efficiency, on-target efficacy, and target range in the genome. Some other modifications in the PAM-determining domain of Cas9 have been made which expand the PAM recognition sites as well as targeting range. Researchers have been working to enhance the specificity of Cas9 protein, and it is still in progress, which included significant improvements in gRNA selection, off-target identification assays, use of novel enzymes, engineering of gRNA or protein, coupling Cas9 to artificial inhibitory domains, orthologs of Cas9 and its variants, inhibition of non-homologous end joining to increase Cas9 specificity, and chemical modifications in single-guide RNAs to reduce off-target effects in the genome. Additionally, Cas9 orthologs such as StCas9 and short-length SaCas9 with expanded PAM recognition site work effectively in the targeted genome. Further improvements and developments of knock-in, knockout, precise genome editing, or replacement of genes in the targeted organism remain a challenge. Although CRISPR/Cas9 nucleases have some off-target effects, off-target detection assays such as Digenome-seq, GUIDE-seq, targeted deep sequencing, BLESS, HTGTS, and ChIP have been used to evaluate off-target sites in the genome. Hence, some precise and sensitive off-target detection assays need to be developed which can detect off-target mutations with indel frequency range $> 0.01\%$. Moreover, CRISPR/Cas9 tool has been developed and modified with increased specificity and a wide range of applications such as targeted RNA cleavage, epigenomic modifications, and chromatin remodeling and imaging in the targeted organisms. Multiplex CRISPR/Cas9 system provides ease in the introduction of heterologous genes in the targeted loci as well as metabolic pathway engineering of organisms. Due to its simplicity, high fidelity, sensitivity, and effectiveness, CRISPR/Cas9 has become a versatile tool in contrast to previously used genome manipulation technologies. So, there is a great potential and hope that CRISPR/Cas9 and many other evolving systems of genome editing will revolutionize life sciences and molecular crop breeding.

References

Adiego-Pérez B, Randazzo P, Daran JM, Verwaal R, Roubos JA, Daran-Lapujade P, Van Der Oost J (2019) Multiplex genome editing of microorganisms using CRISPR-Cas. FEMS Microbiol Lett 366(8):fnz086

Alipanahi B, Delong A, Weirauch MT, Frey BJ (2015) Predicting the sequence specificities of DNA-and RNA-binding proteins by deep learning. Nat Biotechnol 33(8):831–838

Anders C, Niewoehner O, Duerst A, Jinek M (2014) Structural basis of PAM-dependent target DNA recognition by the Cas9 endonuclease. Nature 513(7519):569–573

Andersson M, Turesson H, Nicolia A, Fält AS, Samuelsson M, Hofvander P (2017) Efficient targeted multiallelic mutagenesis in tetraploid potato (Solanum tuberosum) by transient CRISPR-Cas9 expression in protoplasts. Plant Cell Rep 36:117–128

Aschenbrenner S, Kallenberger SM, Hoffmann MD, Huck A, Eils R, Niopek D (2020) Coupling Cas9 to artificial inhibitory domains enhances CRISPR-Cas9 target specificity. Sci Adv 6(6): eaay0187

Bae S, Park J, Kim JS (2014) Cas-OFFinder: a fast and versatile algorithm that searches for potential off-target sites of Cas9 RNA-guided endonucleases. Bioinformatics 30(10):1473–1475

Baek K, Kim DH, Jeong J, Sim SJ, Melis A, Kim JS, Jin E, Bae S (2016) DNA-free two-gene knockout in Chlamydomonas reinhardtii via CRISPR-Cas9 ribonucleoproteins. Sci Rep 6: 30620

Bell RT, Fu BX, Fire AZ (2016) Cas9 variants expand the target repertoire in Caenorhabditis elegans. Genetics 202(2):381–388

Bischoff N, Wimberger S, Maresca M, Brakebusch C (2020) Improving precise CRISPR genome editing by small molecules: is there a magic potion? Cells 9(5):1318

Bolukbasi MF, Gupta A, Oikemus S, Derr AG, Garber M, Brodsky MH, Zhu LJ, Wolfe SA (2015) DNA-binding-domain fusions enhance the targeting range and precision of Cas9. Nat Methods 12(12):1150–1156

Briner AE, Donohoue PD, Gomaa AA, Selle K, Slorach EM, Nye CH, Haurwitz RE, Beisel CL, May AP, Barrangou R (2014) Guide RNA functional modules direct Cas9 activity and orthogonality. Mol Cell 56(2):333–339

Brooks C, Nekrasov V, Lippman ZB, Van Eck J (2014) Efficient gene editing in tomato in the first generation using the clustered regularly interspaced short palindromic repeats/ CRISPRassociated9 system. Plant Physiol 166:1292–1297

Brunet E, Simsek D, Tomishima M, DeKelver R, Choi VM, Gregory P, Urnov F, Weinstock DM, Jasin M (2009) Chromosomal translocations induced at specified loci in human stem cells. Proc Natl Acad Sci 106(26):10620–10625

Cao M, Gao M, Ploessl D, Song C, Shao Z (2018) CRISPR–mediated genome editing and gene repression in *Scheffersomyces stipitis*. Biotechnol J 13(9):1700598

Cebrian-Serrano A, Davies B (2017) CRISPR-Cas orthologues and variants: optimizing the repertoire, specificity, and delivery of genome engineering tools. Mamm Genome 28(7–8):247–261

Cencic R, Miura H, Malina A, Robert F, Ethier S, Schmeing TM, Dostie J, Pelletier J (2014) Protospacer adjacent motif (PAM)-distal sequences engage CRISPR Cas9 DNA target cleavage. PLoS One 9(10):e109213

Čermák T, Baltes NJ, Čegan R, Zhang Y, Voytas DF (2015) High-frequency, precise modification of the tomato genome. Genome Biol 16(1):232

Chandrasekaran J, Brumin M, Wolf D, Leibman D, Klap C, Pearlsman M, Sherman A, Arazi T, Gal-On A (2016) Development of broad virus resistance in non-transgenic cucumber using CRISPR/Cas9 technology. Mol Plant Pathol 17(7):1140–1153

Chen B, Gilbert LA, Cimini BA, Schnitzbauer J, Zhang W, Li GW, Park J, Blackburn EH, Weissman JS, Qi LS, Huang B (2013) Dynamic imaging of genomic loci in living human cells by an optimized CRISPR/Cas system. Cell 155(7):1479–1491

Chen Y, Cao J, Xiong M, Petersen AJ, Dong Y, Tao Y, Huang CTL, Du Z, Zhang SC (2015) Engineering human stem cell lines with inducible gene knockout using CRISPR/Cas9. Cell Stem Cell 17(2):233–244

Cho SW, Kim S, Kim Y, Kweon J, Kim HS, Bae S, Kim JS (2014) Analysis of off-target effects of CRISPR/Cas-derived RNA-guided endonucleases and nickases. Genome Res 24(1):132–141

Cobb RE, Wang Y, Zhao H (2015) High-efficiency multiplex genome editing of Streptomyces species using an engineered CRISPR/Cas system. ACS Synth Biol 4(6):723–728

Cong L, Ran FA, Cox D, Lin S, Barretto R, Habib N, Hsu PD, Wu X, Jiang W, Marraffini LA, Zhang F (2013) Multiplex genome engineering using CRISPR/Cas systems. Science 339(6121):819–823

Crispo M, Mulet AP, Tesson L, Barrera N, Cuadro F, dos Santos-Neto PC, Nguyen TH, Crénéguy A, Brusselle L, Anegón I, Menchaca A (2015) Efficient generation of myostatin knock-out sheep using CRISPR/Cas9 technology and microinjection into zygotes. PLoS One 10(8):e0136690

Dang XT, Tran DH, Hirose O, Satou K (2015) SPY: a novel resampling method for improving classification performance in imbalanced data. In: 2015 seventh international conference on knowledge and systems engineering (KSE), pp 280–285. IEEE

Davis KM, Pattanayak V, Thompson DB, Zuris JA, Liu DR (2015) Small molecule–triggered Cas9 protein with improved genome-editing specificity. Nat Chem Biol 11(5):316–318

Dellinger DJ, Timár Z, Myerson J, Sierzchala AB, Turner J, Ferreira F, Kupihár Z, Dellinger G, Hill KW, Powell JA, Sampson JR (2011) Streamlined process for the chemical synthesis of RNA using 2′-O-thionocarbamate-protected nucleoside phosphoramidites in the solid phase. J Am Chem Soc 133(30):11540–11556

Dickinson DJ, Ward JD, Reiner DJ, Goldstein B (2013) Engineering the *Caenorhabditis elegans* genome using Cas9-triggered homologous recombination. Nat Methods 10(10):1028–1034

Doench JG, Hartenian E, Graham DB, Tothova Z, Hegde M, Smith I, Sullender M, Ebert BL, Xavier RJ, Root DE (2014) Rational design of highly active sgRNAs for CRISPR-Cas9–mediated gene inactivation. Nat Biotechnol 32(12):1262–1267

Doench JG, Fusi N, Sullender M, Hegde M, Vaimberg EW, Donovan KF, Smith I, Tothova Z, Wilen C, Orchard R, Virgin HW (2016) Optimized sgRNA design to maximize activity and minimize off-target effects of CRISPR-Cas9. Nat Biotechnol 34(2):184–191

Duan J, Lu G, Xie Z, Lou M, Luo J, Guo L, Zhang Y (2014) Genome-wide identification of CRISPR/Cas9 off-targets in human genome. Cell Res 24(8):1009–1012

East-Seletsky A, O'Connell MR, Knight SC, Burstein D, Cate JH, Tjian R, Doudna JA (2016) Two distinct RNase activities of CRISPR-C2c2 enable guide-RNA processing and RNA detection. Nature 538(7624):270–273

Eggenschwiler R, Moslem M, Fráguas MS, Galla M, Papp O, Naujock M, Fonfara I, Gensch I, Wähner A, Beh-Pajooh A, Mussolino C (2016) Improved bi-allelic modification of a transcriptionally silent locus in patient-derived iPSC by Cas9 nickase. Sci Rep 6:38198

Endo M, Mikami M, Toki S (2016) Biallelic gene targeting in rice. Plant Physiol 170(2):667–677

Esvelt KM, Mali P, Braff JL, Moosburner M, Yaung SJ, Church GM (2013) Orthogonal Cas9 proteins for RNA-guided gene regulation and editing. Nat Methods 10(11):1116–1121

Ferenczi A, Pyott DE, Xipnitou A, Molnar A (2017) Efficient targeted DNA editing and replacement in Chlamydomonas reinhardtii using Cpf1 ribonucleoproteins and single-stranded DNA. Proc Natl Acad Sci 114(51):13567–13572

Fister AS, Landherr L, Maximova SN, Guiltinan MJ (2018) Transient expression of CRISPR/Cas9 machinery targeting TcNPR3 enhances defense response in Theobroma cacao. Front Plant Sci 9: 268

Friedland AE, Baral R, Singhal P, Loveluck K, Shen S, Sanchez M, Marco E, Gotta GM, Maeder ML, Kennedy EM, Kornepati AV (2015) Characterization of Staphylococcus aureus Cas9: a smaller Cas9 for all-in-one adeno-associated virus delivery and paired nickase applications. Genome Biol 16(1):1–10

Frock RL, Hu J, Meyers RM, Ho YJ, Kii E, Alt FW (2015) Genome-wide detection of DNA double-stranded breaks induced by engineered nucleases. Nat Biotechnol 33(2):179–186

Fu Y, Foden JA, Khayter C, Maeder ML, Reyon D, Joung JK, Sander JD (2013) High-frequency off-target mutagenesis induced by CRISPR-Cas nucleases in human cells. Nat Biotechnol 31(9):822–826

Fu Y, Sander JD, Reyon D, Cascio VM, Joung JK (2014) Improving CRISPR-Cas nuclease specificity using truncated guide RNAs. Nat Biotechnol 32(3):279–284

Gabriel R, Von Kalle C, Schmidt M (2015) Mapping the precision of genome editing. Nat Biotechnol 33(2):150–152

Gao J, Wang G, Ma S, Xie X, Wu X, Zhang X, Wu Y, Zhao P, Xia Q (2015) CRISPR/Cas9-mediated targeted mutagenesis in Nicotiana tabacum. Plant Mol Biol 87(1–2):99–110

Geisinger JM, Turan S, Hernandez S, Spector LP, Calos MP (2016) In vivo blunt-end cloning through CRISPR/Cas9-facilitated non-homologous end-joining. Nucleic Acids Res 44(8):e76

Gomez MA, Lin ZD, Moll T, Luebbert C, Chauhan RD, Vijayaraghavan A, Kelley R, Beyene G, Taylor NJ, Carrington J, Staskawicz B (2017) Simultaneous CRISPR/Cas9-mediated editing of cassava eIF4E isoforms nCBP-1 and nCBP-2 confers elevated resistance to cassava brown streak disease. bioRxiv: 209874

Guilinger JP, Thompson DB, Liu DR (2014) Fusion of catalytically inactive Cas9 to FokI nuclease improves the specificity of genome modification. Nat Biotechnol 32(6):577

Heigwer F, Kerr G, Boutros M (2014) E-CRISP: fast CRISPR target site identification. Nat Methods 11(2):122–123

Hendel A, Bak RO, Clark JT, Kennedy AB, Ryan DE, Roy S, Steinfeld I, Lunstad BD, Kaiser RJ, Wilkens AB, Bacchetta R (2015) Chemically modified guide RNAs enhance CRISPR-Cas genome editing in human primary cells. Nat Biotechnol 33(9):nbt-3290

Heo YT, Quan X, Xu YN, Baek S, Choi H, Kim NH, Kim J (2015) CRISPR/Cas9 nuclease-mediated gene knock-in in bovine-induced pluripotent cells. Stem Cells Dev 24(3):393–402

Horvath P, Romero DA, Coûté-Monvoisin AC, Richards M, Deveau H, Moineau S, Boyaval P, Fremaux C, Barrangou R (2008) Diversity, activity, and evolution of CRISPR loci in Streptococcus thermophilus. J Bacterial 190(4):1401–1412

Hou Z, Zhang Y, Propson NE, Howden SE, Chu LF, Sontheimer EJ, Thomson JA (2013) Efficient genome engineering in human pluripotent stem cells using Cas9 from Neisseria meningitidis. Proc Natl Acad Sci 110(39):15644–15649

Hsu PD, Scott DA, Weinstein JA, Ran FA, Konermann S, Agarwala V, Li Y, Fine EJ, Wu X, Shalem O, Cradick TJ (2013) DNA targeting specificity of RNA-guided Cas9 nucleases. Nat Biotechnol 31(9):827–832

Hsu PD, Lander ES, Zhang F (2014) Development and applications of CRISPR-Cas9 for genome engineering. Cell 157(6):1262–1278

Hu X, Chang N, Wang X, Zhou F, Zhou X, Zhu X, Xiong JW (2013) Heritable gene-targeting with gRNA/Cas9 in rats. Cell Res 23(11):1322–1325

Hwang WY, Fu Y, Reyon D, Maeder ML, Kaini P, Sander JD, Joung JK, Peterson RT, Yeh JRJ (2013) Heritable and precise zebrafish genome editing using a CRISPR-Cas system. PLoS One 8(7):e68708

Iqbal Z, Sattar MN, Shafiq M (2016) CRISPR/Cas9: a tool to circumscribe cotton leaf curl disease. Front Plant Sci 7:475

Inoue YU, Morimoto Y, Hoshino M, Inoue T (2018) Generation of Pax6-IRES-EGFP knock-in mouse via the cloning-free CRISPR/Cas9 system to reliably visualize neurodevelopmental dynamics. Neurosci Res 132:1–7

Iseli C, Ambrosini G, Bucher P, Jongeneel CV (2007) Indexing strategies for rapid searches of short words in genome sequences. PLoS One 2(6):e579

Jia H, Orbovic V, Jones JB, Wang N (2016) Modification of the PthA4 effector binding elements in type I Cs LOB 1 promoter using Cas9/sg RNA to produce transgenic Duncan grapefruit alleviating XccΔpthA4: dCs LOB 1.3 infection. Plant Biotechnol J 14(5):1291–1301

Jiang W, Bikard D, Cox D, Zhang F, Marraffini LA (2013a) RNA-guided editing of bacterial genomes using CRISPR-Cas systems. Nat Biotechnol 31(3):233–239

Jiang W, Bikard D, Cox D, Zhang F, Marraffini LA (2013b) CRISPR-assisted editing of bacterial genomes. Nat Biotechnol 31(3):233

Jiang F, Zhou K, Ma L, Gressel S, Doudna JA (2015) A Cas9–guide RNA complex preorganized for target DNA recognition. Science 348(6242):1477–1481

Jiang WZ, Henry IM, Lynagh PG, Comai L, Cahoon EB, Weeks DP (2017) Significant enhancement of fatty acid composition in seeds of the allohexaploid, Camelina sativa, using CRISPR/Cas9 gene editing. Plant Biotechnol J 15(5):648–657

Juergens H, Varela JA, Gorter de Vries AR, Perli T, Gast VJ, Gyurchev NY, Rajkumar AS, Mans R, Pronk JT, Morrissey JP, Daran JMG (2018) Genome editing in Kluyveromyces and Ogataea yeasts using a broad-host-range Cas9/gRNA co-expression plasmid. FEMS Yeast Res 18(3): foy012

Jung YJ, Lee HJ, Yu J et al (2021) Transcriptomic and physiological analysis of OsCAO1 knockout lines using the CRISPR/Cas9 system in rice. Plant Cell Rep 40(6):1013–1024. https://doi.org/10.1007/s00299-020-02607-y

Karvelis T, Gasiunas G, Young J, Bigelyte G, Silanskas A, Cigan M, Siksnys V (2015) Rapid characterization of CRISPR-Cas9 protospacer adjacent motif sequence elements. Genome Biol 16(1):1–13

Kaur N, Alok A, Kaur N, Pandey P, Awasthi P, Tiwari S (2018) CRISPR/Cas9-mediated efficient editing in phytoene desaturase (PDS) demonstrates precise manipulation in banana cv. Rasthali genome. Funct Integr Genomics 18(1):89–99

Khanzadi MN, Khan AA (2020) CRISPR/Cas9: Nat.'s gift to prokaryotes and an auspicious tool in genome editing. J Basic Microbiol 60(2):91–102

Kim HJ, Lee HJ, Kim H, Cho SW, Kim JS (2009) Targeted genome editing in human cells with zinc finger nucleases constructed via modular assembly. Genome Res 19(7):1279–1288

Kim S, Kim D, Cho SW, Kim J, Kim JS (2014) Highly efficient RNA-guided genome editing in human cells via delivery of purified Cas9 ribonucleoproteins. Genome Res 24(6):1012–1019

Kim D, Bae S, Park J, Kim E, Kim S, Yu HR, Hwang J, Kim JI, Kim JS (2015) Digenome-seq: genome-wide profiling of CRISPR-Cas9 off-target effects in human cells. Nat Methods 12(3):237–243

Kim D, Kim J, Hur JK, Been KW, Yoon SH, Kim JS (2016) Genome-wide analysis reveals specificities of Cpf1 endonucleases in human cells. Nat Biotechnol 34(8):863–868

Kleinstiver BP, Prew MS, Tsai SQ, Nguyen NT, Topkar VV, Zheng Z, Joung JK (2015) Broadening the targeting range of Staphylococcus aureus CRISPR-Cas9 by modifying PAM recognition. Nat Biotechnol 33(12):1293–1298

Kleinstiver BP, Pattanayak V, Prew MS, Tsai SQ, Nguyen NT, Zheng Z, Joung JK (2016) High-fidelity CRISPR–Cas9 nucleases with no detectable genome-wide off-target effects. Nature 529(7587):490–495

Koo T, Lee J, Kim JS (2015) Measuring and reducing off-target activities of programmable nucleases including CRISPR-Cas9. Mol Cells 38(6):475

Kuscu C, Arslan S, Singh R, Thorpe J, Adli M (2014) Genome-wide analysis reveals characteristics of off-target sites bound by the Cas9 endonuclease. Nat Biotechnol 32(7):677–683

Lawrenson T, Shorinola O, Stacey N, Li C, Østergaard L, Patron N, Uauy C, Harwood W (2015) Induction of targeted, heritable mutations in barley and Brassica oleracea using RNA-guided Cas9 nuclease. Genome Biol 16(1):258
Lee HJ, Kim E, Kim JS (2010) Targeted chromosomal deletions in human cells using zinc finger nucleases. Genome Res 20(1):81–89
Lee CM, Cradick TJ, Bao G (2016a) The Neisseria meningitidis CRISPR-Cas9 system enables specific genome editing in mammalian cells. Mol Ther 24(3):645–654
Lee N, Shin J, Park JH, Lee GM, Cho S, Cho BK (2016b) Targeted gene deletion using DNA-free RNA-guided Cas9 nuclease accelerates adaptation of CHO cells to suspension culture. ACS Synth Biol 5(11):1211–1219
Lee K, Conboy M, Park HM, Jiang F, Kim HJ, Dewitt MA, Mackley VA, Chang K, Rao A, Skinner C, Shobha T (2017) Nanoparticle delivery of Cas9 ribonucleoprotein and donor DNA in vivo induces homology-directed DNA repair. Nat Biomed Eng 1(11):889–901
Lin S, Staahl BT, Alla RK, Doudna JA (2014) Enhanced homology-directed human genome engineering by controlled timing of CRISPR/Cas9 delivery. elife 3:e04766
Liu JJ, Orlova N, Oakes BL, Ma E, Spinner HB, Baney KL, Chuck J, Tan D, Knott GJ, Harrington LB, Al-Shayeb B (2019) CasX enzymes comprise a distinct family of RNA-guided genome editors. Nature 566(7743):218–223
Ma Y, Ma J, Zhang X, Chen W, Yu L, Lu Y, Bai L, Shen B, Huang X, Zhang L (2014) Generation of e GFP and Cre knockin rats by CRISPR/Cas9. FEBS J 281(17):3779–3790
Ma X, Zhu Q, Chen Y, Liu YG (2016) CRISPR/Cas9 platforms for genome editing in plants: developments and applications. Mol Plant 9(7):961–974
Makarova KS, Koonin EV (2015) Annotation and classification of CRISPR-Cas systems. In: CRISPR. Humana Press, New York, pp 47–75
Makarova KS, Aravind L, Wolf YI, Koonin EV (2011) Unification of Cas protein families and a simple scenario for the origin and evolution of CRISPR-Cas systems. Biol Direct 6(1):38
Mali P, Aach J, Stranges PB, Esvelt KM, Moosburner M, Kosuri S, Yang L, Church GM (2013) CAS9 transcriptional activators for target specificity screening and paired nickases for cooperative genome engineering. Nat Biotechnol 31(9):833–838
Mans R, van Rossum HM, Wijsman M, Backx A, Kuijpers NG, van den Broek M, Daran-Lapujade P, Pronk JT, van Maris AJ, Daran JMG (2015) CRISPR/Cas9: a molecular Swiss army knife for simultaneous introduction of multiple genetic modifications in Saccharomyces cerevisiae. FEMS Yeast Res 15(2):fov004
Mao Y, Zhang H, Xu N, Zhang B, Gou F, Zhu JK (2013) Application of the CRISPR–Cas system for efficient genome engineering in plants. Mol Plant 6(6):2008–2011
McCarty NS, Graham AE, Studená L, Ledesma-Amaro R (2020) Multiplexed CRISPR technologies for gene editing and transcriptional regulation. Nat Commun 11(1):1–13
Mertens S, Gallone B, Steensels J, Herrera-Malaver B, Cortebeek J, Nolmans R, Saels V, Vyas VK, Verstrepen KJ (2019) Reducing phenolic off-flavors through CRISPR-based gene editing of the FDC1 gene in Saccharomyces cerevisiae x Saccharomyces eubayanus hybrid lager beer yeasts. PLoS One 14(1):e0209124
Mojica FJ, García-Martínez J, Soria E (2005) Intervening sequences of regularly spaced prokaryotic repeats derive from foreign genetic elements. J Mol Evol 60(2):174–182
Montague TG, Cruz JM, Gagnon JA, Church GM, Valen E (2014) CHOPCHOP: a CRISPR/Cas9 and TALEN web tool for genome editing. Nucleic Acids Res 42(W1):W401–W407
Moreno-Mateos MA, Fernandez JP, Rouet R, Vejnar CE, Lane MA, Mis E, Khokha MK, Doudna JA, Giraldez AJ (2017) CRISPR-Cpf1 mediates efficient homology-directed repair and temperature-controlled genome editing. Nat Commun 8(1):1–9
Müller M, Lee CM, Gasiunas G, Davis TH, Cradick TJ, Siksnys V, Bao G, Cathomen T, Mussolino C (2016) Streptococcus thermophilus CRISPR-Cas9 systems enable specific editing of the human genome. Mol Ther 24(3):636–644
Naito Y, Hino K, Bono H, Ui-Tei K (2015) CRISPRdirect: software for designing CRISPR/Cas guide RNA with reduced off-target sites. Bioinformatics 31(7):1120–1123

Nihongaki Y, Kawano F, Nakajima T, Sato M (2015) Photoactivatable CRISPR-Cas9 for optogenetic genome editing. Nat Biotechnol 33(7):755–760

Nishimasu H, Ran FA, Hsu PD, Konermann S, Shehata SI, Dohmae N, Ishitani R, Zhang F, Nureki O (2014) Crystal structure of Cas9 in complex with guide RNA and target DNA. Cell 156(5):935–949

Nødvig CS, Hoof JB, Kogle ME, Jarczynska ZD, Lehmbeck J, Klitgaard DK, Mortensen UH (2018) Efficient oligo nucleotide mediated CRISPR-Cas9 gene editing in aspergilli. Fungal Genet Biol 115:78–89

Park PJ (2009) ChIP–seq: advantages and challenges of a maturing technology. Nat Rev Genet 10(10):669–680

Park CY, Kim J, Kweon J, Son JS, Lee JS, Yoo JE, Cho SR, Kim JH, Kim JS, Kim DW (2014) Targeted inversion and reversion of the blood coagulation factor 8 gene in human iPS cells using TALENs. Proc Natl Acad Sci 111(25):9253–9258

Park S, Gianotti-Sommer A, Molina-Estevez FJ, Vanuytsel K, Skvir N, Leung A, Rozelle SS, Shaikho EM, Weir I, Jiang Z, Luo HY (2017) A comprehensive, ethnically diverse library of sickle cell disease-specific induced pluripotent stem cells. Stem Cell Rep 8(4):1076–1085

Peng Y, Clark KJ, Campbell JM, Panetta MR, Guo Y, Ekker SC (2014) Making designer mutants in model organisms. Development 141(21):4042–4054

Peng A, Chen S, Lei T, Xu L, He Y, Wu L, Yao L, Zou X (2017) Engineering canker-resistant plants through CRISPR/Cas9-targeted editing of the susceptibility gene Cs LOB 1 promoter in citrus. Plant Biotechnol J 15(12):1509–1519

Pohl C, Kiel JAKW, Driessen AJM, Bovenberg RAL, Nygard Y (2016) CRISPR/Cas9 based genome editing of Penicillium chrysogenum. ACS Synth Biol 5(7):754–764

Port F, Bullock SL (2016) Augmenting CRISPR applications in drosophila with tRNA-flanked sgRNAs. Nat Methods 13(10):852

Prykhozhij SV, Rajan V, Gaston D, Berman JN (2015) CRISPR Multitargeter: a web tool to find common and unique CRISPR single guide RNA targets in a set of similar sequences. PLoS One 10(3):e0119372

Rahdar M, McMahon MA, Prakash TP, Swayze EE, Bennett CF, Cleveland DW (2015) Synthetic CRISPR RNA-Cas9–guided genome editing in human cells. Proc Natl Acad Sci 112(51): E7110–E7117

Ramakrishna S, Dad ABK, Beloor J, Gopalappa R, Lee SK, Kim H (2014) Gene disruption by cell-penetrating peptide-mediated delivery of Cas9 protein and guide RNA. Genome Res 24(6):1020–1027

Ran FA, Hsu PD, Wright J, Agarwala V, Scott DA, Zhang F (2013) Genome engineering using the CRISPR-Cas9 system. Nat Protoc 8(11):2281–2308

Ran FA, Cong L, Yan WX, Scott DA, Gootenberg JS, Kriz AJ, Zetsche B, Shalem O, Wu X, Makarova KS, Koonin EV (2015) In vivo genome editing using Staphylococcus aureus Cas9. Nature 520(7546):186–191

Raza A, Razzaq A, Mehmood SS, Zou X, Zhang X, Lv Y, Xu J (2019) Impact of climate change on crops adaptation and strategies to tackle its outcome: a review. Plants 8(2):34

Roberson ED (2019) A catalog of CasX genome editing sites in common model organisms. BMC Genomics 20(1):528

Romand S, Jostock T, Fornaro M, Schmidt J, Ritter A, Wilms B, Laux H (2016) Improving expression of recombinant human IGF-1 using IGF-1R knockout CHO cell lines. Biotechnol Bioeng 113(5):1094–1101

Ruan J, Li H, Xu K, Wu T, Wei J, Zhou R, Liu Z, Mu Y, Yang S, Ouyang H, Chen-Tsai RY (2015) Highly efficient CRISPR/Cas9-mediated transgene knockin at the H11 locus in pigs. Sci. Rep 5(1):1–10

Ryan DE, Taussig D, Steinfeld I, Phadnis SM, Lunstad BD, Singh M, Vuong X, Okochi KD, McCaffrey R, Olesiak M, Roy S (2018) Improving CRISPR–Cas specificity with chemical modifications in single-guide RNAs. Nucleic Acids Res 46(2):792–803

Safari F, Zare K, Negahdaripour M, Barekati-Mowahed M, Ghasemi Y (2019) CRISPR Cpf1 proteins: structure, function, and implications for genome editing. Cell BioSci 9(1):1–21

Sakuma T, Masaki K, Abe-Chayama H, Mochida K, Yamamoto T, Chayama K (2016) Highly multiplexed CRISPR-Cas9-nuclease and Cas9-nickase vectors for inactivation of hepatitis B virus. Genes Cells 21(11):1253–1262

Sapranauskas R, Gasiunas G, Fremaux C, Barrangou R, Horvath P, Siksnys V (2011) The Streptococcus thermophilus CRISPR/Cas system provides immunity in Escherichia coli. Nucleic Acids Res 39(21):9275–9282

Sauer NJ, Narváez-Vásquez J, Mozoruk J, Miller RB, Warburg ZJ, Woodward MJ, Mihiret YA, Lincoln TA, Segami RE, Sanders SL, Walker KA (2016) Oligonucleotide-mediated genome editing provides precision and function to engineered nucleases and antibiotics in plants. Plant Physiol 170(4):1917–1928

Schiml S, Fauser F, Puchta H (2014) The CRISPR/C as system can be used as nuclease for in planta gene targeting and as paired nickases for directed mutagenesis in Arabidopsis resulting in heritable progeny. Plant J 80(6):1139–1150

Schumann K, Lin S, Boyer E, Simeonov DR, Subramaniam M, Gate RE, Haliburton GE, Chun JY, Bluestone JA, Doudna JA, Marson A (2015) Generation of knock-in primary human T cells using Cas9 ribonucleoproteins. Proc Natl Acad Sci 112(33):10437–10442

Shen C, Que Z, Xia Y, Tang N, Li D, He R, Cao M (2017) Knock out of the annexin gene OsAnn3 via CRISPR/Cas9-mediated genome editing decreased cold tolerance in rice. J Plant Biol 60(6):539–547

Shi J, Gao H, Wang H, Lafitte HR, Archibald RL, Yang M, Hakimi SM, Mo H, Habben JE (2017) ARGOS 8 variants generated by CRISPR-Cas9 improve maize grain yield under field drought stress conditions. Plant Biotechnol J 15(2):207–216

Singh R, Kuscu C, Quinlan A, Qi Y, Adli M (2015) Cas9-chromatin binding information enables more accurate CRISPR off-target prediction. Nucleic Acids Res 43(18):e118

Slaymaker IM, Gao L, Zetsche B, Scott DA, Yan WX, Zhang F (2016) Rationally engineered Cas9 nucleases with improved specificity. Science 351(6268):84–88

Smith C, Gore A, Yan W, Abalde-Atristain L, Li Z, He C, Wang Y, Brodsky RA, Zhang K, Cheng L, Ye Z (2014) Whole-genome sequencing analysis reveals high specificity of CRISPR/Cas9 and TALEN-based genome editing in human iPSCs. Cell Stem Cell 15(1):12–13

Sonoda E, Hochegger H, Saberi A, Taniguchi Y, Takeda S (2006) Differential usage of non-homologous end-joining and homologous recombination in double strand break repair. DNA Repair 5(9–10):1021–1029

Sontheimer EJ (2019) X-tracting a new CRISPR-Cas genome-editing platform from metagenomic data sets. CRISPR J 2(3):148–150

Stemmer M, Thumberger T, del Sol Keyer M, Wittbrodt J, Mateo JL (2017) Correction: CCTop: an intuitive, flexible, and reliable CRISPR/Cas9 target prediction tool. PLoS One 12(4):e0176619

Svitashev S, Young JK, Schwartz C, Gao H, Falco SC, Cigan AM (2015) Targeted mutagenesis, precise gene editing, and site-specific gene insertion in maize using Cas9 and guide RNA. Plant Physiol 169(2):931–945

Teytelman L, Thurtle DM, Rine J, van Oudenaarden A (2013) Highly expressed loci are vulnerable to misleading ChIP localization of multiple unrelated proteins. Proc Natl Acad Sci 110(46):18602–18607

Tsai SQ, Joung JK (2016) Defining and improving the genome-wide specificities of CRISPR–Cas9 nucleases. Nat Rev Genet 17(5):300–312

Tsai SQ, Wyvekens N, Khayter C, Foden JA, Thapar V, Reyon D, Goodwin MJ, Aryee MJ, Joung JK (2014) Dimeric CRISPR RNA-guided FokI nucleases for highly specific genome editing. Nat Biotechnol 32(6):569–576

Tsai SQ, Zheng Z, Nguyen NT, Liebers M, Topkar VV, Thapar V, Wyvekens N, Khayter C, Iafrate AJ, Le LP, Aryee MJ (2015) GUIDE-seq enables genome-wide profiling of off-target cleavage by CRISPR-Cas nucleases. Nat Biotechnol 33(2):187–197

Tsai CJ, Xu P, Xue LJ, Hu H, Nyamdari B, Naran R, Zhou X, Goeminne G, Gao R, Gjersing E, Dahlen J (2020) Compensatory guaiacyl lignin biosynthesis at the expense of syringyl lignin in 4CL1-knockout poplar. Plant Physiol 183(1):123–136

Tu Z, Yang W, Yan S, Yin A, Gao J, Liu X, Zheng Y, Zheng J, Li Z, Yang S, Li S (2017) Promoting Cas9 degradation reduces mosaic mutations in non-human primate embryos. Sci Rep 7(1):1–1

Tycko J, Myer VE, Hsu PD (2016) Methods for optimizing CRISPR-Cas9 genome editing specificity. Mol Cell 63(3):355–370

Vartak SV, Raghavan SC (2015) Inhibition of nonhomologous end joining to increase the specificity of CRISPR/Cas9 genome editing. FEBS J 282(22):4289–4294

Waltz E (2016) Gene-edited CRISPR mushroom escapes US regulation. Nat News 532(7599):293

Wang Y, Cheng X, Shan Q, Zhang Y, Liu J, Gao C, Qiu JL (2014) Simultaneous editing of three homoeoalleles in hexaploid bread wheat confers heritable resistance to powdery mildew. Nat Biotechnol 32(9):947–951

Wang X, Wang Y, Wu X, Wang J, Wang Y, Qiu Z, Chang T, Huang H, Lin RJ, Yee JK (2015) Unbiased detection of off-target cleavage by CRISPR-Cas9 and TALENs using integrase-defective lentiviral vectors. Nat Biotechnol 33(2):175–178

Wedell N, Price TAR, Lindholm AK (2019) Gene drive: progress and prospects. Proc Biol Sci 286(1917):20192709

Wright AV, Sternberg SH, Taylor DW, Staahl BT, Bardales JA, Kornfeld JE, Doudna JA (2015) Rational design of a split-Cas9 enzyme complex. Proc Natl Acad Sci 112(10):2984–2989

Wu X, Scott DA, Kriz AJ, Chiu AC, Hsu PD, Dadon DB, Cheng AW, Trevino AE, Konermann S, Chen S, Jaenisch R (2014) Genome-wide binding of the CRISPR endonuclease Cas9 in mammalian cells. Nat Biotechnol 32(7):670–676

Wyvekens N, Topkar VV, Khayter C, Joung JK, Tsai SQ (2015) Dimeric CRISPR RNA-guided FokI-dCas9 nucleases directed by truncated gRNAs for highly specific genome editing. Hum Gene Ther 26(7):425–431

Xie S, Shen B, Zhang C, Huang X, Zhang Y (2014) sgRNAcas9: a software package for designing CRISPR sgRNA and evaluating potential off-target cleavage sites. PLoS One 9(6):e100448

Xu Z, Xu X, Gong Q, Li Z, Li Y, Wang S, Yang Y, Ma W, Liu L, Zhu B, Zou L (2019) Engineering broad-spectrum bacterial blight resistance by simultaneously disrupting variable TALE-binding elements of multiple susceptibility genes in rice. Mol Plant 12(11):1434–1446

Yamano T, Nishimasu H, Zetsche B, Hirano H, Slaymaker IM, Li Y, Fedorova I, Nakane T, Makarova KS, Koonin EV, Ishitani R (2016) Crystal structure of Cpf1 in complex with guide RNA and target DNA. Cell 165(4):949–962

Yan WX, Mirzazadeh R, Garnerone S, Scott D, Schneider MW, Kallas T, Custodio J, Wernersson E, Li Y, Gao L, Federova Y (2017) BLISS is a versatile and quantitative method for genome-wide profiling of DNA double-strand breaks. Nat Commun 8(1):1–9

Yang H, Patel DJ (2019) CasX: a new and small CRISPR gene-editing protein. Cell Res 29(5):345–346

Yang L, Grishin D, Wang G, Aach J, Zhang CZ, Chari R, Homsy J, Cai X, Zhao Y, Fan JB, Seidman C (2014) Targeted and genome-wide sequencing reveal single nucleotide variations impacting specificity of Cas9 in human stem cells. Nat Commun 5(1):1–6

Yao R, Liu D, Jia X, Zheng Y, Liu W, Xiao Y (2018) CRISPR-Cas9/Cas12a biotechnology and application in bacteria. Synth Syst Biotechnol 3(3):135–149

Yoshimi K, Kunihiro Y, Kaneko T, Nagahora H, Voigt B, Mashimo T (2016) ssODN-mediated knock-in with CRISPR-Cas for large genomic regions in zygotes. Nat Commun 7(1):1–10

Zetsche B, Gootenberg JS, Abudayyeh OO, Slaymaker IM, Makarova KS, Essletzbichler P, Volz SE, Joung J, Van Der Oost J, Regev A, Koonin EV (2015) Cpf1 is a single RNA-guided endonuclease of a class 2 CRISPR-Cas system. Cell 163(3):759–771

Zhang XH, Tee LY, Wang XG, Huang QS, Yang SH (2015) Off-target effects in CRISPR/Cas9-mediated genome engineering. Mol Ther Nucleic Acids 4:e264

Zhang H, Cheng QX, Liu AM, Zhao GP, Wang J (2017) A novel and efficient method for bacteria genome editing employing both CRISPR/Cas9 and an antibiotic resistance cassette. Front Microbiol 8:812

Zhang Y, Wang J, Wang Z, Zhang Y, Shi S, Nielsen J, Liu Z (2019) A gRNA-tRNA array for CRISPR-Cas9 based rapid multiplexed genome editing in Saccharomyces cerevisiae. Nat Commun 10(1):1053

Zhao Y, Zhang C, Liu W, Gao W, Liu C, Song G, Li WX, Mao L, Chen B, Xu Y, Li X (2016) An alternative strategy for targeted gene replacement in plants using a dual sgRNA/Cas9 design. Sci Rep 6:23890

Zhu LJ, Holmes BR, Aronin N, Brodsky MH (2014) CRISPRseek: a bioconductor package to identify target-specific guide RNAs for CRISPR-Cas9 genome-editing systems. PLoS One 9(9): e108424

Beyond Genome Editing: CRISPR Approaches

6

Qaisar Sultan, Sidra Ashraf, Ahmad Munir, Sultan Habibullah Khan, Nayla Munawar, Kamel A. Abd-Elsalam, and Aftab Ahmad

Contents

Q. Sultan
Department of Botany, Government College University, Faisalabad, Pakistan

S. Ashraf · A. Munir
Department of Biochemistry, University of Agriculture, Faisalabad, Pakistan

S. H. Khan
Center of Agricultural Biochemistry and Biotechnology (CABB)/US-Pakistan Center for Advanced Studies in Agriculture and Food Security, University of Agriculture, Faisalabad, Pakistan

N. Munawar
Department of Chemistry, College of Science, United Arab Emirates University, Al-Ain, UAE

K. A. Abd-Elsalam
Plant Pathology Research Institute, Agricultural Research Center (ARC), Giza, Egypt

A. Ahmad (✉)
Department of Biochemistry/US-Pakistan Center for Advanced Studies in Agriculture and Food Security, University of Agriculture, Faisalabad, Pakistan
e-mail: aftab.ahmad@uaf.edu.pk

A. Ahmad et al. (eds.), *The CRISPR/Cas Tool Kit for Genome Editing*,
https://doi.org/10.1007/978-981-16-6305-5_6

Abstract

CRISPR has become a powerful tool for precise editing of user-defined genomic loci in living organisms. Within mere eight years, CRISPR/Cas has become an indispensable tool in fundamental and applied research of biological sciences. Today CRISPR is no longer just a method to edit a specific locus, but applications of dCas9-based platforms are exceeding the genome editing functions of WT Cas9. Applications of CRISPR technology are extending beyond genome editing such as expression modulation, rewriting epigenetic signatures, and imaging spatiotemporal organization of the genome. In this chapter we discuss, dCas9 and engineered gRNA-based CRISPRi and CRISPRa approach for transcriptional regulation of a gene. In addition, we also describe visualization of spatiotemporal organization of genomic loci through dCas9-based tools. Moreover, we also explain how CRISPR barcodes are exploited to study cell lineage. Furthermore, we also highlight CRISPR-mediated tools such as epigenome regulation, base editing, prime editing, and directed evolution. Finally, we discussed prospects beyond genome editing application of CRISPR/Cas.

Keywords

CRISPR/Cas · Beyond genome editing · Base editing · CRISPRi/CRISPRa · EvolvR · Prime editing · Gene drive · Genome imaging

Abbreviations

ABE	Adenine base editor
BE	Base editor
CBE	Cytosine base editor
CDE	CRISPR-based directed evolution
CRISPR	Clustered regularly interspaced short palindromic repeats
CRISPRa	CRISPR activation
CRISPRi	CRISPR interference
FDA	Food and Drug Administration
GMOs	Genetically modified organisms
gRNA	Guide RNA
HDR	Homology-directed repair
KI	Knock-in
KO	Knockout
KRAB	Kruppel-associated box
MCP	Monocyte chemoattractant protein
NHEJ	Non-homologous end joining
PAM	Protospacer adjacent motif
PE	Prime editor

spCas9	Streptococcus pyogenes Cas9
TALEN	Transcription activator-like effector nucleases
tracrRNA	Trans-activating CRISPR RNA
WGS	Whole genome sequencing
ZFN	Zinc finger nuclease

6.1 Introduction

Eukaryotic genomes are quite complicated and large as they contain billions of bases. Similarly, gene regulation in eukaryotes is significantly complicated compared with prokaryotes. Researchers have been modifying these genomes for desired traits using genetic engineering tools; however, making programmable and precise changes in eukaryotic genomes was not possible before invention of genome editing with engineered nucleases. Engineered nucleases are one of the initial breakthroughs in genome editing in the twenty-first century for site-specific modifications (Adli 2018). Genome editing using engineered nucleases is the technique to introduce the desired changes in the genome by altering the DNA bases at the target site. Precise genome editing at the specific sites with these nucleases holds remarkable value in agriculture (Ricroch 2019), biotechnology (Whelan and Lema 2019), medical (Carroll 2016) and clinical science (Zheng et al. 2020) with therapeutic and translational applications. These nucleases include transcription activator-like effector nucleases (TALENs) (LaFountaine et al. 2015), zinc finger nucleases (ZFNs) (Handel and Cathomen 2011), and clustered regularly interspaced short palindromic repeats (CRISPR) (Barrangou and Doudna 2016). These are powerful tools to tailor genomes of bacteria, plants, and animals for desired modifications. TALENs and ZFNs consist of binding and nuclease domains that work in form of pairs. The binding domain can bind specifically with the target DNA sequence while nuclease domain can cleave the target sequence to create a double-strand break (DSB), subsequently repaired through homology-directed repair (HDR) or non-homology end joining (NHEJ) system. HDR results in the insertion of a DNA template (double stranded or single stranded) at the cleavage site depending upon the provision of donor template, and NHEJ repair system may lead to gene disruption from indels at the target site (Wright et al. 2014). Both ZFNs and TALENs use the *Flavobacterium okranokoties (FokI)* restriction enzyme for cleavage at a target site and possess higher efficiency of genome editing; however they are free of limitations. The difficulty in cloning process, complexity in protein engineering, and requirement of new protein for a new target are major limitations in worldwide adoption of these tools for genome editing in every lab (Adli 2018). In contrast, simple design, high efficiency, modular nature, and RNA-based targeting of CRISPR/Cas have revolutionized genome editing which ultimately leads to Nobel Prize in Chemistry 2020 for this powerful technology.

CRISPR/Cas system is a powerful tool that consists of an endonuclease protein (Cas9 in CRISPR/Cas9 system) whose cleavage activity and targeting specificity is programmed by a short guide RNA (gRNA) and presence of protospacer adjacent motif (PAM). CRISPR arrays are composed of short repeats (repetitive sequence) interspaced by spacer sequences (non-repetitive elements) which are derived from genomic sequences of bacteriophages (Jinek et al. 2012). CRISPR arrays are present as an adaptive immune system in Archaea (90%) and Bacteria (40%) (Mojica et al. 2000). In addition to these CRISPR elements, multiple CRISPR-associated genes are present adjacent to the CRISPR sequences which play specific roles such as processing of CRISPR RNA (crRNA). The mode of action of CRISPR/Cas system in bacteria is further divided into three phases: a) adaptation, b) expression, and c) interference. Cas proteins (Cas1 and Cas2 in CRISPR/Cas9) recognize the invading viral DNA or plasmids in the adaptation phase and integrate it into the CRISPR arrays (as a new spacer DNA) consequently creating a memory or molecular record for future. In the expression phase, CRISPR arrays are transcribed into pre-crRNA which is further processed into mature crRNA and contain a memory sequence complementary to the invading elements. Mature crRNA forms a complex with tracrRNA (Trans-activating RNA) and engages Cas protein to create an active CRISPR/Cas complex (crRNA-tracrRNA-Cas9). In the interference phase, crRNA in the activated complex of ribonucleoprotein pairs with the complementary spacer sequence in the invading element (virus or plasmid) and activates Cas9 to recognize and cleave the target sequence. crRNA and tracrRNA help to recognize target sequence followed by PAM (NGG or NAG for Cas9) sequence; however, sequence and position of PAM may vary with different Cas nucleases. Cas protein in the activated ribonucleoprotein complex acts as a molecular scissor and cleaves the target genomic sequence in invading elements thus defending bacteria against these attacks (Munawar and Ahmad 2021; Rath et al. 2015).

The standard nomenclature of the CRISPR/Cas system is based on the *Cas* genes coding for Cas proteins, organization of effector nucleases, and processing mechanism of pre-crRNA. CRISPR/Cas system is classified into two main classes named Class 1 and Class 2. Class 1 is a more common class present in archaea and bacteria but relies on multi-subunit effector Cas protein complexes for cleavage of DNA or RNA and further divided into three types (Type I, III, and IV). The systems of Class 2 are characterized by single effector protein-containing multi-domains and further divided into three types (Type II, V, and VI). These six types of both classes are further categorized into many subtypes. Compared with Class 1, CRISPR/Cas loci in Class 2 are more uniformly organized. Therefore, the Class 2 CRISPR/Cas systems are well characterized and adopted universally for basic as well as commercial research applications in agriculture and biomedical sciences. Cas9, Cas12, and Cas13 are well-characterized systems of Class 2 CRISPR/Cas systems and considered as most attractive genome editing tools due to their reprogramming ability to any sequence in the genome (Makarova et al. 2018; Makarova and Koonin 2015). In 2017, Shmakov et al. developed a computational pipeline to investigate the whole genome sequence databases (WGS) in National center for biotechnology information (NCBI) to identify unclassified Class 2 CRISPR system members. These members are filtered to identify the loci containing uncategorized proteins (> 500 amino acids). This approach yielded 50 new Cas loci that expanded the number of type V CRISPR/Cas system. This approach also introduced Type VI-A or Class 2 effectors proteins (C2c2/Cas13) to the world (Shmakov et al. 2017).

CRISPR/Cas technology is continuously evolving with new techniques and new Cas proteins with diverse functions. Very recently, Cas13 has been identified which has opened a new chapter enabling RNA editing in addition to usual DNA editing (Cox et al. 2017). This has opened new vistas of CRISPR/Cas to manipulate transcriptional regulation of genes (Charles et al. 2021). Transcriptional regulation of genes was also possible with a modified CRISPR/Cas system called CRISPRi. With CRISPRi using dCas9, transcription of a gene could be blocked by targeting dCas to TATA box in the promoter region (Gilbert et al. 2013). Conversely, dCas could also be used to activate gene expression by recruiting VP64, an activation domain (AD) to promoter sequence, resulting in transcriptional activation of a selected gene (Mali et al. 2013). Moreover, dCas9 can also be fused with green fluorescent protein (GFP) to visualize chromatin organization and dynamics of the targeted gene (Chen et al. 2013). In addition, Cas subunits can be engineered with a rapamycin-binding domain to design an antibiotic inducible CRISPR system. Light-inducible CRISPR/Cas has been used in CRISPR technologies (Dai et al. 2018). Furthermore, subunits of dCas9 can be engineered with various enzymatic activities, e.g., epigenetic modifications for tailoring methylation patterns and deaminase activity for base conversion, changing C to T and A to G independently of Double-strand break (DSBs) (Bandyopadhyay et al. 2020). Prime editing is one of the most recent additions in CRISPR applications by fusing nCas9 with reverse transcriptase (RT), which allows base editing in the target DNA sequence without any provision of donor template (Li et al. 2020; Anzalone et al. 2019). We can expect a future revolution in CRISPR/Cas with much broader applications using Cas13a, Cas13b, Cpf1, and dCas9 for crop improvement. Using multiplexing through CRISPR/Cas, scientists will be able to target multigenic traits and metabolic engineering for future improvement of crops (Jaganathan et al. 2018). In this chapter, we focus on beyond genome editing applications of CRISPR/Cas systems.

6.2 Genome Editing with CRISPR/Cas

The defensive role of the CRISPR/Cas system in bacterial cells paved the way to change innate genetic modifications into precise and targeted genome engineering. Different CRISPR/Cas systems require different components (single or multiple effector nucleases, PAM requirements, optimum sgRNA length, and processing of pre-crRNA) and use distinct mechanisms to target a specific DNA or RNA sequence (Munawar and Ahmad 2021). However, genome editing with CRISPR/Cas system is based on the principle that it creates DSB at the target site in the genome, which is repaired by either error-prone NHEJ or HDR pathways. The repair of DSB through these pathways permits the desired changes in the genome by deletion, insertion, or modifications of nucleotides in the target site of the genome. NHEJ is an error-prone repair mechanism and results in indels (insertion or deletion of variable nucleotides) that can be utilized for gene knockout (KO). In addition, NHEJ is a preferred repair mechanism in cell and widely used to introduce desired modifications in genomes of plants and animals. Different enzymes are used during the NHEJ repair mechanism that involves re-annealing of the ends of DSB (homologous DNA template is not required). In higher eukaryotes, the NHEJ repair pathway occurs throughout the cell cycle and causes alteration at the target site by deletion or addition of nucleotides. Besides, the HDR is an error-free pathway to introduce desired genetic modifications by using donor DNA (extrinsic homologous). HDR mechanism can be used to perform insertion and gene replacement at the target site in the presence of an exogenous donor template. However, HDR repair pathway has low efficiency in eukaryotes than the NHEJ pathway and occurs during the G2 and S phases of the cell cycle (Adli 2018; Sander and Joung 2014; Munawar and Ahmad 2021).

6.2.1 Prominent CRISPR/Cas Systems

Among all the CRISPR/Cas systems, type II CRISPR/Cas9 system is one of the most widely used and best re-engineered tool for genome editing in eukaryotes. CRISPR/Cas9 system (spCas9) from *S. pyogenes* has been one of the earliest characterized CRISPR/Cas systems that has been adopted universally for genome editing applications in every field of biological sciences (Trevino and Zhang 2014). Cas9 is a multi-domain protein that cleaves the viral DNA and foreign plasmid with the help of crRNA and tracrRNA. In Cas9, the RuvC and HNH are the nuclease domains that are responsible for creating DSB. The HNH domain is responsible for the cleavage of the target sequence while the nontarget strand is cleaved by the RuvC domain. The cleavage activity of Cas9 depends upon the presence of an adjacent PAM, a short conserved sequence present downstream of a non-complimentary strand of the target sequence (Fig. 6.1) (Jinek et al. 2012).

Cas9 derived from other species of bacteria such as *Campylobacter jejuni* has been used in humans and mice for genome editing (Kim et al. 2017). Although,

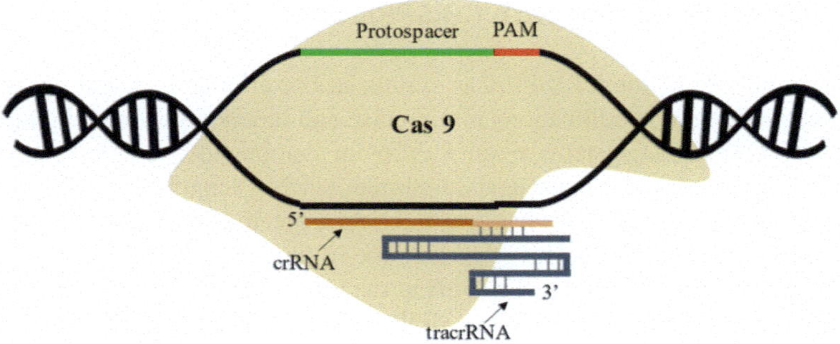

Fig. 6.1 Schematic diagram of CRISPR Cas9

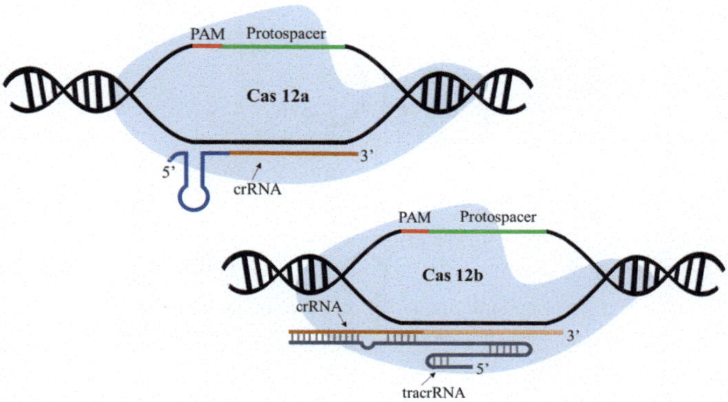

Fig. 6.2 Schematic diagram of CRISPR Cas12a and 12b. CRISPR Cas12 binds and cleave double-stranded DNA upstream of PAM together with protospacer. CRISPR Cas12a only requires crRNA while CRISPR Cas12b requires both crRNA and tracRNA for its activity

CRISPR/Cas9 has been widely used in genome editing; however, it is not limitation-free. For example, large size of Cas9, off-targets, and requirement of PAM sequence for cleavage of target sequence are major limitations. Cas9 proteins from other bacterial species with alternate PAM sequences have been used to address PAM limitation (Hu et al. 2018; Adli 2018). In addition, truncated sgRNA and double nickase have been used to overcome off-targeting in CRISPR/Cas9 (Grünewald et al. 2019). Moreover, different new CRISPR/Cas systems have been identified with superior features to overcome these limitations such as CRISPR/Cas12, CRISPR/Cas13, CRISPR/Cas14, and CRISPR/CasX. For example, CRISPR/Cas12a, a type II CRISPR/Cas system also known as Cpf1, has a smaller Cas nuclease with different PAM requirements. FnCpf1 is derived from *Francisella novicida* and cleaves double-stranded DNA, like the CRISPR/Cas9 system. However, FnCpf1 requires a T-rich PAM sequence (TTN) and produce staggering ends at distal positions for DSB instead of a G-rich PAM (NGG), with blunt ends in CRISPR/Cas9 system (Fig. 6.2). In addition, the efficiency of NHEJ-based gene insertion has also been improved with CRISPR/Cas12. Moreover, Cas12 requires a much smaller crRNA of 42 nucleotides in size while Cas9 requires a longer crRNA of 100 nucleotides. The requirement of much smaller crRNA, T-rich PAM and smaller size of Cas12, makes it a cost-effective and ideal system to target T-rich regions in genomes (Wang et al. 2020; Yan et al. 2019; Wolter and Puchta 2018).

CRISPR/Cas13 is a novel type of type II CRISPR/Cas system with a single effector nuclease which can bind and precisely cleave single-stranded RNA (ssRNA) (Fig. 6.3). Cas13 contains a higher eukaryotes and prokaryotes nucleotide-binding (HEPN) domain that has RNAase activity and can defend against viral RNAs. However, once Cas13 is activated by specific RNA, it can also degrade nonspecific RNA (collateral cleavage), indicating its role in cell death and limiting

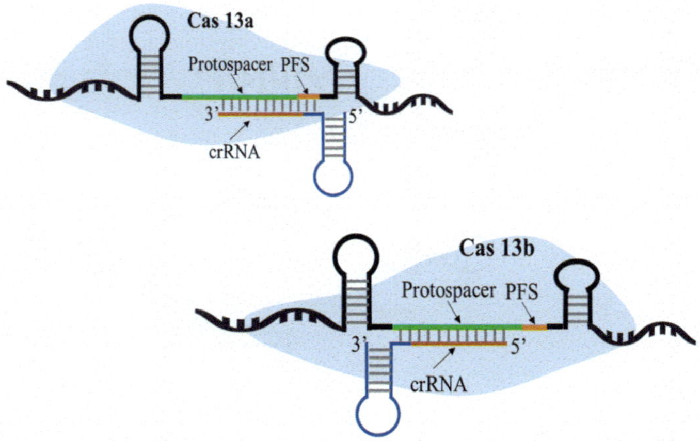

Fig. 6.3 Schematic diagram of CRISPR Cas13a and 13b. CRISPR Cas13a and 13b together with crRNA binds and cleaves single-stranded RNA. Complementarity between crRNA and protospacer flank sequence (PFS) together with protospacer results in the cleavage of RNA. Cas13a carries a direct repeat at 5′ end. Cas13b carries a direct repeat at 3′ end

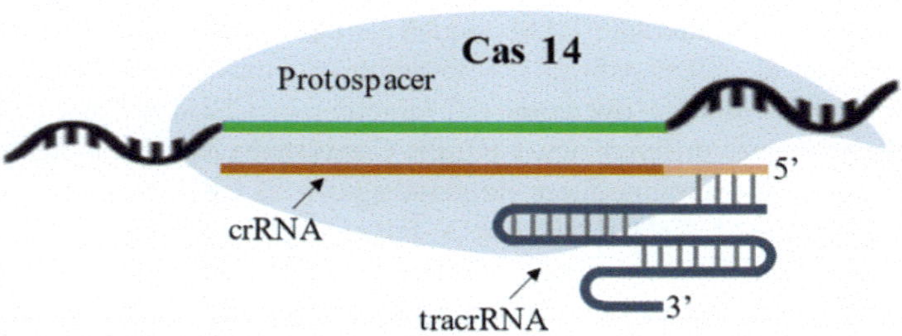

Fig. 6.4 Schematic diagram of CRISPR Cas14. CRISPR Cas14 together with crRNA and tracRNA binds and cleaves single-stranded DNA without PAM recognition

infection. However, nonspecific cleavage was not observed in eukaryotic cells (Abudayyeh et al. 2017; O'Connell 2019).

CRISPR/Cas14 is an emerging CRISPR system that can target single-stranded DNA (ssDNA) (Fig. 6.4). It has been further classified into three subgroups including CRISPR/Cas14a, CRISPR/Cas14b, and CRISPR/Cas14c. The size of Cas14 is almost half of the size of Cas9 and it does not require PAM for site-specific cleavage of ssDNA. CRISPR/Cas14 has been especially useful for genome editing applications against ssDNA viruses such as geminiviruses in plants (Khan et al. 2019; Cana-Quijada et al. 2020).

Another emerging system in CRISPR/Cas family is CRISPR/CasX which is characterized by smaller size and unique PAM requirement and produces staggered ends at DSB (Fig. 6.5). Biochemical studies revealed that CasX is a hybrid nuclease

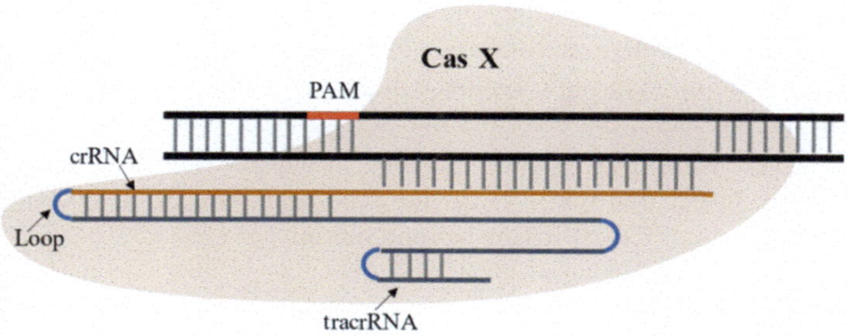

Fig. 6.5 Schematic diagram of CRISPR CasX. RNA-dependent plasmid with two natural RNAs including crRNA and tracRNA binds and cleaves double-stranded DNA

containing features of both Cas9 and Cas12. CRISPR/CasX has been used for genome editing in bacterial as well as human cells (Rees and Liu 2018; Hu et al. 2018).

Several factors influence the specificity and off-targeting in CRISPR/Cas-mediated genome editing. It has been revealed that Cas9 specificity is based on the length of sgRNA. The specificity of Cas9 was increased by increasing the size of the gRNA; however, the optimal length of sgRNA is 20-bp for site-specific cleavage through CRISPR/Cas9. Off-targets are undesirable editings or modifications that occur due to nonspecific binding of sgRNA at nontarget sites in the genome (Wu et al. 2014). Off-targets have been a considerable challenge in CRISPR/Cas-mediated genome editing. It has been reported that CRISPR/Cas9 could produce a considerable number of off-targets because it can tolerate up to 5 mismatches between the target sequence and sgRNA. Therefore, in silico prediction and minimum off-targets are essential for the safer use of the CRISPR/Cas system, particularly for translational research and therapeutic applications. Some online tools have been designed and publicly available to identify off-targets (Munawar and Ahmad 2021).

CRISPR/Cas systems have revolutionized basic as well as applied research in redesigning bacterial, animal, and plant genomes (Zhang et al. 2019; Tu et al. 2015). Although initially CRISPR/Cas9 system was mainly used for genome editing, re-engineering of CRISPR/Cas system such as CRISPR/dCas9 has expanded applications of this system even beyond genome editing (Adli 2018). Therefore, today CRISPR/Cas system is no longer just a tool for genome editing but its applications are extending for screening of genomic libraries for functional genomics (Kampmann 2020), genome imaging (Chen et al. 2013), targeted transcriptional regulation (McCarty et al. 2020), directed evolution (Lee et al. 2018), and base editing (Molla and Yang 2019). In addition, gene drives through CRISPR/Cas has been demonstrated in the laboratory to eradicate vector-borne diseases such as malaria, dengue, and zika (Nateghi Rostami 2020). Moreover, recently, base editing with CRISPR/Cas has been achieved even without providing a donor template called prime editing. Directed evolution is another application to introduce new traits through rewriting of genetic code beyond nature. CRISPR has made genome editing easy, efficient, cost-effective, and more socially acceptable than GMOs. All these applications depend upon the sgRNA that can guide Cas nuclease to the target site to create DSB, nick or recruiting effectors like activators or repressors to the genomic loci (Adli 2018). Multiplex genome editing through CRISPR/Cas has become advantageous to target multiple loci in the genome (Cong et al. 2013). Here we describe applications of CRISPR/Cas beyond editing.

6.3 CRISPRi and CRISPRa for Transcriptional Regulation of Genes

Genomic and transcriptional regulation of genes is important for functional genomic studies as a number of genomes have been sequenced in both plants and animals. For many years, RNAi has ruled the labs to silence the genes at post-transcriptional level without disrupting genomes. RNAi has been used widely to study gene functions in all organisms (Miki and Shimamoto 2004; Maeda et al. 2001); however, off-targets, reduced effect in successive generations and restricted to only transcriptional level, have been major limitations of RNAi (Senthil-Kumar and Mysore 2011; Qiu et al. 2005). With the advent of CRISPR/Cas9 and various techniques based on this novel technology, such as CRISPR/dCas9 (CRISPR interference called as CRISPRi and CRISPR activation called as CRISPRa) and CRISPR/nCas9, it has become a robust and efficient technique for functional genomic studies in almost all fields of biological sciences (Munawar and Ahmad 2021). Compared with RNAi, CRISPR/Cas9 has been simple and precise and have minimum off-targets. Researchers have been using CRISPR/dCas9 for gene regulation (activation as well as interference) soon after the discovery of the CRISPR/Cas9 as a tool for precise genome editing (Boettcher and McManus 2015; Schuster et al. 2019; Kampmann 2018). dCas9 is a catalytically inactive Cas9 that can still bind the targeted region without creating DSB. The catalytic activity of Cas9 was deactivated through substitution of H804A and D10A (Qi et al. 2013; Munawar and Ahmad 2021). dCas9 has been used to recruit various activators, repressors, or fluorescent probes to the targeted site in the genome. CRISPR/dCas9 holds an enormous potential with a broad range of applications in gene regulation, epigenomics, genome imagining, and chromatin studies (Adli 2018). In addition, it has been reported that multiple genes can be regulated (repression and activation) simultaneously through CRISPR/dCas9 approach. CRISPRi has become a versatile tool for downregulation of genes at genomic level. Compared with RNAi, which targets mRNA for silencing of a gene, CRISPRi works at transcriptional level and bind with transcription start site or TATA box consequently blocking RNA polymerase to transcribe a gene. However, RNAi has the advantage that we can use it for organisms whose genomes have not been sequenced yet, and only transcriptomic data is available as RNAi does not work at the genomic level. Moreover, RNAi has no limitation of PAM so we can target conserved regions in mRNA of gene families or splicing variants of a gene. Therefore, one single small interfering RNA (siRNA) can target all transcripts with a

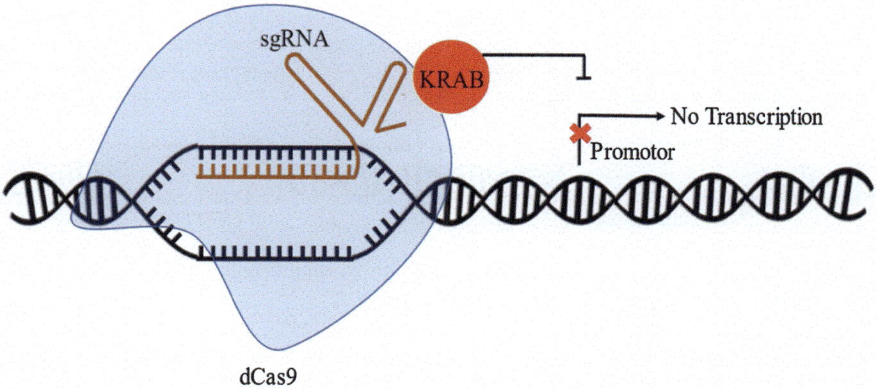

Fig. 6.6 CRISPRi. dCas9 is fused with a transcription repressor. A standard CRISPRi system uses the Krüppel-associated box (KRAB) domain. dCas9-KRAB directs heterochromatin formation at the promoter targeted by the sgRNA, reducing target gene transcription

conserved sequence providing leverage to RNAi over CRISPR/dCas in functional genomics. For CRISPRi, we need genomic data to target dCas9 to transcription start site (TSS) or promoter region of a specific gene (Zhao et al. 2014; Aslam et al. 2021; Schuster et al. 2019).

Transcription in eukaryotes is a complex mechanism regulated through an interplay of DNA regulatory elements called transcription activators and repressors. In addition, epigenetic regulations such as histone acetylation and DNA methylation also play an important role in transcriptional regulation. CRISPR/dCas9 can recruit both transcriptional repressors (KRAB domain, FOG1, and HP1) (Friedman et al. 1996; Adli 2018) or activators (VP64, VP48 and VP16) to robustly repress or activate transcription of a gene without directly editing DNA (Cheng et al. 2013; Mali et al. 2013). Moreover, dCas9 has also been fused with other functional domains such as Tet1 (Choudhury et al. 2016), LSD1 (Zentner and Henikoff 2015), and p300 (Hilton et al. 2015) to modulate the transcriptional regulation. In CRISPRi, dCas9 can be fused with Kruppel-associated Box (KRAB) repressor domain to achieve the transcriptional downregulation of a gene (Fig. 6.6). KRAB is a potent repressor domain, present in human zinc fingers (ZFs) (Birtle and Ponting 2006). In addition, a large family of ZF proteins containing KRAB transcriptional repressor domain is also present in mammals. dCas9-KRAB complex decreases the chromatin accessibility to modulate the gene expression.

Gosh et al. using the CRISPRi have demonstrated that the coexpression of dCas9 and sgRNA resulted in knockdown of the roX1 and roX2 (long non-coding RNAs) in Drosophila. In contrast to the transcriptional repression through CRISPRi, dCas9 can also be fused with transcriptional activators such as VP64, to activate the gene expression (Fig. 6.7). In addition, gene activation by dCas9-VP64 fusion complex was further improved with second-generation CRISPR-based gene activation strategies such as VPr (VP64-p65-rta), a tripartite activator that has been developed

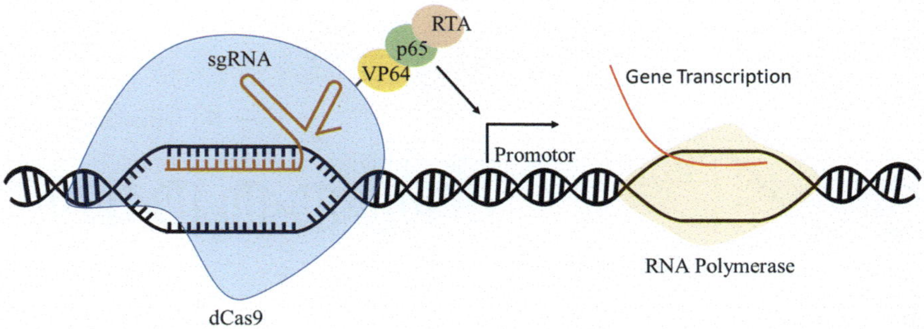

Fig. 6.7 CRISPRa. dCas9 is fused with a transcription activator. A sgRNA programs dCas9 fusion to activate the transcription of the targeted promoter. A commonly used CRISPRa system harnesses four herpes simplex virus VP16 molecules (so-called VP64), the NF-κB transcription factor p65, and the EBV immediate-early transcription activator Rta

to strengthen transcriptional regulation (Adli 2018; Zahoor et al. 2021; Ghosh et al. 2016). In addition to recruiting transcriptional effectors through dCas9, sgRNA can also be used to recruit activation or repression domains as scaffold RNA (scRNA). scRNA contain instructions for both site-specific binding and regulatory functions to perform. scRNA can be programmed for simultaneous activation and repression of multiple and different genes.

For example, Zalatan et al. demonstrated that chimeric sgRNA sequence extended with modular RNA domains such as Com, PP7, and MS2 enabled programmable, parallel, and flexible locus-specific regulation. scRNA with RNA domains like MS2, PP7, and Com can recognize RNA binding proteins such as MCP, PCP, and Com RNA binding proteins fused with VP64 or KRAB domain to regulate the expression of genes. Based on CRISPR/dCas9, third-generation CRISPRa approaches such as SunTag, SAM, and VPR are superior than first-generation activation system (VP64) due to their higher activation (Zalatan et al. 2015). All these strategies have been demonstrated to activate the gene expression in various species such as fly cell lines, mouse, and humans. In addition to these approaches, some inducible CRISPR/dCas systems such as light-inducible and hybrid drug-inducible systems have also been developed. It has been suggested that spatiotemporal control of gene expression through CRISPR/dCas system has remarkable potential, especially in translational and therapeutic applications.

6.4 Epigenome Modulation Through CRISPR

Epigenetics is a dynamic process to control gene expression through changes in chromatin architecture, independent of any permanent change in the primary sequence of DNA. Epigenetic modifications including DNA methylation, histone modifications, and non-coding RNAs are considered to play important roles in gene regulation, genome organization, and cellular processes such as organ development, stress regulation, and disease control (Callinan and Feinberg 2006). CRISPR-mediated site-specific regulation of epigenetic modifications are expected to reveal their functional roles in different cellular processes. In addition, it will also reveal the relationship of epigenetic modifications with regulation of gene expression. Various post-translational modifications of histone proteins such as phosphorylation, methylation, and acetylation lead to induced gene expression while biotinylation and sumoylation are the processes that lead to repressing gene expression. Morever, RNA scaffold produced by small non-coding RNAs can also modify the chromatin architecture and regulate the gene expression (Chinnusamy and Zhu 2009). DNA methylation is the process that occurs in plants and animals distinctly and induces gene expression. Methylation of coding regions and promoters can also modulate the gene expression in tissues and organs during abiotic stress in plants. In the case of plants, cytosine transferase catalyzes the mechanism that inserts methyl group at cytosine residues within CHH dinucleotide, CHG, or CpG islands. While in animals, methylation occurs with the aid of DNMT3B and DNMT3A (DNA methyltransferases) at 5-methyl cytosine of the CpG islands. In addition, endogenous demethylation is catalyzed by ten-eleven translocation (TET) proteins (Adli 2018; Ahmad et al. 2021). In addition to DNA methylation, epigenetic signatures are also characterized by histone modifications. Histone modifications also play an important role in gene regulation. Histone modifications are characterized by mono- and di-methylation at the lysine four positions of Histone H3 (H3K4me1/2) and acetylation at the lysine 27 acetylation position (H3K27ac) of the distal regulatory elements in the genome, while active promoters are characterized by tri-methylation of lysine 4 (H3K4me3) in histone. Methylation and acetylation of histone proteins are associated with gene regulation (Thurman et al. 2012).

Epigenetic readers, writers, and erasers can be used to control the chromatin marks which may influence the gene expression (Fig. 6.8) (Strahl and Allis 2000). Large-scale mapping of epigenetic modifications such as Roadmap Epigenome Mapping Consortium and Encyclopedia of DNA elements have revealed their roles in a cell type-specific gene regulation, histone modifications, and chromatin organization (Bernstein et al. 2010). Although applications of CRISPR/dCas9-mediated epigenome regulation are relatively few, it is expected to greatly empower researchers to reveal functional roles of epigenetic modifications. For example, abnormal methylation pattern in DNA has been associated with cancer, and DNA methylation inhibitors such as 5-azacytidine are approved by FDA as potential therapeutic agents. However, 5-azacytidine targets global methylation of genome; thus it may alter normal pattern methylation at various sites (Kaminskas et al. 2005). CRISPR/dCas9 offers an alternate and an efficient approach to precisely alter the aberrant pattern of methylation in genome. Researchers have demonstrated potential of CRISPR/dCas9 to precisely alter methylation pattern in genome by fusing it with catalytic domains of MQ3 (prokaryotic methyltransferase) and DNMT3A (eukaryotic methyltransferase). In both cases, site-specific deposition of DNA was observed with altered gene expression (Amabile et al. 2016; Stepper et al. 2017; Xu et al. 2016). Similarly, researchers have also demonstrated fusion of CRISPR/dCas9 with TET proteins to specifically erase methylation of DNA up to 90% which resulted in

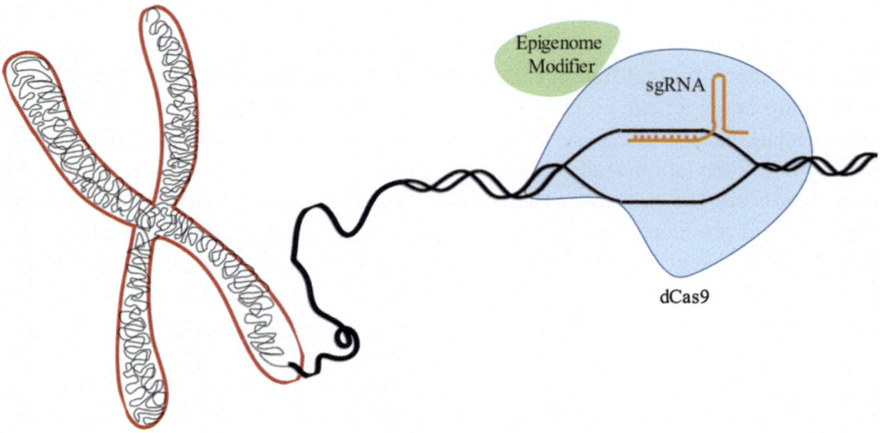

Fig. 6.8 CRISPR and epigenetics. Epigenetic modification with the dCas9 platform involves transcriptional regulation by changing epigenetic signatures. Epigenetic modifiers (writer, reader and erasers) could be used to change methylation pattern, acetylation or histone modifications

an increased level of gene expression (Liu et al. 2016; Xu et al. 2016; Morita et al. 2016). Histone methylation has been precisely altered by recruiting LSD1 with CRISPR/dCas9, resulting in a significant alteration in gene expression through a reduction in enhancer markers H3K4me2 and H3K27ac. In contrast, fusion complex of histone acetyltransferase P300 with CRISPR/dCas9 significantly increased the local level of H3K27ac (Shi et al. 2004). Researchers have also demonstrated the potential of CRISPR/dCas9 system to modify other epigenetic marks. For example, PRDM9 (methyltransferase) was used to influence the level of H3K4me3 marks which resulted in the re-expression of silenced genes in cells (Cano-Rodriguez et al. 2016). Similarly, histone deacetylation was precisely altered by researchers to manipulate chromatin organization (Adli 2018). All these studies demonstrate the potential of CRISPR/dCas9 to precisely manipulate the epigenetic signatures in the genomes. However, all the reports were based on overexpression of fusion complex of dCas9-epigenetic modifier which may also result from nonspecific modifications. Therefore, inducible recruitment of fusion complex may result in increased specificity with higher precision.

6.5 CRISPR-Mediated Site-Specific Base Editing

CRISPR/Cas system has become a versatile and revolutionary technology to accelerate basic and applied developments in agriculture and medical sciences. The basic mechanism of CRISPR/Cas relies on creating DSB at the target site, repaired through NHEJ or HDR repair systems to introduce precise indels. NHEJ is an error-prone mechanism and may lead to indels at the targeted site; however, HDR leads to precise insertion of donor template using homology arms. Although HDR repair mechanism has been demonstrated to introduce site-specific insertions in genome, it is characterized by low efficiency in eukaryotes. In addition, precise insertion through HDR also leads to some undesirable insertions or deletions that limit the potential benefits of this system (Jiang et al. 2020). Moreover, many diseases and several undesirable characters are controlled by single base pair change in the genome. Similarly, several characters in plant genomes are regulated through single base pair alteration in the genome. Therefore, low efficiency and restriction to dividing cells are potential disadvantages of HDR system to precisely replace a single base pair in the genome. CRISPR-mediated base editing is a novel approach to bypass these limitations of HDR and introduces nucleotide substitutions without creating DSB. Base editors (BEs) can safely restore the point mutations and inactivate genes and cis-regulatory regions in hematopoietic cells. CRISPR-mediated precise base editing holds enormous potential for therapeutic applications in human genome and programmable changes in single base pair to introduce elite characters in plant genomes. However, safe and efficient delivery of CRISPR/Cas-based editing system is critical in translational and clinical applications of BEs (Komor et al. 2016; Ahmad et al. 2021).

Two kinds of base editors have been reported thus far: adenine base editors (ABE) and cytosine base editors (CBE). These BEs can be programmed through CRISPR/nCas9 to introduce all four kinds of base substitutions(C $\rightarrow$ T, T $\rightarrow$ C, A $\rightarrow$ G, and G $\rightarrow$ A) in the targeted site. First-generation base editor (CBE1) was developed by fusing dCas9 with apolipoprotein B mRNA editing enzyme, catalytic subunit 1 (APOBEC1). CBE1 deaminates C to U which is recognized by cell replication machinery as T, subsequently resulting in C-G to T-A (Harris et al. 2002). Although CBE1 works perfectly in in vitro applications, its efficiency is very low in human cells because cellular repair mechanism converts U-G intermediate back to C-G pair initiated by uracil N-glycosylate (UGN). Therefore, second-generation base editors CBE2 were developed by fusing uracil glycosylase inhibitor (UGI) with CBE1 to inhibit activity of UGN. Inhibition of UNG by IGI significantly improved activity of CBE2. Third-generation CBE3 was generated by fusing CBE2 with nCas, which further improved the activity of CBE2. As applications of SpCas9-based CBEs were limited by G/C-rich PAM, recent CBEs have been using dCas12-based CBEs. In addition, Cas variants with different PAM requirements have also been developed to address PAM limitations. CBE is limited to substitute C-G to T-A in the targeted site thus urging scientists to expand the window of correctable disease-causing mutations in the genome. The development of ABEs has further improved the base editing capabilities and study of genetic diseases caused by point mutations. ABEs also works similarly to CBEs. CRISPR/dCas9 can be fused with ABE to catalyze conversion of A to I (Inosin) which is recognized by cell replication machinery as G thus substituting A-T with G-C in the target site (Fig. 6.9) (Molla and Yang 2019; Lapinaite et al. 2020). In addition to base editing in DNA, it has also been achieved in RNA using an RNA editing system called REPAIR. Base editing in RNA is useful for reversible gene editing as it does not cause any permanent change in DNA (Mishra et al. 2020).

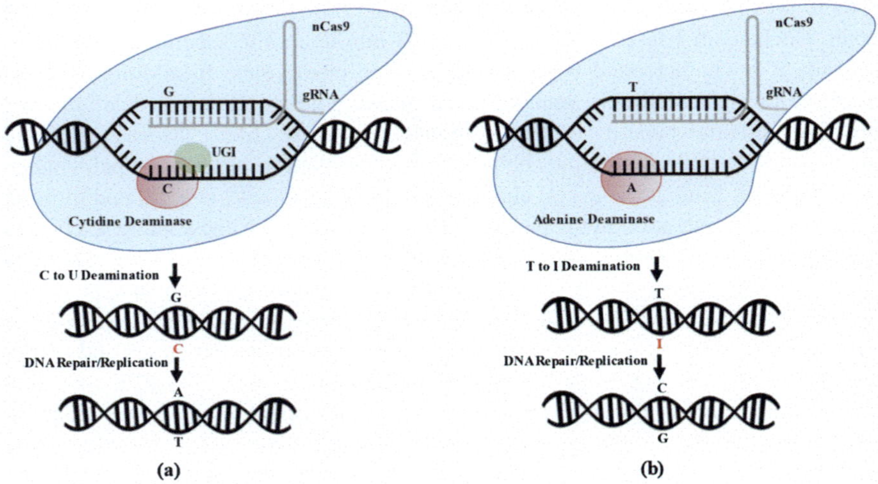

Fig. 6.9 The catalytically impaired nickase Cas9 (nCas9) used for base editing without creating DSB. (**a**) Cytosine base editor (CBE): the CBE consists of a single-stranded cytidine deaminase that linked to Cas9 nickase (nCas9). It catalyzes the transition of cytosine to uridine. The uracil glycosylase inhibitor (UGI) domain prevents the U:G mismatch from being repaired back to a C: G. The nickase clips the opposite non-edited strand, so the U ultimately gets repaired to T. (**b**) Adenine base editor (ABE): the ABE consists of adenosine deaminase that linked to nCas9. It catalyzes the transition of adenosine to guanine. The single-stranded target A is deaminated while the non-edited strand is nicked by nCas9, resulting in the A ultimately being repaired to G

CRISPR-mediated base editing is quite efficient and straightforward than other HDR-based approaches for gene correction as it doesn't require any donor DNA template or DSB to replace particular bases in the target site. Several studies have demonstrated potential of CRISPR-mediated base editing to correct particular bases in the genome. For example, Zong et al. have demonstrated conversion of cytosine into thymine in maize, rice, and wheat with a frequency of 43.48% (Zong et al. 2017). Similarly, Shimatani et al. introduced many herbicide-tolerant point mutations in plants (Shimatani et al. 2017). Mickelbart et al. demonstrated that base substitution in Sub1A-2 gene, sensitive to submergence, could alter the susceptible allele to a tolerant one (Mickelbart et al. 2015).

6.6 Prime Editing (PE)

Prime editing is a recent addition in the CRISPR/Cas toolbox for writing precise and heritable single base pair change in the targeted site without provision of donor template. Although base editing has been used successfully for introducing a single base change in the target site, it is not free of limitations (Anzalone et al. 2019). PAM requirement near base editing site, narrow catalytic window, off-targets, and base editing beyond four transition mutations are major limitations of base editing

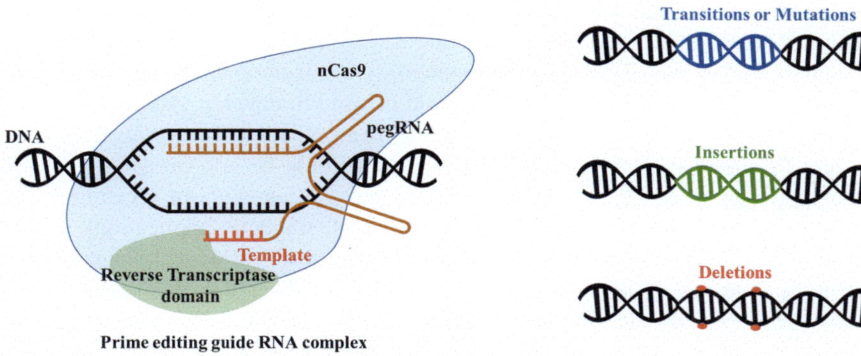

Fig. 6.10 Prime editing. Prime editing requires nCas9 and prime editing guides gRNA (pegRNA) to edit gene sequence without causing double-stranded break. Prime editing uses Cas9 nickase fused with reverse transcriptase which results in transitions, insertions, and deletions at modest editing efficiencies

systems. PE was developed by David Ruchin Liu group from Merkin Institute of Transformative Technologies in Healthcare at Harvard University, USA, to address the limitations of BE system. It employs the same mechanism of CRISPR/Cas9 to install all possible base-to-base conversions and indels without the need for DSB and donor template. PE uses a longer sgRNA known as pegRNA and nCas9 fused with an engineered reverse transcriptase (RT) (Fig. 6.10). The desired sequence changes are programmed in pegRNA as an extension to gRNA, along with a guide sequence for recruiting nCas9 to the target site. The intended sequence spellings in pegRNA are converted to cDNA by RT enzyme thus bypassing the requirement of donor template. The pegRNA guides nCas9 to bind and create a nick in the nontarget strand of DNA, consequently exposing its 3′-hydroxyl group. The nicked strand works as a primer, allowing RT enzyme to extend it using desired changes encoded in pegRNA. The result of this would generate two kinds of flaps, either a 5′-flap containing unedited DNA strand or a 3′-flap containing desired changes. Although hybridization of 5′-flap with non-PAM containing strand is thermodynamically favored, cellular endonuclease such as FEN1 preferably excise 5′-flap thus incorporating intended changes into the host genome (Anzalone et al. 2019; Scholefield and Harrison 2021).

The first-generation primer editors (PE1) were developed by fusing Moloney murine leukemia virus reverse transcriptase (M-MLV RT) with nCas9 and pegRNA; however, their efficiency was low. In the second-generation PEs, Anzalone used different M-MLV RT variants which showed improved binding and increased thermostability with several-fold increase inefficiency. Third-generation PE3 creates a second nick in the unedited strand thus directing replication machinery to repair that strand using the edited strand as a template. PE3s also have improved efficiency of editing with minimum off-targets. PE has been used to introduce site-specific changes in model organisms such as rice, *Arabidopsis*, and mice (Anzalone et al.

2019; Marzec et al. 2020; Lin et al. 2020; Gao et al. 2021). PEs and BEs hold a great promise for medicine, agriculture, and industrial applications. PE offers major advantages as it bypass DSB and requirement of donor template to record precise alterations in the genome. PEs and BEs could be used to develop synthetic devices for a recording of external stimulus, event, or memory. It has been suggested that BE- and PE-based molecular recorder and writers could be approved for medical purposes (Khan et al. 2021). Their template-free editing without creating DSB makes them a suitable candidate for medical and therapeutic applications. DNA recording of events, disease, or external stimuli may result in development of live biosensors. The live biosensors would be helpful for detection of environmental signals such as toxins and diagnosis of diseases. PE is still at developing stage and researchers could expect improved PE-based DNA writers, recorders, and storage systems. PE-based DNA recording system could become a game-changer technology as they can insert arbitrary sequences.

6.7 Genome-Wide Screening

It is important to identify individual genes or entire pathways regulating different phenotypes or disease conditions. Scientists have been using a loss of functions or gain of functions mutagenesis to identify new genes for a particular phenotype. Genome-wide screening aims to develop and screen a population of mutated cells to identify the key genes or a set of genes controlling specific function, pathways, or phenotypes (Sharma and Petsalaki 2018). Recently, CRISPR/Cas has become the most robust and powerful method to perform genome-wide screens in an unbiased manner. Compared with previous methods of genetic screen (T-DNA, transposons, and activation tagging), CRISPR/Cas offers various benefits such as simplicity, easy design, and a broad range of potential targets. In addition, multiplexing, quick gene discovery, targeted mutations, and biallelic mutations are major benefits of CRISPR/Cas approaches. Genome-wide CRISPR/Cas screens are used to reveal a relationship between genotype and a phenotype and identify new genes for different phenotypes by altering gene expression at a global level. Genome-wide gRNA libraries for different organisms such as *E. coli*, mouse, human, and fly have been developed to target every gene in the genome. In addition, genome-wide libraries are not limited to loss of functions (KO), but activation, inhibition, and barcode libraries have also been developed. These libraries contain thousands of plasmids, each containing multiple gRNAs targeting every locus in the genome (Hanna and Doench 2020; Köferle and Stricker 2017; Meltzer et al. 2019). Designing and screening these genome-wide libraries are complex than using CRISPR to edit a single gene. Similarly, position of sgRNAs in the gene also varies according to the applications. Designing and evaluating every single gRNA in is a difficult task; therefore computer softwares are used to design and evaluate all sgRNA in the library. Numerous designing tools including CHOPCHOP (Montague et al. 2014), CRISPR direct (Naito et al. 2015), E-CRISPR (Tarasava et al. 2018), CRISPR-P (Lei et al. 2014), and flyCRISPR (Sangar et al. 2016) have been developed for genome-wide screening. All these tools help to design the sgRNA with low off-target effects. The main objective of all these web tools is to design an optimal sgRNA with low off-target effects. For example, off-target scoring in CHOPCHOP is based on empirical data from multiple studies, while Cas-Finder and E-CRISP evaluate off-targets using user-defined values for mismatch number and position (Bae et al. 2014), CHOPCHOP can design sgRNA for numerous applications such as CRISPRi/CRISPRa, KI, and KO (Afonina et al. 2020; Yang et al. 2021). In addition, some tools are species-specific such as BE-Designer used for the designing of gRNA for base editing and CRISPR-ERA design gRNA for transcriptional regulation (Liu et al. 2015). FlyCRISPR is specialized for designing gRNA of *Drosophila* (Housden et al. 2014), and CRISPR-PLANT develops gRNA for plants (Minkenberg et al. 2019). Zetsche et al. have demonstrated Cas12a and Cas9 as forward genetics platform to screen the mutants against abiotic stress in plants (Zetsche et al. 2015). Moreover, Lu and Zhu used CRISPR-based genome-wide screening in rice against abiotic stress (Lu and Zhu 2017). Sadeek et al. investigated sgRNA libraries cloned into binary vectors to screen the phenotype against abiotic stress in plants (Sedeek et al. 2019).

6.8 Genome Imaging with CRISPR/Cas

Over the time it has been suggested that chromatin structure and its dynamic organization play vital roles in functional output of genome such as expression, replication, and DNA repair (Misteli 2013). In addition, cellular behavior of the genome is also influenced by its interactions with proteins and RNA regulators. To understand the mechanism that how spatiotemporal organizations of genome regulates essential genome functions, DNA and chromatin imaging methods in live cells are indispensable. However, imaging specific genomic loci in the live cells is a challenging task. Historically, fluorescently labelled DNA binding proteins were used to determine position of specific loci in the genome; however this method was limited to specialized sequences such as centromeres and telomeres because of their fixed target sequence (Zink et al. 2003; Bronshtein et al. 2015). Although fluorescence in situ hybridization (FISH) remained flexible and fundamental to determine precise positioning of arbitrary, endogenous, and genomic loci, fixation and denaturation limited its applications in live cell imaging (Tsuchiya 2011; Langer-Safer et al. 1982; Schwarzacher and Heslop-Harrison 1994). With the advent of site-specific DNA binding proteins such as zinc finger proteins and TALEs, scientist used these proteins to recruit fluorescent proteins to the centromeres, telomeres, and genomic loci. However, targeting multiple sequences with these approaches was challenging (Miller et al. 1985; Boch et al. 2009).

Rapid developments in CRISPR/dCas9 have substantially improved specificity, efficiency, and scope of chromatin modeling in live cells. dCas9 fused with fluorescent proteins such as GFP could become a customizable DNA labeller for chromatin imaging in living cells. Fluorescently labelled dCas9 could be recruited to bind any genomic loci depending on the presence of PAM. Alternately, gRNA could also be engineered to bind RNA binding proteins tagged with fluorescent proteins for imaging genomic loci. CRISPR holds several benefits over other imaging techniques including simple design, modular nature, simultaneous programmability to multiple loci in the genome, and compatibility with live cell imaging (Anton et al. 2018). Similarly, multicolor labelling with CRISPR/dCas9 system is also possible which offers simultaneous tracking of multiple loci in the living cells. In this system, one method is characterized by using orthogonal dCas9 fused with different fluorescent proteins, while other method uses engineered sgRNA which recruits orthogonal RNA binding proteins tagged with different fluorescent proteins. It has been demonstrated that sgRNA with eight aptamers provides better imaging of genomic loci. More than 3 genomic loci have been visualized simultaneously, by using three dCas9 orthologs, each derived from different bacterial species and having the ability to recognize the sequence of PAM and sgRNA scaffold (Ma et al. 2015; Takei et al. 2017). Along with chromatin imaging, CRISPR/dCas9 system has also been demonstrated for chromosomal painting. Imaging specific loci require recruitment of multiple copies of fluorescent proteins at the targeted loci which can be achieved by co-delivering multiple sgRNA. However, chromosomal painting requires delivery of hundreds of sgRNA against target sites present throughout chromosome. Researchers have also used CRISPR/dCas system for imaging genomic loci with fluorescent proteins, synthetic dyes, or luminescent nanocrystal particles (Anders et al. 2014). Organic dyes are sensitive, photostable, and smaller in size than fluorescent proteins. Currently, three dyes including Halo-tag, the RNA-aptamers based systems, and molecular beacon-based system have been demonstrated for imaging loci in living cells (Fig. 6.11). Compared with fluorescent proteins and organic dyes, quantum dots represent an excellent system for genome imaging due to

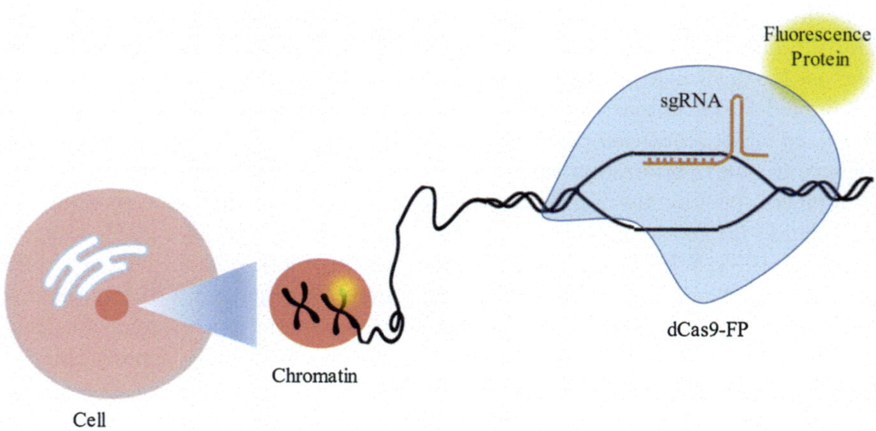

Fig. 6.11 Live cell imaging of chromatin by CRISPR. dCas9 system can bind target DNA under physiological conditions, thereby enabling real-time tracking of chromatin loci in living cells. These CRISPR-based imaging strategies typically use fluorescent proteins (FPs) to generate fluorescence signals

their smaller size and sensitivity. Despite rapid progress in chromatin imaging in live cells with CRISPR/dCas9, several challenges persist. For example, target accessibility, off-targeting, availability of target, and background fluorescence are the major challenges in CRISPR/dCas-based genome imaging (Wu et al. 2019).

6.9 Tracking Cell Lineage Through CRISPR/Cas

Tracking cell lineage holds key importance to answer basic questions in biology. A single totipotent cell of zygote transforms into a complete multicellular organism through remarkable series of developmental and differentiation events, in all sexually producing eukaryotes (Kaufman 1992; McKenna and Gagnon 2019). Elucidating early patterns of cell division and lineage among different cells types is very important to reveal underlying mechanism of tissue and organism development. Methodologies to map detailed mechanism and reconstruction of cellular hierarchies that govern the transformation of a single primordial cell into a multicellular organism would have an enormous impact on our understanding of human development. In addition, it will also improve our understanding of developmental diseases, disease diagnostics, cancer biology, and organ transplantation and critically impacts our abilities to restore normal tissue functioning (Sulston et al. 1983; Riddle et al. 1997). Tracking cell lineage relies on two main conditions: (1) a heritable genetic marker that transmits with cell division and (2) easy detection of that marker. Historically, researchers have been using various methods to construct cell lineage. For example, labelling cells with dyes was one of the most common methods; however dye was diluted with every division of a cell. Traditionally, heritable markers such as fluorescent proteins, transposable elements, and viral DNA barcodes have been used to track cell lineage. In addition, researchers have also used naturally occurring markers such as microsatellite repeats and epigenetic markers to reveal lineage relationship. Although these methods have provided valuable information, their applications were limited by a small number of markers and lack of coupled gene expression (Muñoz-López and García-Pérez 2010). Rapid advancement in RNA sequencing has made it possible to profile single-cell transcriptomic (scRNA-seq) of thousands of individual cells. Large-scale sc-RNA seq datasets could be helpful in expression profiling of diverse cells types and manage their identities across tissues. While these scRNA-seq datasets have been used to track developmental lineage, sometimes lack of expression coherence in developmental stages makes it difficult to construct the lineage (Haque et al. 2017). Recently scRNA-seq and CRISPR/Cas9 barcode editing have been combined to study cell lineage. CRISPR has emerged as a potent tool to track cell lineage in a way, never possible before. In CRISPR-based DNA barcodes, first barcodes (DNA fragments) are inserted into genome and then these barcodes serve as a target for CRISPR/Cas to introduce indels. These CRISPR-based indels can be monitored to track cell lineage at different stages of development (Raj et al. 2018). For example, Kalhor et al. (2018) have demonstrated CRISPR-based in vivo barcoding technique for tracking the developmental lineages in mammalian models. Kalhor applied a self-targeting version of CRISPR/Cas9 using homing gRNA (hgRNA) to introduce a unique barcode that enabled spatiotemporal tracking of cells. They used multiple homing guide RNAs, to study axis development in mouse brain. Similarly, Bowling et al. (2020) developed CRISPR array repair lineage tracing (CARLIN) mouse line to track the lineage and transcriptomic profiling in single cells in vivo. They have exploited CRISPR technology to generate 44,000 transcribed barcodes, in an inducible manner to investigate intrinsic biasness in activity of fetal liver hematopoietic cell (HSC) clones. CRISPR barcoding has also been used to study cancer biology. For example, CRISPR barcodes were used to determine intra-tumor genetic heterogeneity in cancer cells that help tumor cells to evolve.

6.10 CRISPR-Based Gene Drives

Gene drives are the DNA sequences with biased inheritance during sexual reproduction, thus accelerating the spread of a trait throughout the population over generations (Burt 2003). The idea of gene drive was originally developed during the 1960s, and it represents the potential solution to important issues such as vector-borne diseases, pesticide resistance, and controlling invasive species. Earlier gene drives were based on natural gene drives such as homing endonuclease genes present in some bacteria, fungi, and plants. Although scientists have long recognized the power of gene drives, their applications were limited because it was difficult to control their genomic location (James et al. 2018). The engineered CRISPR/Cas9-based gene drive has made this natural process simple, faster, and precise. CRISPR/Cas9 has opened new possibilities for precise and engineered drives. In CRISPR-based gene drive, "drive allele" contains CRISPR/Cas cassette, gRNA, and the desired variant to be propagated in population. CRISPR components of the drive allele cleave on the specific position in the wild-type chromosome, and to repair that break, cell uses an HDR system using variant-containing chromosome as a template, thus copying drive allele into wild-type chromosome. Under normal inheritance, an altered allele would have 50% chances to pass to offsprings; thus altered allele would not spread to entire population. However, if the altered allele would be linked to drive allele, it would spread to the entire population (Fig. 6.12) (Phelps et al. 2020). The process of engineering a CRISPR-based gene drive into a wild population starts with creating a genetically engineered by replacing wild-type sequence on a chromosome with a drive allele containing Cas9, gRNA, and an altered allele.

One or more transgenic organisms containing drive allele can be released in wild to spread to the entire population. Engineered synthetic gene drives hold great potential to alter, reduce, and eliminate the entire population from the environment. CRISPR-based gene drives have been demonstrated at laboratory scale to eradicate vector-borne diseases such as malaria, dengue, and zika. In addition, CRISPR-based gene drive could offer solutions for real problems such as insect pest management in agriculture, managing vector-borne disease in animals and humans, eradicating invasive species, protecting endangered species, and tagging an entire population (Enzmann 2018). Gene drives offer great benefits for food security challenges and

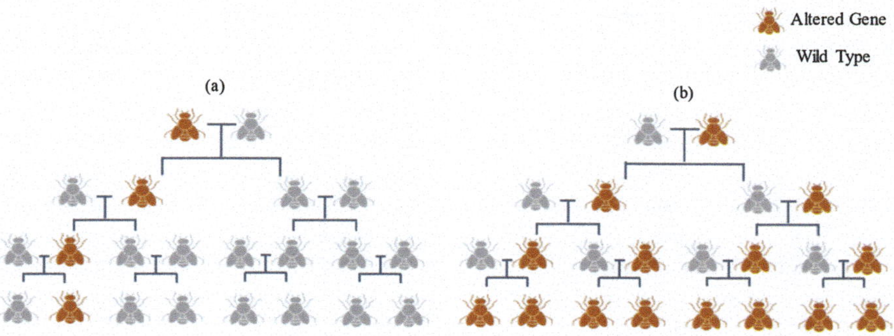

Fig. 6.12 Gene drive by through CRISPR Cas9: (**a**) in normal inheritance, there is only a 50% chance of inheritance of altered gene offspring. (**b**) In gene drive inheritance, there is a nearly 100% chance of inheritance of altered gene in offspring

human health if used responsibly. However, CRISPR-based gene drives need strict biosafety regulations because once a drive is released it is difficult to recall it. Scientists have cautioned about release of gene drives in environment because of possible off-targets, possible spread to non-target species, and environmental concerns (Collins 2018).

6.11 CRISPR-Based Directed Evolution (CDE)

Proteins, the functional workhorses of cells, are composed of amino acids, which dictate their three-dimensional structure and functions. Evolution is the basis of all beneficial changes in protein structure and functions. Although evolution is thought to be a slow process, it allows life to respond and adapt to environmental changes. Directed evolution is a process to increases the rate of change and produces a protein with desirable functions. It is a powerful method to improve functionality of proteins beyond nature. Using traditional methods of directed evolution like phage-assisted continuous evolution (PACE), numerous engineered proteins of pharmaceutical importance have been generated (Harms and Thornton 2010). Although several methods of directed evolution have existed, none was able to continuously diversify all nucleotide within user-defined genomic loci. Combining directed evolution with CRISPR represents a powerful method to produce tailor-made proteins important for cell biology, medicine, and agriculture. CRISPR can be programmed for targeted mutagenesis of genomic loci to perform directed evolution (Jakočiūnas et al. 2018). Recently, a CRISPR-based directed evolution platform (EvolvR) has been developed by combining CRISPR/nCas9 with error-prone DNA polymerase (Poll3M). nCas9-Poll3M fusion complex is directed by gRNA to the target sequence in the genome, where nCas9 creates a nick and disassociates from the strand. Poll3M extends the 3′-end of the nicked DNA strand and degrades the replaced strand. New versions of EvolvR were also created to increase the rate of mutation and expand the editing window. Applications of EvolvR would have an enormous impact on agriculture, medicine, and health (Lee et al. 2018). In addition, it will help scientist to map protein-protein interaction, discover new proteins with customized functions, and study intergenic regions of genome and diversification of bacterial genomes at user-defined loci. EvolvR has been used to mutate mammalian genomes for in-frame mutations in essential genes (Khademi et al. 2019). In addition, CRISPR-based directed evolution was demonstrated as a proof of concept to produce a new variant of OsSF3B1 in rice (Zhang and Qi 2019). The generated variant of OsSF3B1 showed tolerance to herbicide GEX1A. Moreover, CRISPR-based directed evolution could also be used in plant to generate new variants helpful for biotic and abiotic stress tolerance in plants. Further detailed applications of CRISPR-based directed evolution to rewrite genetic code have been discussed in Chap. 8.

6.12 CRISPR/Cas13 for RNA Editing

DNA editing can cause nonspecific mutations in the genome (called as off-targets), which may cause problems. RNA editing has several advantages over genome editing as it does not cause permanent changes in DNA and it is reversible as well (Xu and Li 2020). In addition, RNA editing does not require HDR mechanism so it could be useful for non-dividing cells. CRISPR/Cas13, a type II CRISPR/Cas system, is rapidly becoming a major player for making precise editing of ssRNA. Cas13 has been characterized with two higher eukaryotes and prokaryotes nucleotide-binding (HEPN) RNase domains that can precisely cleave RNA with a preference of post flanking sequence (PFS) (Abudayyeh et al. 2016; East-Seletsky et al. 2016; Smargon et al. 2017; Yan et al. 2018). Four Cas13 families have been characterized thus far: Cas13a, Cas13b, Cas13c, and Cas13d. All these variants of Cas13 are single-effector RNases with functions in RNA processing and programmed cleavage. CRISPR/Cas13a and CRISPR/Cas13b have demonstrated their programmed RNA editing abilities in mammals as well as plants (Abudayyeh et al. 2017; Yang and Chen 2017; Konermann et al. 2018). Compared with other RNA editing techniques such as RNAi, CRISPR/Cas13 offers several advantages. Its simple design, RNA guide module, and large scalability make it a versatile RNA targeting system, opening new horizons for RNA world. In addition, dCas13 module could recruit RNA binding proteins to specific RNAs in a programmed manner. In contrast to RNAi, Cas13 could also precisely target nuclear RNAs, non-coding nuclear RNAs, and pre-mRNAs, by adding nuclear localization signal. Moreover, Cas13 and its mutant versions enable different RNA manipulations such as targeting specific pathogenic splicing isoforms, RNA imaging, post-transcriptional and reversible gene regulation, combating RNA viruses, RNA tracking, and precise editing in RNA. Based on Cas13, RNA base editing system has also been developed for precise base editing in RNA such as RNA Editing Programmable A to I Replacement (REPAIR) and RNA editing for specific C-to-U exchange (RESCUE). However, multiplexing ability of REPAIR and RESCUE is not yet confirmed.

Similarly, ultrasensitive diagnostic tools like Specific high-sensitivity Enzymatic Reporter unLOCKing (SHERLOCK) and (PACMAN) have been developed based on collateral cleavage activity of Cas13. Detailed applications of CRISPR/Cas13 have been discussed in Chap. 7.

6.13 Prospects

Progress in science is linked with new techniques and discoveries. CRISPR was a breakthrough discovery of the twenty-first century that has reshaped the genomic landscape of medical, health, and agricultural sciences. Within the last decade, we have witnessed stunning progress in development and applications of CRISPR/Cas technology. Continuous developments in CRISPR technologies include dCas9-based CRISPRi/a, base editing, prime editing, directed evolution, CRISPR-based epigenetic modulation, genome imaging, cell lineage, and data storage in DNA. das9 is an enzymatically inactive Cas9, which cannot create DSB but still can bind site-specific location in a genome using gRNA. Initially, dCas9 was developed to recruit transcriptional effectors (repressors and activators) to the targeted loci without causing DSB in the host genome. Subsequent applications used this property of dCas9 to recruit reporter proteins and modifying enzymes. Similarly, gRNA was also engineered to recruit RNA binding proteins fused with reporter proteins. Recent developments in dCas-based platforms enabled researchers to study molecular pathways in a precise manner. In addition, applications of CRISPR technologies are expanding our vision of genome organization, functional genomics, epigenetics, and gene regulation at whole-genome level.

CRISPR technology is rapidly evolving with a broad range of applications in genome editing and beyond. However, despite its great potential and developments, CRISPR has been facing technical, ethical, and regulatory challenges. CRISPR offers great potential in therapeutics but delivery of such tools in living organisms represents a major challenge. Viral delivery vectors are most used for delivering CRISPR reagents in vivo; however their carcinogenesis, immunogenicity, and small packaging capacity are potential concerns associated with them. Details of CRISPR reagents and delivery methods have been discussed in Chap. 4. In addition, regulatory and ethical concern are also growing about CRISPR/Cas. For example, world community is divided about regulatory framework of CRISPR crops. The USA has been deregulating SDN1 and SDN2 CRISPR crops, while EU countries consider all CRISPR edited crops as genetically engineered crops, imposing strict regulation on their commercialization. Similarly, scientific community is concerned about release of gene drives in the environment. Although gene drive represents a great potential to eradicate vector-borne diseases, additional regulatory procedures must be considered before their practical use.

In conclusion, thanks to CRISPR, today researchers have been manipulating genomes in ways hardly imaginable before. CRISPR will continue to impact our food, medicine, and also our understanding of natural world all around us. Researchers will continue to harness the core capabilities of CRISPR for ambitious applications such as human therapeutics, agricultural improvement, and eliminating infectious diseases.

References

Abudayyeh O, Gootenberg JS, Konermann S, Joung J, Slaymaker IM, Cox DB, Zhang F (2016) C2c2 is a single-component programmable RNA-guided RNA-targeting CRISPR effector. Science 353(6299):aaf5573

Abudayyeh OO, Gootenberg JS, Essletzbichler P, Han S, Joung J, Belanto JJ, Zhang F (2017) RNA targeting with CRISPR–Cas13. Nature 550(7675):280–284

Adli M (2018) The CRISPR tool kit for genome editing and beyond. Nat Commun 9(1):1–13

Afonina I, Ong J, Chua J, Lu TK, Kline KA (2020) Multiplex CRISPRi-Cas9 silencing of planktonic and stage-specific biofilm genes in *Enterococcus faecalis*. bioRxiv

Ahmad A, Ashraf S, Munawar N, Jamil A, Ghaffar A, Shahbaz M (2021) CRISPR/Cas-mediated abiotic stress tolerance in crops. In: CRISPR crops. Springer, Berlin, pp 177–211

Amabile A, Migliara A, Capasso P, Biffi M, Cittaro D, Naldini L, Lombardo A (2016) Inheritable silencing of endogenous genes by hit-and-run targeted epigenetic editing. Cell 167(1):219–232. e214

Anders C, Niewoehner O, Duerst A, Jinek M (2014) Structural basis of PAM-dependent target DNA recognition by the Cas9 endonuclease. Nature 513(7519):569–573

Anton T, Karg E, Bultmann S (2018) Applications of the CRISPR/Cas system beyond gene editing. Biol Methods Protoc 3(1):bpy002

Anzalone AV, Randolph PB, Davis JR, Sousa AA, Koblan LW, Levy JM, Chen PJ, Wilson C, Newby GA, Raguram A (2019) Search-and-replace genome editing without double-strand breaks or donor DNA. Nature 576(7785):149–157

Aslam MA, Hammad M, Ahmad A, Altenbuchner J, Ali H (2021) Delivery methods, resources and design tools in CRISPR/Cas. In: CRISPR crops. Springer, Berlin, pp 63–116

Bae S, Park J, Kim J-S (2014) Cas-OFFinder: a fast and versatile algorithm that searches for potential off-target sites of Cas9 RNA-guided endonucleases. Bioinformatics 30(10):1473–1475

Bandyopadhyay A, Kancharla N, Javalkote VS, Dasgupta S, Brutnell TP (2020) CRISPR-Cas12a (Cpf1): a versatile tool in the plant genome editing toolbox for agricultural advancement. Front Plant Sci 11:584151

Barrangou R, Doudna JA (2016) Applications of CRISPR technologies in research and beyond. Nat Biotechnol 34(9):933–941

Bernstein BE, Stamatoyannopoulos JA, Costello JF, Ren B, Milosavljevic A, Meissner A, Kellis M, Marra MA, Beaudet AL, Ecker JR (2010) The NIH roadmap epigenomics mapping consortium. Nat Biotechnol 28(10):1045–1048

Birtle Z, Ponting CP (2006) Meisetz and the birth of the KRAB motif. Bioinformatics 22(23):2841–2845

Boch J, Scholze H, Schornack S, Landgraf A, Hahn S, Kay S et al (2009) Breaking the code of DNA binding specificity of TAL-type III effectors. Science 326(5959):1509–1512

Boettcher M, McManus MT (2015) Choosing the right tool for the job: RNAi, TALEN, or CRISPR. Mol Cell 58(4):575–585

Bowling S, Sritharan D, Osorio FG, Nguyen M, Cheung P, Rodriguez-Fraticelli A, Camargo FD (2020) An engineered CRISPR-Cas9 mouse line for simultaneous readout of lineage histories and gene expression profiles in single cells. Cell 18(6):1410–1422

Bronshtein I, Kepten E, Kanter I, Berezin S, Lindner M, Redwood AB, Garini Y (2015) Loss of Lamin a function increases chromatin dynamics in the nuclear interior. Nat Commun 6(1):1–9

Burt A (2003) Site-specific selfish genes as tools for the control and genetic engineering of natural populations. Proc R Soc Lond B Biol Sci 270:921–928

Callinan PA, Feinberg AP (2006) The emerging science of epigenomics. Hum Mol Genet 15 (suppl_1):R95–R101

Cana-Quijada P, Romero-Rodríguez B, Vallejo PG, Castillo AG, Bejarano ER (2020) Cutting-edge technology to generate plant immunity against geminiviruses. Curr Opin Virol 42:58–64

Cano-Rodriguez D, Gjaltema RAF, Jilderda LJ, Jellema P, Dokter-Fokkens J, Ruiters MHJ, Rots MG (2016) Writing of H3K4Me3 overcomes epigenetic silencing in a sustained but context-dependent manner. Nat Commun 7(1):1–11

Carroll D (2016) Genome editing: progress and challenges for medical applications. Genome Med 8(1):1–3

Charles EJ, Kim SE, Knott GJ, Smock D, Doudna J, Savage DF (2021) Engineering improved Cas13 effectors for targeted post-transcriptional regulation of gene expression. bioRxiv

Chen B, Gilbert LA, Cimini BA, Schnitzbauer J, Zhang W, Li G-W, Park J, Blackburn EH, Weissman JS, Qi LS (2013) Dynamic imaging of genomic loci in living human cells by an optimized CRISPR/Cas system. Cell 155(7):1479–1491

Cheng AW, Wang H, Yang H, Shi L, Katz Y, Theunissen TW, Rangarajan S, Shivalila CS, Dadon DB, Jaenisch R (2013) Multiplexed activation of endogenous genes by CRISPR-on, an RNA-guided transcriptional activator system. Cell Res 23(10):1163–1171

Chinnusamy V, Zhu J-K (2009) Epigenetic regulation of stress responses in plants. Curr Opin Plant Biol 12(2):133–139

Choudhury SR, Cui Y, Lubecka K, Stefanska B, Irudayaraj J (2016) CRISPR-dCas9 mediated TET1 targeting for selective DNA demethylation at BRCA1 promoter. Oncotarget 7(29):46545

Collins JP (2018) Gene drives in our future: challenges of and opportunities for using a self-sustaining technology in pest and vector management. BMC Proc 12(8):37–41

Cong L, Ran FA, Cox D, Lin S, Barretto R, Habib N, Hsu PD, Wu X, Jiang W, Marraffini LA (2013) Multiplex genome engineering using CRISPR/Cas systems. Science 339(6121):819–823

Cox DB, Gootenberg JS, Abudayyeh OO, Franklin B, Kellner MJ, Joung J, Zhang F (2017) RNA editing with CRISPR-Cas13. Science 358(6366):1019–1027

Dai X, Chen X, Fang Q, Li J, Bai Z (2018) Inducible CRISPR genome-editing tool: classifications and future trends. Crit Rev Biotechnol 38(4):573–586

East-Seletsky A, O'Connell MR, Knight SC, Burstein D, Cate JH, Tjian R, Doudna JA (2016) Two distinct RNase activities of CRISPR-C2c2 enable guide-RNA processing and RNA detection. Nature 538(7624):270–273

Enzmann B (2018) Gene drives explained: how to solve problems with CRISPR. synthego.com, https://www.synthego.com/blog/gene- drive-crispr

Friedman JR, Fredericks WJ, Jensen DE, Speicher DW, Huang X-P, Neilson EG, Rauscher FJ (1996) KAP-1, a novel corepressor for the highly conserved KRAB repression domain. Genes Dev 10(16):2067–2078

Gao P, Lyu Q, Ghanam AR, Lazzarotto CR, Newby GA, Zhang W, Choi M, Slivano OJ, Holden K, Walker JA (2021) Prime editing in mice reveals the essentiality of a single base in driving tissue-specific gene expression. Genome Biol 22(1):1–21

Ghosh S, Tibbit C, Liu J-L (2016) Effective knockdown of drosophila long non-coding RNAs by CRISPR interference. Nucleic Acids Res 44(9):e84–e84

Gilbert LA, Larson MH, Morsut L, Liu Z, Brar GA, Torres SE, Stern-Ginossar N, Brandman O, Whitehead EH, Doudna JA (2013) CRISPR-mediated modular RNA-guided regulation of transcription in eukaryotes. Cell 154(2):442–451

Grünewald J, Zhou R, Iyer S, Lareau CA, Garcia SP, Aryee MJ, Joung JK (2019) CRISPR DNA base editors with reduced RNA off-target and self-editing activities. Nat Biotechnol 37(9):1041–1048

Handel E-M, Cathomen T (2011) Zinc-finger nuclease based genome surgery: it's all about specificity. Curr Gene Ther 11(1):28–37

Hanna RE, Doench JG (2020) Design and analysis of CRISPR–Cas experiments. Nat Biotechnol 38(7):813–823

Haque A, Engel J, Teichmann SA, Lonnberg T (2017) A practical guide to single-cell RNA-sequencing for biomedical research and clinical applications. Genome Med 9(1):75

Harms MJ, Thornton JW (2010) Analyzing protein structure and function using ancestral gene reconstruction. Curr Opin Struct Biol 20(3):360–366

Harris RS, Petersen-Mahrt SK, Neuberger MS (2002) RNA editing enzyme APOBEC1 and some of its homologs can act as DNA mutators. Mol Cell 10(5):1247–1253

Hilton IB, D'ippolito AM, Vockley CM, Thakore PI, Crawford GE, Reddy TE, Gersbach CA (2015) Epigenome editing by a CRISPR-Cas9-based acetyltransferase activates genes from promoters and enhancers. Nat Biotechnol 33(5):510–517

Housden BE, Lin S, Perrimon N (2014) Cas9-based genome editing in drosophila. Methods Enzymol 546:415–439

Hu JH, Miller SM, Geurts MH, Tang W, Chen L, Sun N, Zeina CM, Gao X, Rees HA, Lin Z (2018) Evolved Cas9 variants with broad PAM compatibility and high DNA specificity. Nature 556(7699):57–63

Jaganathan D, Ramasamy K, Sellamuthu G, Jayabalan S, Venkataraman G (2018) CRISPR for crop improvement: an update review. Front Plant Sci 9:985

Jakočiūnas T, Pedersen LE, Lis AV, Jensen MK, Keasling JD (2018) CasPER, a method for directed evolution in genomic contexts using mutagenesis and CRISPR/Cas9. Metab Eng 48: 288–296

James S, Collins FH, Welkhoff PA, Emerson C, HCJ G, Gottlieb M, Touré YT (2018) Pathway to deployment of gene drive mosquitoes as a potential bio control tool for elimination of malaria in sub-Saharan Africa: recommendations of a scientific working group. Am J Trop Med 98(6_Suppl):1–49

Jiang Z, Hong X, Zhang S, Yao R, Xiao Y (2020) CRISPR base editing and prime editing: DSB and template-free editing systems for bacteria and plants. Synth Syst Biotechnol 5(4):277–292

Jinek M, Chylinski K, Fonfara I, Hauer M, Doudna JA, Charpentier E (2012) A programmable dual-RNA–guided DNA endonuclease in adaptive bacterial immunity. Science 337(6096):816–821

Kalhor R, Kalhor K, Mejia L, Leeper K, Graveline A, Mali P, Church GM (2018) Developmental barcoding of whole mouse via homing CRISPR. Science 361(6405):eaat9804

Kaminskas E, Farrell AT, Wang YC, Sridhara R, Pazdur R (2005) FDA drug approval summary: Azacitidine (5-azacytidine, Vidaza™) for injectable suspension. Oncologist 10(3):176–182

Kampmann M (2018) CRISPRi and CRISPRa screens in mammalian cells for precision biology and medicine. ACS Chem Biol 13(2):406–416

Kampmann M (2020) CRISPR-based functional genomics for neurological disease. Nat Rev Neurol 16(9):465–480

Kaufman MH (1992) The atlas of mouse development. Elsevier, Amsterdam

Khademi SH, Sazinas P, Jelsbak L (2019) Within-host adaptation mediated by intergenic evolution in Pseudomonas aeruginosa. Genome Biol Evol 11(5):1385–1397

Khan MZ, Haider S, Mansoor S, Amin I (2019) Targeting plant ssDNA viruses with engineered miniature CRISPR-Cas14a. Trends Biotechnol 37(8):800–804

Khan Z, Khan SH, Ahmad A (2021) Challenges and future perspective of CRISPR/Cas technology for crop improvement. In: CRISPR crops. Springer, Berlin, pp 289–306

Kim E, Koo T, Park SW, Kim D, Kim K, Cho H-Y, Song DW, Lee KJ, Jung MH, Kim S (2017) In vivo genome editing with a small Cas9 orthologue derived from campylobacter jejuni. Nat Commun 8(1):1–12

Köferle A, Stricker SH (2017) A universal protocol for large-scale gRNA library production from any DNA source. JoVE 130:e56264

Komor AC, Kim YB, Packer MS, Zuris JA, Liu DR (2016) Programmable editing of a target base in genomic DNA without double-stranded DNA cleavage. Nature 533(7603):420–424

Konermann S, Lotfy P, Brideau NJ, Oki J, Shokhirev MN, Hsu PD (2018) Transcriptome engineering with RNA-targeting type VI-D CRISPR effectors. Cell 173(3):665–676

LaFountaine JS, Fathe K, Smyth HD (2015) Delivery and therapeutic applications of gene editing technologies ZFNs, TALENs, and CRISPR/Cas9. Int J Pharm 494(1):180–194

Langer-Safer PR, Levine M, Ward DC (1982) Immunological method for mapping genes on drosophila polytene chromosomes. PNAS 79(14):4381–4385

Lapinaite A, Knott GJ, Palumbo CM, Lin-Shiao E, Richter MF, Zhao KT, Doudna JA (2020) DNA capture by a CRISPR-Cas9–guided adenine base editor. Science 369(6503):566–571

Lee JK, Jeong E, Lee J, Jung M, Shin E, Y-h K, Lee K, Jung I, Kim D, Kim S (2018) Directed evolution of CRISPR-Cas9 to increase its specificity. Nat Commun 9(1):1–10

Lei Y, Lu L, Liu H-Y, Li S, Xing F, Chen L-L (2014) CRISPR-P: a web tool for synthetic single-guide RNA design of CRISPR-system in plants. Mol Plant 7(9):1494–1496

Li H, Li J, Chen J, Yan L, Xia L (2020) Precise modifications of both exogenous and endogenous genes in rice by prime editing. Mol Plant 13(5):671–674

Lin Q, Zong Y, Xue C, Wang S, Jin S, Zhu Z, Wang Y, Anzalone AV, Raguram A, Doman JL (2020) Prime genome editing in rice and wheat. Nat Biotechnol 38(5):582–585

Liu H, Wei Z, Dominguez A, Li Y, Wang X, Qi LS (2015) CRISPR-ERA: a comprehensive design tool for CRISPR-mediated gene editing, repression and activation. Bioinformatics 31(22):3676–3678

Liu XS, Wu H, Ji X, Stelzer Y, Wu X, Czauderna S, Shu J, Dadon D, Young RA, Jaenisch R (2016) Editing DNA methylation in the mammalian genome. Cell 167(1):233–247.e217

Lu Y, Zhu J-K (2017) Precise editing of a target base in the rice genome using a modified CRISPR/Cas9 system. Mol Plant 10(3):523–525

Ma X, Zhang Q, Zhu Q, Liu W, Chen Y, Qiu R, Liu YG (2015) A robust CRISPR/Cas9 system for convenient, high-efficiency multiplex genome editing in monocot and dicot plants. Mol Plant 8(8):1274–1284

Maeda I, Kohara Y, Yamamoto M, Sugimoto A (2001) Large-scale analysis of gene function in Caenorhabditis elegans by high-throughput RNAi. Curr Biol 11(3):171–176

Makarova KS, Koonin EV (2015) Annotation and classification of CRISPR-Cas systems. Methods Mol Biol 1311:47–75

Makarova KS, Wolf YI, Koonin EV (2018) Classification and nomenclature of CRISPR-Cas systems: where from here? CRISPR J 1(5):325–336

Mali P, Aach J, Stranges PB, Esvelt KM, Moosburner M, Kosuri S, Yang L, Church GM (2013) CAS9 transcriptional activators for target specificity screening and paired nickases for cooperative genome engineering. Nat Biotechnol 31(9):833–838

Marzec M, Brąszewska-Zalewska A, Hensel G (2020) Prime editing: a new way for genome editing. Trends Cell Biol 30(4):257–259

McCarty NS, Graham AE, Studená L, Ledesma-Amaro R (2020) Multiplexed CRISPR technologies for gene editing and transcriptional regulation. Nat Commun 11(1):1–13

McKenna A, Gagnon JA (2019) Recording development with single-cell dynamic lineage tracing. Development 146(12):dev169730

Meltzer H, Marom E, Alyagor I, Mayseless O, Berkun V, Segal-Gilboa N, Unger T, Luginbuhl D, Schuldiner O (2019) Tissue-specific (ts) CRISPR as an efficient strategy for in vivo screening in drosophila. Nat Commun 10(1):1–9

Mickelbart MV, Hasegawa PM, Bailey-Serres J (2015) Genetic mechanisms of abiotic stress tolerance that translate to crop yield stability. Nat Rev Genet 16(4):237–251

Miki D, Shimamoto K (2004) Simple RNAi vectors for stable and transient suppression of gene function in rice. Plant Cell Physiol 45(4):490–495

Miller J, McLachlan AD, Klug A (1985) Repetitive zinc-binding domains in the protein transcription factor IIIA from Xenopus oocytes. EMBO J 4(6):1609–1614

Minkenberg B, Zhang J, Xie K, Yang Y (2019) CRISPR-PLANT v2: an online resource for highly specific guide RNA spacers based on improved off-target analysis. Plant Biotechnol J 17(1):5

Mishra R, Raj KJ, Kaijun Z (2020) Base editing in crops: current advances, limitations and future implications. Plant Biotechnol J 18(1):20–31

Misteli T (2013) The cell biology of genomes: bringing the double helix to life. Cell 152(6):1209–1212
Mojica FJ, Díez-Villaseñor C, Soria E, Juez G (2000) Biological significance of a family of regularly spaced repeats in the genomes of archaea, bacteria and mitochondria. Mol Microbiol 36(1):244–246
Molla KA, Yang Y (2019) CRISPR/Cas-mediated base editing: technical considerations and practical applications. Trends Biotechnol 37(10):1121–1142
Montague TG, Cruz JM, Gagnon JA, Church GM, Valen E (2014) CHOPCHOP: a CRISPR/Cas9 and TALEN web tool for genome editing. Nucleic Acids Res 42(W1):W401–W407
Morita S, Noguchi H, Horii T, Nakabayashi K, Kimura M, Okamura K, Sakai A, Nakashima H, Hata K, Nakashima K (2016) Targeted DNA demethylation in vivo using dCas9–peptide repeat and scFv–TET1 catalytic domain fusions. Nat Biotechnol 34(10):1060–1065
Munawar N, Ahmad A (2021) CRISPR/Cas system: an introduction. In: CRISPR crops. Springer, Berlin, pp 1–35
Muñoz-López M, García-Pérez JL (2010) DNA transposons: nature and applications in genomics. Curr Genomics 11(2):115–128
Naito Y, Hino K, Bono H, Ui-Tei K (2015) CRISPRdirect: software for designing CRISPR/Cas guide RNA with reduced off-target sites. Bioinformatics 31(7):1120–1123
Nateghi Rostami M (2020) CRISPR/Cas9 gene drive technology to control transmission of vector-borne parasitic infections. Parasite Immunol 42(9):e12762
O'Connell MR (2019) Molecular mechanisms of RNA targeting by Cas13-containing type VI CRISPR–Cas systems. J Mol Biol 431(1):66–87
Phelps MP, Seeb LW, Seeb JE (2020) Transforming ecology and conservation biology through genome editing. Conserv Biol 34(1):54–65
Qi LS, Larson MH, Gilbert LA, Doudna JA, Weissman JS, Arkin AP, Lim WA (2013) Repurposing CRISPR as an RNA-guided platform for sequence-specific control of gene expression. Cell 152(5):1173–1183
Qiu S, Adema CM, Lane T (2005) A computational study of off-target effects of RNA interference. Nucleic Acids Res 33(6):1834–1847
Raj B, Gagnon JA, Schier AF (2018) Large-scale reconstruction of cell lineages using single-cell readout of transcriptomes and CRISPR–Cas9 barcodes by scGESTALT. Nat Prot 13(11):2685–2713
Rath D, Amlinger L, Rath A, Lundgren M (2015) The CRISPR-Cas immune system: biology, mechanisms and applications. Biochimie 117(201):119–128
Rees HA, Liu DR (2018) Base editing: precision chemistry on the genome and transcriptome of living cells. Nat Rev Genet 19(12):770–788
Ricroch A (2019) Global developments of genome editing in agriculture. In: Transgenic research, vol 2. Springer, Berlin, pp 45–52
Riddle DL, Blumenthal T, Meyer BJ, Priess JR (1997) Organelle morphogenesis during spermatogenesis. In: *C. elegans* II, 2nd edn. CSHLP, Cold Spring Harbor, NY
Sander JD, Joung JK (2014) CRISPR-Cas systems for editing, regulating and targeting genomes. Nat Biotechnol 32(4):347–355
Sangar V, Samant L, Pawar S, Dhawale P, Chowdhary A (2016) Online and offline tools: Crispr/Cas off-target detection. Int J Pharm Sci Res 7(5):1889
Scholefield J, Harrison PT (2021) Prime editing—an update on the field. Gene Ther 28:369–401
Schuster A, Erasimus H, Fritah S, Nazarov PV, van Dyck E, Niclou SP, Golebiewska A (2019) RNAi/CRISPR screens: from a pool to a valid hit. Trends Biotechnol 37(1):38–55
Schwarzacher T, Heslop-Harrison JP (1994) Direct fluorochrome-labeled DNA probes for direct fluorescent in situ hybridization to chromosomes. In: Protocols for nucleic acid analysis by non-radioactive probes. Humana Press, Totowa, NJ, pp 167–176
Sedeek KE, Mahas A, Mahfouz M (2019) Plant genome engineering for targeted improvement of crop traits. Front Plant Sci 10:114

Senthil-Kumar M, Mysore KS (2011) Caveat of RNAi in plants: the off-target effect. In: RNAi and plant gene function analysis, pp 13–25
Sharma S, Petsalaki E (2018) Application of CRISPR-Cas9 based genome-wide screening approaches to study cellular signalling mechanisms. Int J Mol Sci 19(4):933
Shi Y, Lan F, Matson C, Mulligan P, Whetstine JR, Cole PA, Casero RA, Shi Y (2004) Histone demethylation mediated by the nuclear amine oxidase homolog LSD1. Cell 119(7):941–953
Shimatani Z, Kashojiya S, Takayama M, Terada R, Arazoe T, Ishii H, Teramura H, Yamamoto T, Komatsu H, Miura K (2017) Targeted base editing in rice and tomato using a CRISPR-Cas9 cytidine deaminase fusion. Nat Biotechnol 35(5):441–443
Shmakov S, Smargon A, Scott D, Cox D, Pyzocha N, Yan W, Abudayyeh OO, Gootenberg JS, Makarova KS, Wolf YI (2017) Diversity and evolution of class 2 CRISPR–Cas systems. Nat Rev Microbiol 15(3):169–182
Smargon AA, Cox DB, Pyzocha NK, Zheng K, Slaymaker IM, Gootenberg JS, Zhang F (2017) Cas13b is a type VI-B CRISPR-associated RNA-guided RNase differentially regulated by accessory proteins Csx27 and Csx28. Mol Cell 65(4):618–630
Stepper P, Kungulovski G, Jurkowska RZ, Chandra T, Krueger F, Reinhardt R, Reik W, Jeltsch A, Jurkowski TP (2017) Efficient targeted DNA methylation with chimeric dCas9–Dnmt3a–Dnmt3L methyltransferase. Nucleic Acids Res 45(4):1703–1713
Strahl BD, Allis CD (2000) The language of covalent histone modifications. Nature 403(6765):41–45
Sulston JE, Schierenberg E, White JG, Thomson JN (1983) The embryonic cell lineage of the nematode *Caenorhabditis elegans*. Dev Biol 100:64–119
Takei Y, Shah S, Harvey S, Qi LS, Cai L (2017) Multiplexed dynamic imaging of genomic loci by combined CRISPR imaging and DNA sequential FISH. Biophys J 112(9):1773–1776
Tarasava K, Liu R, Garst A, Gill RT (2018) Combinatorial pathway engineering using type I-E CRISPR interference. Biotechnol Bioeng 115(7):1878–1883
Thurman RE, Rynes E, Humbert R, Vierstra J, Maurano MT, Haugen E, Sheffield NC, Stergachis AB, Wang H, Vernot B (2012) The accessible chromatin landscape of the human genome. Nature 489(7414):75–82
Trevino AE, Zhang F (2014) Genome editing using Cas9 nickases. Methods Enzymol 546:161–174
Tsuchiya KD (2011) Fluorescence *in situ* hybridization. Clin Lab Med 31:525–542
Tu Z, Yang W, Yan S, Guo X, Li X-J (2015) CRISPR/Cas9: a powerful genetic engineering tool for establishing large animal models of neurodegenerative diseases. Mol Neurodegener 10(1):1–8
Wang J, Zhang C, Feng B (2020) The rapidly advancing class 2 CRISPR-Cas technologies: a customizable toolbox for molecular manipulations. J Cell Mol Med 24(6):3256–3270
Whelan AI, Lema MA (2019) Regulation of genome editing in plant biotechnology: Argentina. In: Regulation of genome editing in plant biotechnology. Springer, Berlin, pp 19–62
Wolter F, Puchta H (2018) The CRISPR/Cas revolution reaches the RNA world: Cas13, a new Swiss Army knife for plant biologists. Plant J 94(5):767–775
Wright DA, Li T, Yang B, Spalding MH (2014) TALEN-mediated genome editing: prospects and perspectives. Biochem J 462(1):15–24
Wu X, Kriz AJ, Sharp PA (2014) Target specificity of the CRISPR-Cas9 system. Quant Biol 2(2):59–70
Wu X, Mao S, Ying Y, Krueger CJ, Chen AK (2019) Progress and challenges for live-cell imaging of genomic loci using CRISPR-based platforms. Genom Proteom Bioinform 17(2):119–128
Xu Y, Li Z (2020) CRISPR-Cas systems: overview, innovations and applications in human disease research and gene therapy. Comput Struct Biotechnol 18:2401–2415
Xu X, Tao Y, Gao X, Zhang L, Li X, Zou W, Ruan K, Wang F, Xu G-l HR (2016) A CRISPR-based approach for targeted DNA demethylation. Cell Discov 2(1):1–12
Yan WX, Chong S, Zhang H, Makarova KS, Koonin EV, Cheng DR, Scott DA (2018) Cas13d is a compact RNA-targeting type VI CRISPR effector positively modulated by a WYL-domain-containing accessory protein. Mol Cell 70(2):327–339

Yan F, Wang W, Zhang J (2019) CRISPR-Cas12 and Cas13: the lesser known siblings of CRISPR-Cas9. Springer, Berlin

Yang L, Chen LL (2017) Enhancing the RNA engineering toolkit. Science 358(6366):996–997

Yang Y, Feng Q, Hu K, Cheng F (2021) Using CRISPRa and CRISPRi technologies to study the biological functions of ITGB5, TIMP1, and TMEM176B in prostate cancer cells. Front Mol Biosci 8:485

Zahoor MK, Ahmad A, Zahoor MA, Majeed HN, Zulhussnain M, Ranian K (2021) CRISPR/Cas-based insect resistance in crops. In: CRISPR crops. Springer, Berlin, pp 117–149

Zalatan JG, Lee ME, Almeida R, Gilbert LA, Whitehead EH, La Russa M, Tsai JC, Weissman JS, Dueber JE, Qi LS (2015) Engineering complex synthetic transcriptional programs with CRISPR RNA scaffolds. Cell 160(1–2):339–350

Zentner GE, Henikoff S (2015) Epigenome editing made easy. Nat Biotechnol 33(6):606–607

Zetsche B, Gootenberg JS, Abudayyeh OO, Slaymaker IM, Makarova KS, Essletzbichler P, Volz SE, Joung J, Van Der Oost J, Regev A (2015) Cpf1 is a single RNA-guided endonuclease of a class 2 CRISPR-Cas system. Cell 163(3):759–771

Zhang Y, Qi Y (2019) CRISPR enables directed evolution in plants. Genome Biol 20(1):1–4

Zhang Y, Malzahn AA, Sretenovic S, Qi Y (2019) The emerging and uncultivated potential of CRISPR technology in plant science. Nat Plants 5(8):778–794

Zhao Y, Dai Z, Liang Y, Yin M, Ma K, He M, Ouyang H, Teng C-B (2014) Sequence-specific inhibition of microRNA via CRISPR/CRISPRi system. Sci Rep 4(1):1–5

Zheng N, Li L, Wang X (2020) Molecular mechanisms, off-target activities, and clinical potentials of genome editing systems. Clin Transl Med 10(1):412–426

Zink D, Sadoni N, Stelzer E (2003) Visualizing chromatin and chromosomes in living cells. Methods 29(1):42–50

Zong Y, Wang Y, Li C, Zhang R, Chen K, Ran Y, Qiu J-L, Wang D, Gao C (2017) Precise base editing in rice, wheat and maize with a Cas9-cytidine deaminase fusion. Nat Biotechnol 35(5):438–440

RNA Editing with CRISPR/Cas13 7

Sidra Ashraf, Muhammad Zubair Ghouri, Muhammad Asif Javed, Hassan Zafar, Hazrat Ali, Sameer H. Qari, and Aftab Ahmad

Contents

Sidra Ashraf, Muhammad Zubair Ghouri and Muhammad Asif Javed have contributed equally to this chapter.

S. Ashraf
Department of Biochemistry, University of Agriculture, Faisalabad, Pakistan

M. Z. Ghouri
Center for Advanced Studies in Agriculture and Food Security (CAS-AFS), University of Agriculture, Faisalabad, Pakistan

M. A. Javed
Department of Forestry and Range Management, University of Agriculture, Faisalabad, Pakistan

H. Zafar
Department of Structural Biology, Central European Institute of Technology, Masaryk University, Brno, Czech Republic

H. Ali
National Institute for Biotechnology and Genetic Engineering (NIBGE), Faisalabad, Pakistan

S. H. Qari
Department of Biology/Genetics and Molecular Biology Central Laboratory, Aljumum University College, Umm Al-Qura University, Makkah, Saudi Arabia

A. Ahmad (✉)
Department of Biochemistry/US-Pakistan Center for Advanced Studies in Agriculture and Food Security, University of Agriculture, Faisalabad, Pakistan
e-mail: aftab.ahmad@uaf.edu.pk

A. Ahmad et al. (eds.), *The CRISPR/Cas Tool Kit for Genome Editing*,
https://doi.org/10.1007/978-981-16-6305-5_7

Abstract

CRISPR/Cas system has emerged as a powerful tool for precise genome editing with numerous applications such as knockout, knock-in, base editing, prime editing and rewriting the genetic code. Cas9 and Cas12 both belong to Class 2 CRISPR/Cas system and among the most widely used Cas nucleases for manipulation of DNA. Although both Cas9 could be used for transcriptional control (CRISPRi and CRISPRa), targeting RNA at posttranscriptional level with Cas9 nuclease was not possible. So far, RNAi was the best tool for posttranscriptional regulation of gene; however, identification of Cas13, an RNA targeting Cas nuclease, provides an alternate to RNAi for precise editing of RNA. Like Cas9 and Cas12, Cas13 also belongs to class 2 CRISPR/Cas system and a single effector Cas protein; however, it does not require PAM but relies on PFS region to cleave ssRNA. In this chapter, we discuss different approaches for posttranscriptional control of RNA like RNAi, CRISPR/Cas III system, CRISPR/Cas II, and CRISPR/Cas VI system and provide a detailed comparison between all these RNA editing approaches. We also discuss CRISPR/Cas13 system in detail such as processing mechanism of CRISPR/Cas13 and different Cas13 proteins like Cas13a. In addition, we discuss detailed applications of CRISPR/Cas13 system such as RNA editing (REPAIR), the diagnostic use of CRISPR/Cas13 (SHERLOCK and CARVER), RNA imagining, alternate splicing, RNA virus interferences such as Zika and COVID-19, and specific isolation of RNA. These highlights just provide a glimpse about the potential of CRISPR/Cas13 to manipulate RNA and further advancement of the CRISPR/Cas system. Finally, we highlight future prospects as RNA editing through CRISPR/Cas13 provides a safer alternative to CRISPR/Cas9-mediated genome editing.

Keywords

CRISPR/Cas · CRISPR/Cas13 · RNA editing · RNAi · RNA imaging · SHERLOCK · REPAIR · PAC-MAN · CARVER

Abbreviations

ADAR	Adenosine deaminase RNA specific
BLP	Bacterial lipoprotein
CARF	CRISPR-associated Rossmann fold
HDR	Homology-directed repair
HEPN domain	Higher eukaryotes and prokaryotes nucleotide-binding domain
KEN	Kinase extension domain
miRNA	microRNA
NHEJ	Non-homologous end joining
NTD	N-terminal domain
NUC	Nuclease lobe
PAC-MAN	Prophylactic Antiviral CRISPR in huMAN cells
PCD	Programmed cell death
PFS	Protospacer flanking sequence
REC	Recognition lobe
RISC	RNA-induced silencing complex
RPA	Recombinase polymerase amplification
SHERLOCK	Specific high sensitivity enzymatic reporter UnLOCKing
shRNA	Short hairpin RNA

7.1 Introduction

The CRISPR/Cas9 is a revolutionary toolbox in genome engineering to precisely manipulate almost every gene or genetic elements in the genome. CRISPR/Cas is an adaptive immune system in prokaryotic cells (Bacteria and Archaea) which protect them from invading foreign nucleic acids (viruses and plasmids). CRISPR/Cas represents a diverse antiviral defense mechanism comprising both DNA and RNA editing systems (Fineran and Dy 2014). The CRISPR genomic loci contain, Cas effector proteins, and viral DNA sequences also called as protospacers, flanked by the conserved palindromic repeats (Makarova et al. 2017; Wright et al. 2016). The spacer sequences express as CRISPR RNA (crRNA), which along with transactivating RNA (tracrRNA) recruit Cas endonuclease protein to produce a surveillance complex which recognizes specific DNA sequence depending upon the presence of PAM. Once the presence of PAM adjacent to the DNA/RNA duplex is recognized by Cas endonuclease, it creates double-stranded break (DSB) consequently repaired through either non-homologous end joining (NHEJ) repair system or homology-directed repair (HDR) if repair template is available. The mechanism of natural CRISPR/Cas system in bacteria consists of three general steps: (a) an adaptation, i.e., the foreign viral DNA is cleaved and incorporates as spacer DNA in the CRISPR repeats arrays; (b) transcription of CRISPR arrays to produce mature crRNA; and (c) interference, i.e., the crRNA directs the Cas endonuclease towards the target sequence to cause DSBs (Mojica et al. 2009).

The common nomenclature of CRISPR/Cas system is based on associated *Cas* genes coding Cas proteins (Makarova et al. 2017), organization of effector proteins, such as multi-subunit effector protein complex or a single effector protein. Based on these features, CRISPR/Cas systems have been classified into two main classes and six types. The detailed classification of CRISPR/Cas system has been discussed in Chap. 2. Class 1 systems (Type I, III, and IV) consist of multi-subunit effector protein complexes for processing of crRNA and create double-stranded break in DNA. Due to their requirement of multiple subunits to cleave DNA, Class 1 systems are difficult to program for genome editing. In contrast, Class 2 CRISPR/Cas system (Type II, V, and VI) relies on single and large protein with multiple domains to cleave DNA or RNA. Class 2 CRISPR/Cas effectors like Cas9, Cas12, Cas13, and Cas14 are most widely adopted systems for genome editing (Fig. 7.1).

In addition, Class II systems are considered more attractive genome editing tools due to their reprogramming ability to target any sequence in the genome. Computational tools are always used to identify new variants of CRISPR/Cas systems. Reengineering of Class II effector Cas proteins for structural and functional properties has resulted in several new versatile tools such as CRISPRi, CRISPRa, base editing, prime editing and epigenome regulation for genome engineering and regulation (Koonin et al. 2017). Different types of CRISPR/Cas systems recognize and cleave different nucleic acid substrates such as DNA, ssRNA, or ssDNA. In addition, PAM requirements, position of DSB, and optimal length of sgRNA to make DSB also vary among CRISPR/Cas systems. For example, type I, type II, type IV (likely), and type V specifically target DNA, while type III can target both DNA

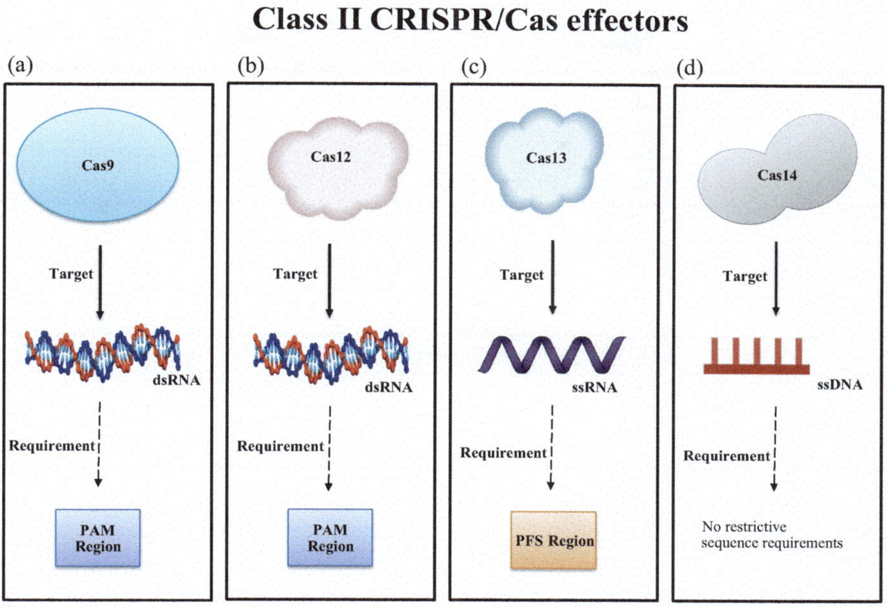

Fig. 7.1 Representation of Class II effector proteins: (**a**) Cas9 is an effector endonuclease that is widely used in genome editing and targets dsDNA to create DSB. (**b**) Cas12 also targets the dsDNA but its nuclease domain is different from Cas9 domain. Both Cas9 and Cas12 require the PAM region to target the specific DNA. (**c**) The newly discovered type VI (Cas13) may target the ssRNA and some orthologue of Cas13 may require PFS region to target specific ssRNA. (**d**) Cas14 targets ssDNA and does not require restrictive sequence for target

and RNA (Table 7.1). However, type VI exclusively cleaves RNA. DNA targeting systems (CRISPR/Cas9, CRISPR/Cas12) specifically recognize PAM sequence in the genome for binding and cleavage activity; however RNA targeting system recognizes protospacer flanking sequence (PFS) to cleave the RNA. Historically, the most widely used type II CRISPR/Cas system (CRISPR/Cas9) cleaves dsDNA with NGG as PAM requirement and cleaves precisely 3 bp away from PAM site, producing blunt ends. In contrast, another type II CRISPR/Cas system (CRISPR/Cas12 or Cpf1) recognizes TTTN as PAM and creates DSB with staggered ends (Shmakov et al. 2017). Moreover, CRISPR/Cas9 system has been reengineered and programmed to just bind DNA, recruit enhancer or activators, and insert tags or epigenetic modifiers to create diverse modifications in the genome. Although, Cas9 system has been programed into various platforms for genetic modifications, their use was limited to manipulations at DNA level; however posttranscriptional regulations and studying non-coding RNAs were not possible with this system. For posttranscriptional regulation without disrupting genes at DNA level, Cas13 has emerged as a powerful system with enormous potential to functionally characterize coding RNA (mRNA) and non-coding RNA (siRNA, miRNA, lncRNA, snRNA). Novel tools based on CRISPR/Cas13 system have been developed for

Table 7.1 Classification of CRISPR/Cas system

Class	Type	Subtype	Target	Effector nuclease	Nuclease domain	References
Class1	Type I	I-A, I-B, I-C, I-D, I-E, I-F, I-G	DNA	Cas3	HD	Makarova et al. (2018) and Koonin and Makarova (2019)
	Type III	III-A, III-B, III-C, III-D	DNA/ RNA	Cas10, Csx1, Csm6	HD/ HEPN	Makarova et al. (2018) and Koonin and Makarova (2019)
	Type IV	IV-A, IV-B	DNA	Csf1		Makarova et al. (2018) and Koonin and Makarova (2019)
Class 2	Type II	II-A, II-B, II-C	DNA	Cas9	RuvC +HNH	Makarova et al. (2018) and Koonin and Makarova (2019)
	Type V	V-A, V-B, V-C	DNA	Cas12/ Cpf1	RuvC	Makarova et al. (2018)
	Type VI	VI-A, VI-B, VI-C, VI-D	RNA	Cas13/ C_2c_2	HEPN	Makarova et al. (2018) and Koonin and Makarova (2019)

molecular diagnostic and viral control such as SHERLOCK and CARVER. In addition, CRISPR/dCas13 can be modulated to study imaging, trafficking, site-specific base editing, and therapeutic applications. The focus of this chapter is to highlight functional features of CRISPR/Cas13 for RNA editing. In addition, we summarize comparison of CRISPR/Cas13 with other techniques and also discuss potential applications of CRISPR/Cas13 in RNA world.

7.2 RNA Editing Systems: An Overview

CRISPR/dCas9-based CRISPRi and CRISPRa are efficient and robust systems for transcriptional activation and repression by recruiting transcriptional activator and repressors to the promoter region of a gene; however these systems must overcome several barriers to function properly (Zheng et al. 2019). For example, binding of dCas9 to the target DNA sequence may be influenced by DNA-bound proteins as well as chromatin structure of eukaryotic genomes. In addition, transcriptional gene repression through CRISPRi may eliminate all splice variants of the target gene leading to unpredictable results and making it difficult to study particular splicing isoforms. Therefore, it is very important to manipulate endogenous RNA at post-transcriptional level for tissue, organ, and process-specific studies (Hutvagner 2005). Until recently, RNAi was the only choice for regulating RNA at posttranscriptional level; however, RNAi has several limitations such as off-targeting, limited to cytoplasmic RNA only, specificity, and reduced effect in successive generations. Therefore, an effective platform for targeting and controlling the activity of cellular transcripts at the posttranscriptional level was highly desired. Discovery of CRISPR/Cas13 platform provides an alternate to RNAi and holds an enormous potential to transform the field of RNA biology. Here, we discuss different systems for posttranscriptional regulation of RNA.

7.2.1 RNA Interference (RNAi)

RNAi was first established in 1990, when Fire and Craig Mello revealed that in *Caenorhabditis elegans* (*C. elegans*), double-stranded RNA (dsRNA) triggers gene silencing in a sequence-specific manner (Sugimoto 2004). This discovery was followed by a number of studies to reveal the mechanism behind sequence-specific gene regulation through dsRNA. Before the emergence of CRISPR/Cas9, RNAi was one of the most widely used reverse genetic approaches for functional genomics, characterization of genes, and gene silencing both in plants and animals. Cellular mechanism of gene regulation through RNAi relies on endogenous small RNAs such as small interfering RNA (siRNA), short hairpin RNA (shRNA), and microRNA (miRNA) (Pushparaj et al. 2008). Although RNAi has been demonstrated using endogenous small RNA (miRNA, shRNA, or siRNA) mechanism, cellular machinery can also be triggered using exogenous double-stranded synthetic siRNAs or shRNAs. Endogenous pre-miRNA or exogenous dsRNA will be cleaved into smaller RNAs (21 nt in length) by endonuclease called as dicer. Following cleavage with dicer endonuclease, small RNAs will be loaded into RNA-induced silencing complex (RISC), consequently directed to complementary mRNA. Once, miRNA or siRNA makes complex with perfectly matched target mRNA, Argonaute, an integral protein will cleave the targeted complementary mRNA, thus inhibiting expression of mRNA into protein (Tijsterman and Plasterk 2004). If a perfect match does not occur between small RNAs and mRNA, RISC complex can still bind mRNA leading to blockage of translation without cleavage.

Although, RNAi have been extensively used for functional genomics and post-transcriptional gene regulation studies, certain limitations may hamper the successful applications of RNAi to target specific mRNA. For example, RNAi is very effective against cytoplasmic mRNA; however, targeting nuclear mRNA with RNAi is challenging. Similarly, off-targets are one of the major concerns in applications of RNAi as partial matches between target mRNA and small RNAs may occur. Non-specific interaction between small RNAs and nontarget mRNA may result in off-targets, and off-target phenotypes sometimes dominate on-target effects (Buchholz et al. 2006). These limitations, especially off-targets, urged the scientists to develop alternate approaches to manipulate RNA with precise modifications and high specificity.

7.2.2 Type II CRISPR/Cas Systems for RNA Editing

Cas9 is a dual RNA-guided DNA endonuclease that is required for interference and immunity in type II systems. Based on Cas9 genes and subtype-specific genes, the type II CRISPR/Cas system is further classified into II-A, II-B, and II-C systems. Type II CRISPR/Cas systems comprise three components, i.e., a Cas9 endonuclease, CRISPR arrays, and a tracr-RNA. Pre CRISPR RNA (Pre-crRNA) in type II CRISPR/Cas system is processed to a mature crRNA, through RNase III. Tracr-RNA in type II systems is required for interaction between crRNA and Cas9 and 5′ fragment of tracr-RNA binds with CRISPR repeat sequence to make a duplex, which is subsequently cleaved with RNase III (Makarova et al. 2017). Intermediate crRNAs are further processed through nucleases to a mature crRNA; however, the underlying mechanism and the nucleases are not fully understood. The complex of mature crRNA and tracrRNA is loaded into Cas9 followed by conformational changes in Cas9. These conformational changes drive the Cas9 to trace the target sequence in the genome recognized by PAM region at 5′-end. A complementary pairing between crRNA and target sequence activates nuclease activity of Cas9, leading to double-stranded break in DNA by HNH and RuvC domains (Jinek et al. 2012). Although Cas9 possesses different nucleolytic activities; however, targeting and cleavage of dsDNA is the most common one and extensively used in genome editing. In contrast to DNase activity by type II systems, they can also target RNA in a precise manner, such as type II system in *Francisella novicida*, which downregulates lipoprotein gene by targeting its mRNA in a posttranscriptional mechanism, along with usual nucleolytic activity against DNA (Sampson et al. 2013). Here we discuss RNA targeting using type II CRISPR/Cas systems.

7.2.2.1 RCas9 System for Targeting RNA

CRISPR/Cas9, a type II CRISPR/Cas system from *Streptococcus pyogenes*, is one of the most widely adopted genome editing systems in the scientific community for a broad range of applications in basic research to industrial applications (Marraffini 2016). In addition, applications of CRISPR/Cas9 are not limited to genome editing, but reengineering of CRISPR/Cas9 to create single-stranded break (nCas9) or just bind DNA site specifically (dCas9) has also led to its applications in precise transcriptional regulations (Braun et al. 2016), epigenome engineering (O'Geen et al. 2017), base editing (Ma et al. 2018), and chromatin imaging (Duan et al. 2018). Although, majorities of studies on CRISPR/Cas9 are limited to DNA, its simplicity, ease in design, and modular nature have urged scientists to use it for RNA manipulation as well. During 2014, O'Connell for the first time used SpCas9 for targeting single-stranded RNA (ssRNA) in vitro using RCas9 system. As the presence of PAM (NGG – just next to the target site and on opposite strand) is an essential requirement for recognition and cleavage of DNA by SpCas9, when PAM was provided exogenously using ssDNA oligonucleotides called as PAMmers, SpCas9 was able to bind and cleave ssRNA very precisely. By providing exogenous PAM sequence, the system could specifically target ssRNA rather than DNA (O'Connell et al. 2014). In addition, RCas9 system specifically recognized and cleaved

DNA/RNA hybrid but remained ineffective against dsRNA. RCas9 system is not only used for RNA cleavage but also have been used for RNA isolation and RNA imaging (Nelles et al. 2015). For example, O'Connell also used dCas9 with PAMmers to isolate and pull down endogenous mRNA from cell lysate (O'Connell et al. 2014). Moreover, RCas9 system was programed to track and image endogenous mRNA under in vivo conditions. These applications highlight the exciting, unexpected, and yet new reprogramming ability of CRISPR/Cas9 system for RNA regulation. However, along with successful reports of RCas9 system, certain concerns are also present regarding RNA target through RCas9 which may restrict widespread applications of this system. For example, firstly, efficient delivery of PAMmers, sgRNA, and Cas9 to the target cells is a limiting factor in applications of RCas9 system. Secondly, PAMmers must be chemically modified to protect the internal cell degradation system in living organisms, which makes them costly thus remaining a limiting factor in large-scale applications of RCas9 system. Finally, although RCas9 system can be programmed to target nuclear RNAs using U6 or U3 promoter-driven sgRNA, native DNA cleavage activity of SpCas9 may result in off-targets.

7.2.2.2 FnCRISPR/Cas9 System for RNA Editing

Similar to Cas9 endonuclease of *Streptococcus pyogenes*, FnCas9 is a type II CRISPR/Cas endonuclease from bacterial species *Francisella novicida*. In addition, like other Cas9 effectors such as SpCas9 and SaCas9, FnCas9 works in a similar pattern to cleave double-stranded DNA (Chen et al. 2017). For example, like SpCas9, it recognizes NGG as PAM sequence, and in association with tracrRNA and crRNA, it creates DSB in the target DNA sequence. Similarly, on provision of sgRNA, it can also target dsDNA in an exactly similar mechanism to SpCas9 and SaCas9. Application of FnCas9 has been reported for genome editing in mouse. However, in contrast to DNA cleavage activity of FnCRISPR/Cas9 system, it also downregulates bacterial lipoprotein (BLP) gene in a posttranscriptional mechanism of gene regulation. Downregulation of BLP through CRISPR/Cas9 system helps *F. novicida* to escape BLP-induced host immune response in bacteria and increase its virulence in eukaryotes (Chen et al. 2018). Sampson and Weiss (2014), first time reported that *F. novicida* uses CRISPR/Cas9 system to cleave its lipoprotein mRNA leading to suppression of lipoprotein gene expression.

Apart from tracrRNA and crRNA, CRISPR/Cas9 system in *F. novicida* encodes an additional small CRISPR/Cas-associated RNA (scaRNA), with 5′-end complementary to tracrRNA while 3′-end complementary to mRNA of lipoprotein gene. scaRNA can replace crRNA and make heteroduplex with tracrRNA using complementary base pairing. Heteroduplex between scaRNA and tracrRNA, in association with FnCas9, targets BLP mRNA. However, crRNA, HNH, and RuvC domains do not take part in RNA cleavage through FnCas9. In contrast to RCas9 system using PAMmers, RNA cleavage through FnCas9 is independent of PAM sequence. The exact mechanism of FnCas9-based RNA targeting is not yet clearly understood. These findings highlight the potential of type II CRISPR/Cas9 system to target RNA and novel activity of FnCas9, which has been successfully used for targeting RNA viruses in eukaryotes. For example, Price et al. (2015) engineered tracrRNA and scaRNA into single RNA targeting gRNA (rgRNA) and used the engineered FnCas9/rgRNA system to target hepatitis C virus (HCV) in human hepatocarcinoma cells leading to significant (60%) reduction in viral protein. As described above that RNA targeting through FnCas9 system does not rely on PAM sequence, the exact mechanism of RNA suppression and viral inhibition through FnCas9 is unknown. Price et al. (2015) suggested that the viral inhibition of HCV by FnCas9 was not due to mRNA cleavage instead mRNA binding consequently inhibiting viral translation and replication, because similar results were obtained using FndCas9 system. Therefore, it was concluded that HCV inhibition was due to the binding of FnCas9 with viral mRNA (Price et al. 2015). This study further advances the use and applications of FnCas9-based RNA targeting in eukaryotes; however like RCas9 system, DNase activity of FnCas9 is one of the major limitations of this system as it may result in off-targets.

7.2.3 Type III CRISPR/Cas System (CRISPR/Cas3)

In contrast to the most widely used Class 2 type II CRISPR/Cas system which relies on single multi-domain effector protein (Cas9) for DNA cleavage, class 1 type III CRISPR/Cas system (CRISPR/Cas3) relies on multiple Cas proteins for targeting nucleic acids. In addition, while CRISPR/Cas9 system is specific for genome editing, CRISPR/Cas3 system can target both DNA and RNA. So, due to its DNA and RNA cleavage activities along with requirement of multiple Cas proteins for cleavage, reengineering and reprogramming of type III CRISPR/Cas system for genome editing is quite difficult (Terns 2018). Effector proteins of type III CRISPR/Cas system have been characterized from *Thermus thermophilus (*Staals et al. 2013*)* and *Pyrococcus furiosus (*Hale et al. 2009*)*.

7.2.3.1 Types of Type III CRISPR/Cas System

Type III CRISPR/Cas system has been further classified into four subtypes, type III-A, III-B, III-C, and III-D. The first two subtypes were identified earlier but type III-C and III-D have been characterized recently. Cas10 is commonly found in all subtypes and have two domains: an HD domain and a palm domain. Both these domains play an important role in recognition and nuclease activity of type III CRISPR/Cas system (Wang et al. 2019b).

7.2.3.2 RNP Complex and Processing of crRNA

The backbone of RNP complex in type III CRISPR/Cas system comprises two multi-subunit filaments, in which the first filament is composed of six subunits of Cas7, while the second filament is composed of three subunits of Cas11 (Csm2 or Cmr5). crRNA in type III CRISPR/Cas systems stretches along filaments of RNP complex and is bound at 5′-end by Cas5 (Csm4 or Csm3) and Cas10 (Csm1 or Cmr2) proteins. Processing of pre-crRNA in most of the type III systems is a two-step process, in which during the first step, Cas6 cleaves the larger pre-crRNA into single spacer units, while in the second step, trimming of crRNA takes place to generate two populations of mature crRNA of different lengths. In the first step, Cas6 homologs from other CRISPR/Cas systems can be used, while the exact mechanism of trimming in the second step is not clearly known.

7.2.3.3 Nuclease Activities of Type III System

The interesting feature of type III CRISPR/Cas system is that it contains three different nuclease activities. Here we discuss nuclease activities possessed by type III systems.

7.2.3.4 Sequence-Specific RNA Cleavage

Targeted and sequence-specific cleavage of RNA is common in all type III CRISPR/Cas systems. Sequence-specific RNA cleavage by type III CRISPR/Cas system relies on Cas7 effector nuclease. Complementary pairing between crRNA and target RNA in the RNP complex is precisely cleaved by Cas7 at 6 nt intervals (Fig. 7.4) (Terns 2018). Perfect Watson and Crick pairing between crRNA and target RNA triggers cleavage of target RNA; however, mismatches in the complementary pairing of crRNA and target RNA can be tolerated without any effect on nuclease activity. The differentiation between self and non-self RNA by type III systems is ensured by one directed transcription which does not produce transcript complementary to crRNA.

7.2.3.5 Non-specific Cleavage of ssDNA

In addition to sequence-specific cleavage of RNA, type III systems also non-specifically cleave ssDNA which is based on activity of HD domain of Cas10 and the presence of a protospacer sequence. For transcription, RNA polymerase unwinds both DNA strands, leaving antisense strand exposed to HD domain of Cas10, and complementary pairing between crRNA-spacer and target RNA activates Cas10, a divalent metal-activated DNase. Protection of self CRISPR loci relies on complementary pairing between handle of crRNA present at 5′ end and 3′ protospacer region of target RNA, which inhibits Cas10 DNase.

7.2.3.6 Non-specific RNA Cleavage

Type III CRISPR/Cas system also shows a non-specific cleavage of RNA through Cas10 nuclease activity; however in contrast to non-specific cleavage of ssDNA, which is based on HD domain of Cas10, non-specific cleavage of RNA relies on palm domain of Cas10. Palm domain of Cas10 also catalyzes non-constitutive conversion of ATP into cyclic oligoadenylates which activates Csm6 or Csx1 proteins. However, these proteins are not part of RNP complex of CRISPR/Cas III system. Like Cas10, these proteins contain two domains, a C-terminal HEPN domain and CRISPR-associated Rossmann fold (CARF) present at N-terminal. It is believed that CARF is responsible for detection of cyclic oligoadenylate, while HEPN is involved in RNA cleavage (Terns 2018). Like specific RNA cleavage, non-specific RNA cleavage can also tolerate mismatches in spacer region, leading to cleavage of foreign RNA.

7.2.4 Type VI CRISPR/Cas Systems (CRISPR/Cas13)

Type VI CRISPR/Cas systems belong to Class II systems of CRISPR/Cas, which use single effector nuclease for nuclease activity. Effector proteins of type VI contain *RxxxxH* motif which are commonly found in HEPN domains of RNases. In contrast to other Class II systems which target DNA, type VI systems target ssRNA (East-Seletsky et al. 2017) instead of dsDNA, ssDNA, and dsRNA. Target binding site is recognized by the presence of post flanking site (PFS), and binding of RNP complex (Cas13 and crRNA-spacer) with target RNA induces conformational changes in Cas13, leading to activation of HEPN nuclease activity. Once activated, Cas13 cleaves the target RNA and non-specific RNA resulting in collateral cleavage or global RNA cleavage. Here we discuss detailed molecular architecture of type VI CRISPR/Cas system. Like other effector nucleases (Cas9 and Cas12) of Class II CRISPR/Cas systems, Cas13 possesses bilobed globular protein structure, in which one lobe recognizes crRNA, thus called as recognition lobe (REC), while other lobe is called as nuclease lobe (NUC) (Fig. 7.2).

While Cas9 and Cas12 possess RuvC nuclease domain, Cas13 comprises the HEPN nuclease domain which is a common feature of RNA processing systems in higher eukaryotes and kinase extension domain (KEN) containing nucleases such as RNase L and IRE I in eukaryotes. Detailed structural studies have further revealed that multiple domains are present in REC and NUC lobes of Cas13 (Fig. 7.2). REC lobe possesses an N-terminal domain (NTD) and a helical-1 domain, while NUC lobe is comprised of a split HEPN domain (HEPN1) with a helical-2 domain inserted between HEPN1-I and HEPN1-II. Next to HEPN1 domain, there is helical-3 domain followed by the second HEPN domain (HEPN2). Interestingly, helical-1 domain of Cas13a was absent in Cas13d indicating that Cas13d relies on only NTD domain to recognize crRNA direct repeats (Konermann et al. 2018).

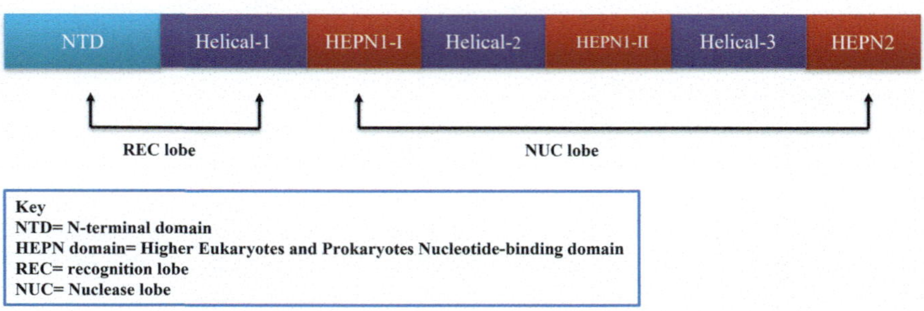

Fig. 7.2 Bilobed architecture of Cas13: both lobes, the REC lobe and NUC lobe, contain multidomains. REC lobe consists of N-terminal domain and helical-1 domain, and the other lobe (NUC) contains 4 domains. The HEPN1 domain is further splitted into HEPN1 I and HEPN1 II domains

7.3 Structure of crRNA in CRISPR/Cas13 Systems

Processing mechanism of crRNA in CRISPR/Cas13 systems is the same as other CRISPR/Cas systems in which crRNA is expressed as pre-crRNA which is further processed into mature RNA (Fig. 7.3).

For example, in type II CRISPR/Cas systems, RNase III recognizes a complex of crRNA-tracrRNA and Cas9 cleaves the duplex between crRNA and tracrRNA to produce a mature crRNA. Processing of crRNA is present in type II, type Vb, and type Ve CRISPR/Cas systems which rely on RNase III, while type Va and type VI CRISPR/Cas systems use an RNA nuclease present within Cas12a and Cas13, respectively. The structure of crRNA in type VI systems is quite simple and consists of two parts, a direct repeat hairpin loop handle and an adjacent guide sequence of 20–30 nt in length present either at 5′ end or 3′ end of repeat handle.

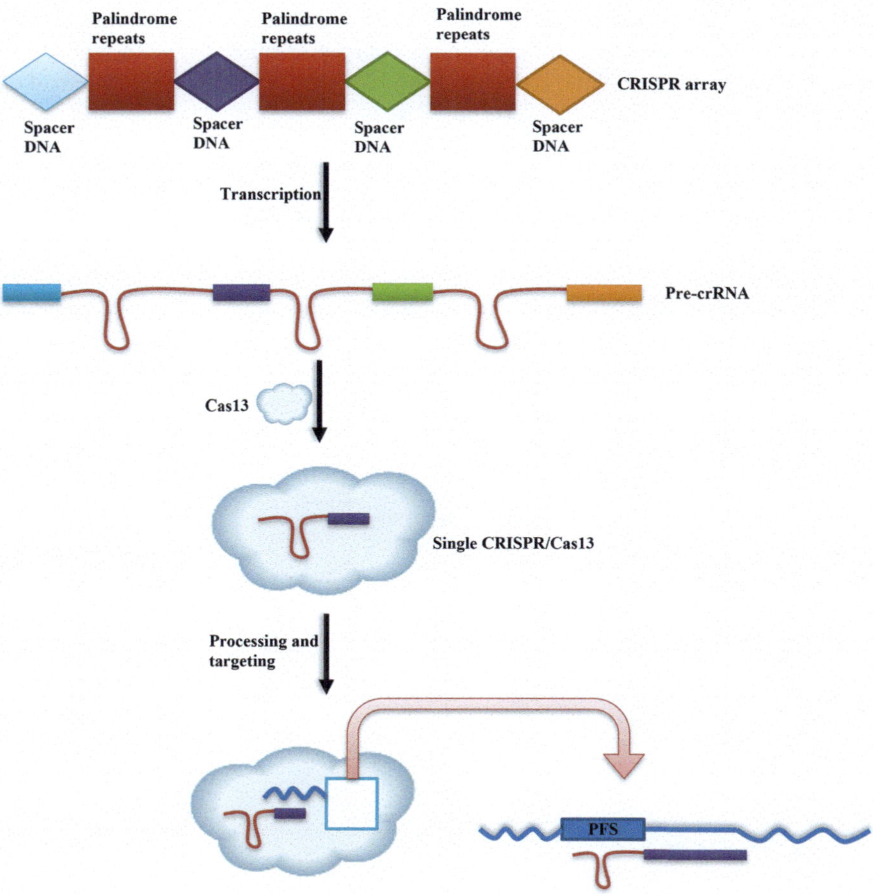

Fig. 7.3 Processing and targeting of CRISPR/Cas13: pre-crRNA is formed by the transcription of CRISPR array. Processing of pre-crRNA to mature crRNA is done by Cas13 that has distinct domains for processing and targeting. Cas13 recognizes PFS region for targeting

7.3.1 Type VI-A (Cas13a)

In Cas13a, direct repeat region of crRNA is composed of 5–6 bp stem using Watson and Crick base pairing with an unpaired 2 nt bulge (AA or AC) present at 3′ end of stem, while loop region of stem contains 7–9 nt depending upon homologs of Cas13a. In addition, an 8–12 nt single-stranded region is present at 5′-end of the stem-loop structure. This single-stranded region is important for the recognition of crRNA by Cas13a (Liu et al. 2017a) (Fig. 7.4).

7.3.2 Type VI-B (Cas13b)

Although detailed structure of crRNA for type VI-B does not exist, predictive models of secondary structure showed that compared with other types (VI-A, VI-B, and VI-C), stem structure of DR in VI-B is longer (9–14 bp), with bulges and a smaller loop structure. In addition, in contrast to other types, in which DR repeat is present at 5′ end of the crRNA-spacer, VI-B contains DR region at 3′-end of the crRNA-spacer (O'Connell 2019) (Fig. 7.4).

7.3.3 Type VI-C (Cas13c) and Type VI-D (Cas13d)

Compared with type VI-A and VI-B, DR region in type VI-C is much smaller in length comprising 30 nt (Figure). The stem-loop structure in type VI-C crRNA comprises a stem of 9 bp with 4–5 nt loop. In addition, a 3 nt short flanking sequence

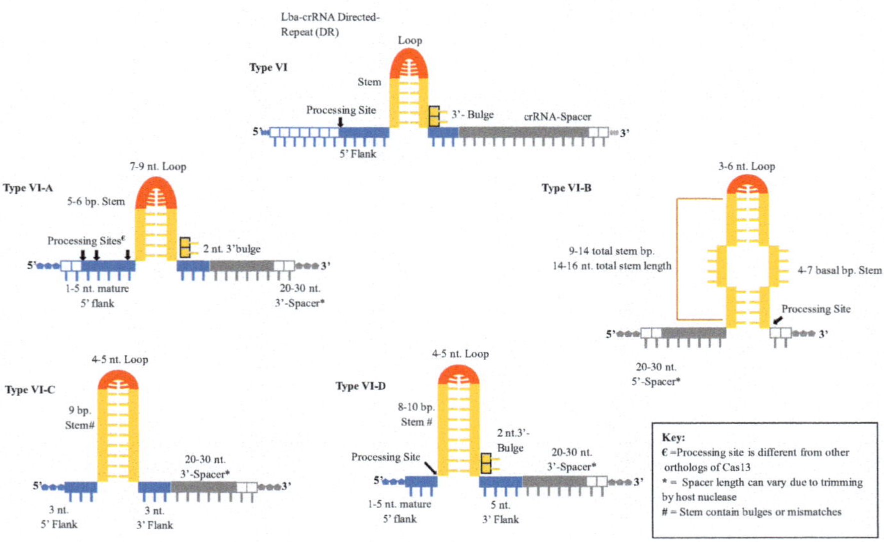

Fig. 7.4 Different subtypes of type VI system: the important similarities and differences between crRNAs from each of the different type VI system subtypes

is present on both sides (5′ and 3′) of stem-loop structure. DR region in type VI-D consists of 36 nt in length with a stem-loop structure formed by 10 bp stem and a 4 to 6 nt loop (Fig. 7.4). A detailed structural analysis of Cas13d in *Eubacterium siraeum* (EsCas13d) showed that stem-loop structure in type VI-D comprises a 9 bp stem and a 4 nt loop region. In addition, stem also contains a 2 nt bulge, commonly present in type VI-A and VI-C. In addition, stem-loop structure in VI-D is flanked by single-stranded region of 5–10 nt at 5′-end and a 5–7 nt with conserved motif AAAAC at 3′-end (O'Connell 2019; Yan et al. 2018) (Fig. 7.4).

7.4 Processing of Pre-crRNA

Like other Class II CRISPR/Cas systems (CRISPR/Cas9 and CRISPR/Cas12), type VI CRISPR/Cas systems also process pre-crRNA into mature crRNA. Similarly, type VI systems also rely on single effector nuclease which is recruited by crRNA to the target RNA. However, in contrast to other CRISPR/Cas systems which are sensitive to processing of pre-crRNA, processing of pre-crRNA is not necessarily required for cleavage of ssRNA by Cas13a. All Cas13 effector nucleases possess two nuclease activities: one is involved in processing of pre-crRNA to mature crRNA, while the other is required for cleavage of target RNA. In the following section, we discuss processing of crRNA in different type VI CRISPR/Cas systems.

7.4.1 Cas13a

Preprocessing mechanism in Cas13a was first described by East-Seletsky et al. (2017) and they have shown that Cas13a processes pre-crRNA and it does not require HEPN domains. However, mutagenesis studies with *Leptotrichia buccalis* Cas13a (Lbu-Cas13a) revealed that a single mutation of arginine to alanine (R1079A) in HEPN domain may completely abolish processing of crRNA without having any effect on HEPN nuclease activity. Moreover, it was also revealed that the processing mechanism of crRNA is based on sequence-specific recognition of processing site but does not require divalent metal ions. In another study with Lsh-Cas13a, it further showed that active site for processing of pre-crRNA lies within helical-1 domain rather than HEPN2 domain. To resolve these contradictions, East-Seletsky performed mutagenesis studies with *Leptotrichia shahii* Cas13a (Lbu-Cas13a) and found that specific amino acids in helical-1 and HEPN2 domain participate in processing of crRNA. In addition, the authors also observed that cleavage site in pre-crRNA lies 4–5 nt upstream of stem-loop structure (Fig. 7.5) (East-Seletsky et al. 2017).

7.4.2 Cas13b

Biogenesis of crRNA in Cas13b showed that pre-crRNA is processed into 66 nt mature crRNA, containing a 30 nt spacer at 5′ end followed by a 36 nt DR region at

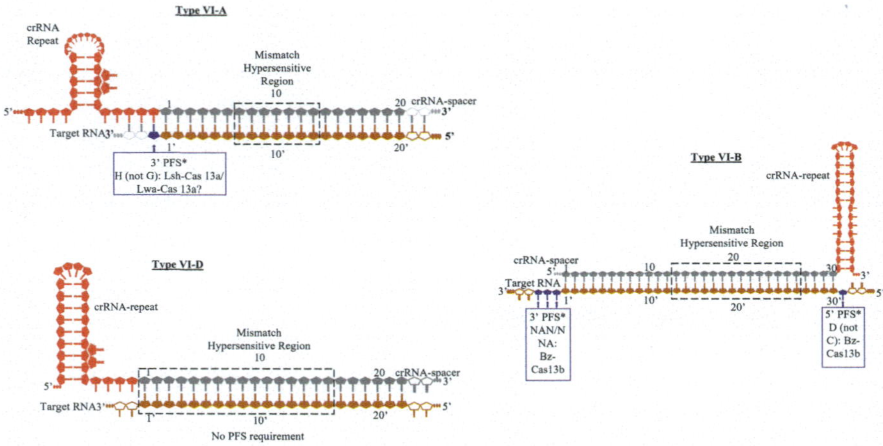

Fig. 7.5 A schematic representation of the interaction between target RNA and crRNA. Emphasizing the significance of the components needed for stable RNA targeting and nuclease activity in the target RNA. It has been shown that the PFS, or Protospacer Flanking Sequence, indicated in blue bordered box, regulates nuclease activity. The need for PFS is not always seen as it is in Type VI-D. Mismatched hypersensitive regions, shown by a black dotted box, are the locations where mismatches between target RNA and crRNA spacer result in a high number of mutations in target RNA within each subtype

3′-end. In addition, cleavage of pre-crRNA takes place at the junction of single- and double-stranded region of crRNA at the base of stem-loop structure (Fig. 7.5) (O'Connell 2019).

7.4.3 Cas13c and Cas13d

Although processing mechanism Cas13c is not clearly understood but due to conserved mechanism of pre-crRNA processing in Cas13a, b, and d, it is believed that processing of crRNA in Cas13c follows the same pattern. Processing of pre-crRNA in Cas13d results in a mature crRNA with a repeat-spacer structure like Cas13a. Moreover, like Cas13a, cleavage of pre-crRNA in Cas13d takes place at the junction of DR-hairpin structure. However, in contrast to Cas13a, which does not depend upon divalent metal ion for cleavage, Cas13d requires divalent metal ion to process the pre-crRNA (O'Connell 2019).

7.4.4 Processing of crRNA-Spacer in Type VI Systems

In addition to initial processing of pre-crRNA by Cas13, crRNA-spacer further undergoes trimming process in vivo. Initially, crRNA-spacer is longer in length which is trimmed by host RNases to an optimal length which varies with different types of Cas13. Using heterologous expression of crRNA-arrays in *E. coli* showed that optimal length for crRNA-spacer in Lsh-Cas13a was 14–20 nt, while optimal length of crRNA-spacer in *Eubacterium siraeum* Cas13d (Es-Cas13d) and Ur-Cas13d was 20–30 nt and 14–26 nt, respectively (Fig. 7.5) (O'Connell 2019).

7.4.5 crRNA-Spacer Organization and Target Search by Cas13

In most of the well-characterized CRISPR/Cas systems, crRNA (sgRNA) sequence possesses a seed region and a distal region. Seed region of sgRNA helps Cas effector nuclease to scan target regions and stabilize sgRNA-target strand complexes. In addition, mismatches in the seed region may affect the stability of Cas nuclease-sgRNA complex and subsequently decrease the cleavage efficiency of the Cas effector nuclease. For example, in CRISPR/Cas9, the first 12 nucleotides of sgRNA in close proximity to PAM serve as seed region and any mismatches in this region may decrease or even completely abolish the nuclease activity of Cas9. In order to explore whether mismatches in the crRNA-spacer also effect Cas13, Abudayyeh et al. evaluated the effect of mismatches between crRNA-spacer for Lsh-Cas13a and target RNA on nuclease activity of HEPN domain. The study showed that mismatches in the middle region crRNA-spacer resulted in the decrease of HEPN-mediated nuclease activity. However, it was not clear whether it was due to defective binding of Cas13 or inactivated HEPN nuclease. HEPN nuclease activity remains in an active autoinhibitory mode in the absence of target RNA, thus preventing non-specific cleavage of RNA. Once Cas13a-crRNA-spacer complex recognizes and binds target RNA, both Cas13 and crRNA spacer undergo conformational changes leading to activation of HEPN nuclease domains (Abudayyeh et al. 2017). Activation of HEPN domains cleaves target RNA as well as non-specific RNA (also called as collateral cleavage—explained in Sect. 7.4.7).

7.4.6 The Post-Flanking Sequence (PFS)

Most of the well-characterized CRISPR/Cas systems for DNA cleavage recognize PAM sequence (NGG for Cas9, TTTN for Cas12), flanking target site at 3′-end or 5′-end. PAM sequence is the safety switch of CRISPR/Cas9 to recognize self and non-self DNA, thus preventing interference complex to cleave its own CRISPR/Cas loci. Abudayyeh et al. (2017) studied the sequence preference flanking target site in Cas13a and showed that Lsh-Cas13a prefers A, U, or C (H) but not G at first position adjacent to 3′-end of the crRNA. These findings were further supported by in vitro experiments which confirmed that the presence of G just next to 3′-end of crRNA significantly reduces HEPN nuclease activity, while in the presence of A, U, or C, HEPN-mediated nuclease activity was maximum. To avoid any confusion with PAM, authors named this sequence preference for Cas13a as post-flanking sequence (PFS). Further studies by the same authors revealed that *Leptotrichia wadei* (Lwa-Cas13) does not show any sequence preference for cleavage of target DNA and the presence of G-PFS did not show any significant reduction in HEPN nuclease activity indicating different PFS preferences for different Cas13 homologs (Abudayyeh et al. 2017).

In contrast to Cas13a, Cas13b from *Bergeyella zoohelcum* (Zhang et al. 2018) and *Prevotella buccae* (Slaymaker et al. 2019) showed PFS preference at both 5′- and 3′-end of the target site. The authors observed that PFS preference at 5′-end was as A, U, or G, while 3′-end requires NAN or NNA as PFS for optimal targeting of RNA. The additional requirement of PFS for Cas13b was attributed to inverted orientation of crRNA-spacer-repeat in Cas13b systems. Some studies have suggested that for Cas13b only shows 5′-end PFS preferences but not 3′-end preferences. Compared with Cas13a and Cas13b, Cas13d did not show any PFS requirement, while PFS preferences for Cas13c are not yet known. Taken together, these studies suggest that sequence preferences of flanking target site in ssRNA do exist; however these preferences vary among different Cas13 effector nucleases.

7.4.7 Collateral Cleavage/Global RNA Degradation

Interestingly, in vitro studies with Cas13 showed that crRNA-spacer-target RNA duplex leads to cleavage of multiple RNA targets including specific as well as non-specific targets. This observation contrasted with editing outcomes from DNA targeting CRISPR/Cas systems. In addition, it was also observed that as cleavage proceeds, smaller RNA cleavage products increase as compared to larger RNA cleavage products suggesting that larger cleavage products are further cleaved into smaller fragments. To explore this unusual cleavage activity of Cas13 in detail, nontarget RNA fragments were added along with specific targets, and it was observed that once HEPN nuclease is activated by specific RNA targets, it cleaves both specific and non-specific (unbound as well as noncomplementary) RNA targets. Further, mutagenesis studies of HEPN nuclease confirmed that cleavage of cis (specific) and trans (non-specific) RNA targets was property of HEPN nuclease. This non-specific cleavage of RNA by HEPN nuclease was named as collateral cleavage. Although, precise roles of collateral cleavage or global RNA degradation are not known, it is believed that it induces dormancy or cell death to limit the phage infection. Interestingly, collateral cleavage was not observed in studies of Cas13 with human or plants cells (Ali et al. 2018; Meeske et al. 2019) .

7.4.8 Specificity of Cas13

Regarding specificity of Cas13, it has been described earlier that cleavage activity of Lsh-Cas13a is sensitive to mismatches in the middle region of crRNA-spacer and target RNA. In addition, it was further confirmed with MS2 phage infections that single mismatches in the crRNA-spacer and target RNA (activator RNA) pose no effect on cleavage activity of Lsh-Cas13a (Abudayyeh et al. 2016). However, mismatches between activator RNA and crRNA-spacer at position 5 and 17 showed maximum effect on cleavage by Lsh-Cas13a. Similarly, double mismatches in the middle region of crRNA spacer showed maximum effect on cleavage activity of

Cas13, while double mismatches towards the ends of crRNA spacer had least effect on cleavage activity by Cas13. Similarly, studies with Lwa-Cas13a and Psp-Cas13b confirmed that these proteins are sensitive to mismatches in the central region of crRNA-spacer. In addition, these studies further showed that sequence composition in crRNA-spacer and activator RNA also affect the specificity of Cas13 effector nuclease. Plasmid library expressing activator RNA with mismatches to crRNA-spacer for Psp-Cas13b showed that mismatch sensitive region for Cas13b lies between 12 and 26 nt. This observation further indicates that while central region is very sensitive, the proximal end is more sensitive than distal end (Cox et al. 2017). Similarly, in vitro studies confirmed that mismatch sensitive region for Psp-Cas13d exists between position 1 and 16 of crRNA-spacer. HEPN cleavage specificity was also studied by targeting RNA in human cells. For example, Cwa-Cas13a and Psp-Cas13b were used to target Gluc transcript in human cells with high specificity and negligible off-targets (O'Connell 2019).

Further studies confirmed the importance of length of crRNA-spacer for specificity and efficiency of Cas13. For example, using human cells, it was revealed that by reducing length of crRNA below 28 nt decreases cleavage efficiency of Cas13a indicating that the length of crRNA also plays an important role in specificity of Cas13a in cells. It contrasts with the CRISPR/Cas9 system, in which cleavage efficiency increases by using truncated sgRNA of 18–19 nt. In addition, Tambe et al. (2018) showed that binding specificity and HEPN nuclease activity are differentially regulated depending on position and number of mismatches between activator RNA and crRNA-spacer. Mismatches at certain positions enhance binding but reduce cleavage while mismatches at other positions may enhance cleavage but minimize binding (Tambe et al. 2018).

7.5 Comparison of CRISPR/Cas13 with Other RNA Editing Systems

Although, all RNA targeting systems such as RNAi (Martinez et al. 2002), RCas9 (Batra et al. 2017) and CRISPR/Cas13 (Abudayyeh et al. 2017) have been used for RNA editing with variable degree of success, they are based on different principles with different effector nucleases. For example, RNAi is based on posttranscriptional mechanism of RNA targeting and have been limited to cytoplasmic RNA degradation. Nuclear RNAs are difficult to target with RNAi. Similarly, CRISPR/Cas9 can be used to regulate RNA, but both work at genomic or transcriptional level and any change in the genome will be permanent and inheritable. In contrast to CRISPR/Cas9, CRISPR/Cas13 works at posttranscriptional level; however, it also possesses non-specific cleavage or collateral cleavage in bacteria. Here we illustrate similarities and differences of CRISPR/Cas13 with other RNA targeting systems.

7.5.1 CRISPR/Cas13 Vs. RNAi

RNAi was the first RNA targeting system developed during the last decade of the twentieth century and has been successfully used to manipulate expression of various genes in animals (Zhang et al. 2007), plants (Nakatsuka et al. 2008), and bacteria (Wang et al. 2010). However, applications of RNAi were hampered by concerns regarding specificity, reproducibility, and stability. Recently, CRISPR/Cas13 emerged as a new tool for targeting RNA and works on posttranscriptional level like RNAi. A common similarity between both approaches is their mechanism which is based on small non-coding RNA and ribonucleoprotein complex for targeting RNA. RNAi relies on siRNA, shRNA, and AGO complex for interference of RNA (Aagaard and Rossi 2007), while CRISPR/Cas13 is based on crRNA and Cas13 effector nuclease for targeting RNA (Garcia-Doval and Jinek 2017). Both approaches share many other similarities such as applications, efficiency, and time required to significantly reduce transcript level of a gene. For example, RNAi and CRISPR/Cas13 can be used to knock down a gene at posttranscriptional level without causing a permanent change at genomic level. In addition, in both approaches, significant reduction in transcript level can be observed 24 h post delivery of targeting reagents even that targeting reagents were present enough, in the cellular compartments. Moreover, both approaches show high efficiency to limit the transcript level. Finally, although both techniques have comparable efficiency during in vitro experiments, both are limited by delivery of reagents for in vivo applications. In addition, to these similarities both have many differences such as requirement of PFS in CRISPR/Cas13 and RNA binding and imaging by CRISPR/Cas13 system which is not possible with RNAi. In contrast to CRISPR/Cas13, RNAi does not rely on PFS, and RNA binding, tracking, and imaging is not possible with RNAi. In addition, off-targeting remains a concern with RNAi, while CRISPR/Cas13 enjoys minimal off-targets; however collateral cleavage with CRISPR/Cas13 in prokaryotes also remains a concern. Similarities and differences between RNAi, CRISPR/Cas9, and CRISPR/Cas13 have been explained in Table 7.2.

7.5.2 CRISPR/Cas13 Vs. CRISPR/Cas9

Although CRISPR/Cas9 is mainly a DNA targeting system, CRISPR/dCas9 offers an efficient and robust method of gene silencing however at transcriptional level. Compared with CRISPR/Cas9 which works at DNA level, CRISPR/Cas13 emerged as a first naturally occurring CRISPR/Cas system to target ss-RNA (Freije et al. 2019), relies on single effector nuclease, and works at posttranscriptional level. In addition, both systems have different applications with their own advantages and disadvantages. For example, CRISPR/dCas9 can be used to track endogenous RNA in living cells (Nelles et al. 2016). Similarly, CRISPR/dCas13 has been programmed to track RNA in living cells (Yang et al. 2019). In addition, CRISPR/Cas13 has been programed for RNA binding and trafficking and for diagnostic applications. CRISPR/Cas9 recognizes NGG as PAM to recognize target DNA; however

Table 7.2 Comparison of RNAi, Cas9, and Cas13

Properties	RNAi	CRISPR/ Cas9	CRISPR/ Cas13	References
Effector protein	AGO2	Cas9	Cas13	Koonin and Makarova (2019) and Granados-Riveron and Aquino-Jarquin (2018)
Enzymatic domain	PIWI	RuvC +HNH	HEPN	Koonin and Makarova (2019) and Granados-Riveron and Aquino-Jarquin (2018)
Natural target	RNA	DNA	RNA	Koonin and Makarova (2019) and Granados-Riveron and Aquino-Jarquin (2018)
Target substrate preference	None	PAM 5′NGG-3′	PFS 3′A, U or C	Granados-Riveron and Aquino-Jarquin (2018)
Off-target effects	Yes	Slight	No detectable off-target	Granados-Riveron and Aquino-Jarquin (2018)

CRISPR/Cas13 recognizes PFS preferences for cleavage of target RNA (Abudayyeh et al. 2017). Optimal length of sgRNA in CRISR/Cas9 is 20 nt, while optimal length for crRNA-spacer varies among different types CRISPR/Cas13 systems. Similarities and differences between both systems have been summarized in Table 7.2.

7.6 Types of Cas13 Effector Nuclease

Unlike other CRISPR/Cas system, type VI effector nucleases exclusively cleave the ssRNA target (Zetsche et al. 2015). As described earlier, the effector proteins of type VI contain two active sites that are distinct from each other, performing different functions. One active site of Cas13 is involved in processing of pre-crRNA, while the other one is responsible for degradation of targeted ssRNA. Type VI CRISPR/Cas proteins have four subtypes: VI-A (C_2c_2/Cas13a), VI-B (Cas13b which includes Cas13b1 and Cas13b2), VI-C (Cas13c), and VI-D (Cas13d) (Table 7.3). These subtypes of Cas13 share low sequence similarities and their classification is based on the location of two HEPN domains. Although both HEPN domains are conserved in all types, Cas13 features and their spacing within the effecter protein are unique for each variant (Shmakov et al. 2015).

7.6.1 Type VI-A (CRISPR/Cas13a)

Cas13a effector protein is a programmable RNA-guided ssRNA nuclease containing two HEPN domains, which are associated with RNAase activity (Nethery and Barrangou 2019). Abudayyeh et al. explored the activity of LshCas13a that defends *E. coli* against the ssRNA virus. The LshCas13a recognizes the PFS, present within

Table 7.3 Subtypes of Cas13

Subtype	Cas13	Characterized species	PFS	Target	Accessory protein	References
VI-A	Cas13a	*Leptotrichia shahii* (LshCas13a) *Leptotrichia wadei* (LwdCas13a)	Required	ssRNA	Cas1 and Cas2	Molla et al. (2020) and Abudayyeh et al. (2017)
VI-B	Cas13b	*Prevotella sp.* (PspCas13b)	Not required	ssRNA	Csx27 and Csx28	Chaudhary (2018)
VI-C	Cas13c	Unknown	Unknown	Unknown	Unknown	–
VI-D	Cas13d	*Ruminococcus* and *Eubacterium*	Not required	ssRNA	WYL1	O'Connell (2019)

target sequence to cleave the target RNA. In vitro, Cas13a targets the exposed regions of specific ssRNA and preferentially cleaves uracil residues at various sites. Furthermore, mutating the putative arginine and histidine residues within HEPN domains obliterate the cleavage activity of Cas13, generating an inactive version of protein, i.e., dCas13. This indicates that the HEPN domains are responsible for catalytic activity and cleavage of ssRNA. However, dCas13 behaves like dCas9, i.e., binds to ssRNA resulting in a complex protein which is known as RNA-guided RNA binding protein. In the middle of protospacer base pairing, Cas13a cannot tolerate the double nucleotide mismatches, thus indicating the existence of seed sequence. In vivo, the reprogramming ability of Cas13a system was successfully used to target the non-phage RNA; i.e., mRNA of red flourescent protein (RFP) (Abudayyeh et al. 2016). In 2016, East-Seletsky et al. investigated the enzymatic activities of LshCas13a and other homologs of Cas13a such as LseCas13a (*Listeria seeligeri*) and LbuCas13a *(Leptotrichia buccalis*) and suggested that Cas13a is a dual ribonuclease, which processes the pre-crRNA to generate mature crRNA (East-Seletsky et al. 2017). In 2016, Lie et al. investigated the crystal structure of LshCas13a and revealed that distinct catalytic sites are used for ssRNA cleavage and pre-crRNA processing. The processing of pre-crRNA is a function of positively charged group of residues, located in N-terminal of helical-1 domain, present within REC lobe. However, cleavage of ssRNA is a function of catalytic site, present between the two HEPN domains (Liu et al. 2017b). East-Seletsky et al. found that the processing activity of pre-crRNA is highly conserved among type VI-A (Cas13a) proteins and processing improves the targeting of Cas13a to ssRNA. The biochemical and functional characterization of Cas13a proteins suggested that Cas13a can be classified into two subfamilies with different substrate preference (East-Seletsky et al. 2017). A non-specific trans-cleavage activity of ssRNA was observed upon activation of Cas13a protein by binding to the crRNA, complementary to ssRNA (Abudayyeh et al. 2016). Conformational change occurs in Cas13a when it binds with the ssRNA, leading to activation of catalytic sites of HEPN domains. Cas13a binds with target RNA but binding changes are far from HEPN catalytic site due to which short target RNA cannot be cleaved by catalytic site in *cis*; however there is possibility that longer target RNA can be cleaved in *cis*. Liu et al. analyzed the structure of Cas13a and suggested that the HEPN sites are exposed to the surface due to which non-specific and collateral cleavage of ssRNA occur by catalytic sites in *trans* position (Liu et al. 2017b). This process suggests a mechanism to sense viral RNA and stimulate the dormancy or programmed cell death (PCD) of the host cell that prevents the spreading of viral infection. The recent study of East-Seletsky et al. demonstrated the collateral cleavage activity of Cas13a used to detect and sense the presence of specific transcripts (East-Seletsky et al. 2017). Gootenberg and colleagues developed a diagnostic tool for the detection of RNA and DNA in vitro with a single base mismatch and atto-molar sensitivity, demonstrating the potential utility of Cas13a in a wide range of diagnostic applications (Gootenberg et al. 2017). Abudayyeh and colleagues identified the interference activity of LwaCas13a, and its heterologous expression resulted in knockdown of an endogenous or reporter transcript in plant and mammalian cells (Abudayyeh et al. 2016).

7.6.2 Type VI-B1 and B2 (CRISPR/Cas13b)

Cas13b is another member of Class 2, type VI CRISPR/Cas systems, with significant RNA targeting capability. The effector protein of Cas13b is considerably different from Cas13a, having novel protein sequence that shows two HEPN domains responsible for cleavage of ssRNA. In 2017, Smargon et al. functionally characterized Cas13b and suggested that the effector protein cleaves the ssRNA both in vivo and in vitro. The targeting ability of Cas13b is like Cas13a in many ways including the ability to target ssRNA but not dsRNA, no specific collateral cleavage activity and generating mature crRNA by processing of its own pre-crRNA. In addition, inactive Cas13b, i.e., dCas13b, binds specifically to a target RNA. However, Cas13b is distinct from Cas13a, i.e., Cas13b requires PFS regions on the both sides of crRNA-protospacer with nucleotide sequence different from Cas13a. The Cas13b also need accessory proteins Csx27 and Csx28 present within the CRISPR loci and required for modulating activity of Cas13b. Based on the presence Csx28 and Csx27, Cas13b is further categorized into VI-B1 and VI-B2, respectively (Smargon et al. 2017). In 2017, Cox et al. investigated that the RNAase activity of Cas13b was enhanced by the presence of Csx28 because it dimerizes the two HEPN domain together thus making RNAase active site more composite. Csx28 significantly boosts the efficacy of Cas13b, due to which the HEPN domains show non-specific RNA nuclease activity, while Csx27 tightly associated with the Cas13b and inhibit its nuclease activity. While infection occurs, Csx27 release its inhibitory response on Cas13b. Cox et al. suggested that the ribonuclease activity of Cas13b can specifically knock down the endogenous transcript in mammalian cells, and Abudayyeh et al. identified the more specific and robust cleavage activity of Cas13b (Cox et al. 2017; Abudayyeh et al. 2016). Cox et al. evaluated the interference activity of several orthologs of Cas13 nuclease, including 15 of Cas13b, 21 of Cas13a, and 7 of Cas13c, and identified that the most efficient ortholog of Cas enzyme is PspCas13b (from *Prevotella*) which has maximum interference activity and knockdown ability (Cox et al. 2017). Cas13b shows robust targeting of ssRNA, thus expanding the potential of CRISPR/Cas system to manipulate RNA with a range of applications.

7.6.3 Type VI-C (CRISPR/Cas13c)

Type VI-C is not a well-characterized system and detailed architecture of effector nuclease, and crRNA-spacer are not yet explored.

7.6.4 Type VI-D (CRISPR/Cas13d)

Predominately, Cas13d is found in two genus of bacteria that are *Ruminococcus* and *Eubacterium*. The protein structure of Cas13d consists of one or multiple WYL domain which is potently associated within the prokaryotes as a defensive system (Yan et al. 2018). WYL domain contains 170 amino acids which are frequently found at C-terminal of ribbon-helix-helix (RHH) or helix-turn-helix (HTH) DNA binding domains (Makarova et al. 2017). Hein et al. revealed that CRISPR/Cas system of type I-D contains WYL domain, which acted as transcriptional repressor of crRNA in *Synechococcus*. While Yan et al. illustrated that the Cas13d contain WYL1 domain in *Ruminococcus* (RspWYL1) that stimulate the collateral cleavage of ssRNA both in vitro and in bacterial negative screens (Yan et al. 2018). The efficient collateral activity of Cas13d is due to WYL1 domain that makes this system particularly important and attractive for the manipulation of RNA. The CRISPR locus of Cas13d lacks the acquisition protein Cas1 and the length of direct repeats (DR) of CRISPR/Cas13d is highly conserved with 36 nucleotides.

7.6.5 Type VI-E (Cas13e) and VI-F (Cas13f)

Recently, two new types of Cas13 have been identified as Cas13e and Cas13f. Cas13e includes two members (Cas13e.1 and Cas13e.2), while Cas13f includes five members (Cas13f.1 to Cas13f.5), respectively. Architecture of DR and crRNA spacer in Cas13e showed that it contains DR region at 3′ end of crRNA spacer similar to Cas13b. Compared with Cas13a and Cas13b, Cas13e.1 and Cas13.f1 showed higher knockdown efficiency in HEK293T cells. Cas13e.1 was also used to target RdRP mRNA as an antiviral strategy against COVID-19 in HEK293T cells. Moreover, it was also showed that Cas13e.1 can tolerate mismatches like other previously identified Cas effector nucleases. Compared with other Cas13 effectors such as Cas13a/b/d, Cas13e.1 is very compact in size and Cas13e.1-based system has been suggested as an efficient system to target COVID-19 for therapeutic purpose (Xu et al. 2020).

7.7 Applications of CRISPR/Cas in RNA Editing

In biological research, CRISPR/Cas is becoming an important tool. Today, CRISPR/Cas9 is not only used as a gene-editing tool but application of catalytically inactive Cas9 (dCas9) and Cas9 nickase (nCas9) in epigenetic editing, genome imaging, prime editing, base editing, gene regulation, and chromatin engineering are now exceeding the gene-editing functionality of Cas9 (Adli 2018). However, researchers are still actively exploring other CRISPR systems to identify CRISPR/Cas9 like effector proteins. As the Cas9 can target the dsDNA but not ssRNA, now Cas13 is used to target ssRNA. Cas13 is a newly discovered CRISPR/Cas system, which has applications in plant biology, biotechnology, and therapeutics. Cas13 is a powerful tool for RNA imaging (Yang et al. 2019), RNA regulation (Ali et al. 2018), and RNA detection (Freije et al. 2019). The RNAase activity of Cas13 provides the diagnostic applications (Khambhati et al. 2019). dCas13 is a binding protein which is used for RNA imaging and regulates the specific transcript. In plants Cas13 is used for defensive mechanism against viruses (Aman et al. 2018b). Here, we discuss applications of CRISPR/Cas13 for RNA editing.

7.7.1 SHERLOCK: A CRISPR/Cas13-Based Viral Diagnostic Tool

The conformational changes in Cas13 leading to its activation and collateral cleavage have been programmed as diagnostic tool to detect the nucleic acids in vitro. The technique is termed as "specific high sensitivity enzymatic reporter UnLOCKing (SHERLOCK)," in which the DNA or cDNA (for RNA genome viruses) are amplified by RPA (recombinase polymerase amplification) or (reverse transcriptase RPA for RNA viruses) under isothermal conditions. A T7 promoter is added during this amplification step for transcription. Amplified DNA and cDNA are transcribed into RNA that is cleaved by Cas13a in the presence of crRNA and RNA sensors, which release fluorescence signals on cleavage (Fig. 7.6) (Gootenberg et al. 2018). SHERLOCK has the ability to detect the viral particles and clearly discriminate between flavivirus Dengue (DENV) and Zika virus (ZIKV). All the components of SHERLOCK can be lyophilized and rehydrated which enables this procedure to be used in the field with paper spotting. SHERLOCK is a rapid detection method with high mismatch sensitivity. It is an affordable and more efficient countermeasure to minimize the spread of virus in plants (Nicaise 2014). This technique discriminates the different species of bacteria, detecting the mutations, which are related to cancer, and detecting the antibiotic resistance genes.

7.7.2 CARVER

7.7.2.1 A Cas13-Based Detection and Prevention Platform Against RNA Viruses

RNA viruses such as Ebola, Zika, and COVID-19 contain ssRNA as a genome and cause infectious diseases in human beings leading to health and economic crises at a

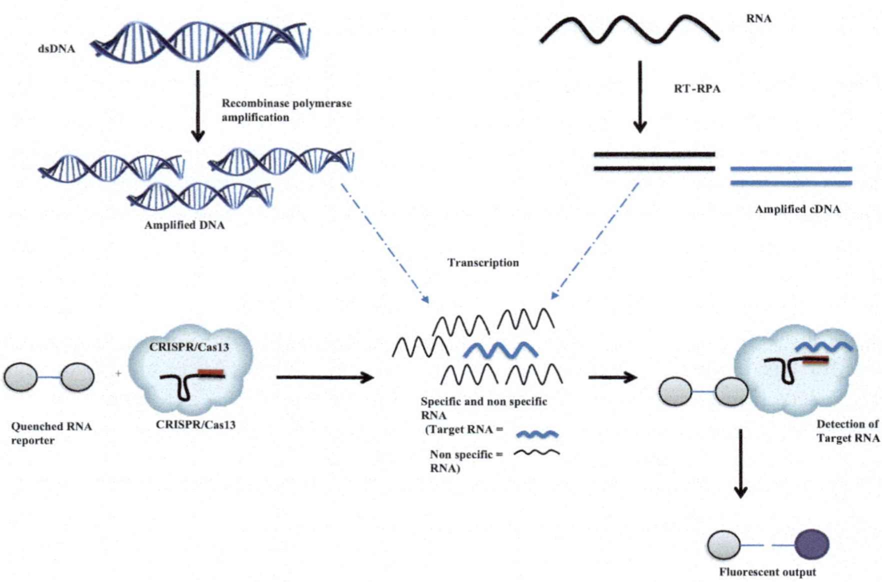

Fig. 7.6 SHERLOCK technique: dsDNA or ssRNA are amplified by RPA enzyme under isothermal condition. CRISPR/Cas13 with quenched RNA reporter targets the specific RNA and reporter activation occurs when target RNA is detected

global level. So far, there is no approved treatment against these viruses. As Cas13 targets ssRNA, recently, scientists from MIT, Harvard University, USA, have developed a Cas13-based, combined detection and prevention method called as CARVER (Cas13-assisted restriction of viral expression and readout) against RNA viruses using human cells. This study was reported as one of the first CRISPR/Cas13-based antiviral approaches against ssRNA viruses. So, CARVER could be transformed as a potential diagnostic and therapeutic platform against RNA viruses in the future (Jia et al. 2020; Freije et al. 2019).

7.7.3 CRISPR/Cas13-Based Antiviral Strategy Against COVID-19 (PAC-MAN)

Corona virus disease-2019 (COVID-19, caused by SARS-CoV-2 (severe acute respiratory syndrome coronavirus-2), represents one of the greatest challenges to human health in the twenty-first century. First time reported in Wuhan, China, during 2019, it was rapidly transmitted to almost all countries in the world, causing respiratory problem in human beings with a death rate of more than 2.5% in some countries. So far there is no approved treatment however, vaccines against COVID-19 have been developed to control this pandemic which has seriously impacted public health and economies. From the very beginning, CRISPR/Cas system was perceived as a promising tool against this highly transmissible and deadliest outbreak. Recently, Abbott et al. (2020) have developed a CRISPR/Cas13-based platform known as PAC-MAN (prophylactic antiviral CRISPR in human cells) to combat COVID-19 virus. In PAC-MAN, Rf-Cas-13d (*Ruminococcus flavefaciens*-Cas13d) was programed to target RdRP (RNA-dependent RNA polymerase) and nucleocapsid gene of COVID-19 in human cells. PAC-MAN offers an efficient and remarkable approach to control COVID-19 by limiting its mRNA abundance in cells.

7.7.4 Knockdown of RNA

Cas13 can target a specific transcript for downregulation of gene expression (Fig. 7.7). Abudayyeh et al. identified that Cas13a from *Leptotrichia wadei* does not require PFS and used this effector nuclease for successful RNA knockdown in human cells. The activity of LwCas13a depends on stabilization domain, i.e., msGFP, responsible for cellular and nuclear localization. LwCas13a was also used for RNA knockdown in plants and rapidly reducing the cytoplasmic RNA pool (Abudayyeh et al. 2016). The knockdown ability of Cas13a is superior than RNAi in terms of specificity. In addition, Cas13b has very efficient RNA cleavage activity

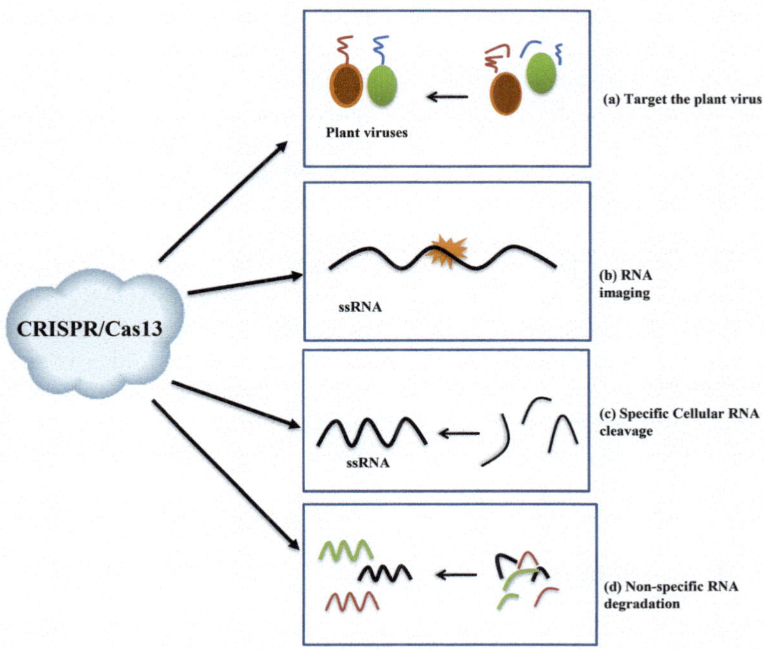

Fig. 7.7 Potential applications of CRISPR/Cas13 in plant biology. (**a**) Targeting the plant RNA viruses at multiple sites with Cas13. (**b**) RNA imaging. (**c**) Downregulation of gene expression by targeting specific RNA transcript. (**d**) Collateral degradation of non-specific RNA to promote PCD

and does not require msGFP for stabilization. Cox et al. identified that the RNA targeting ability of Cas13b ortholog in eukaryotic cells is more efficient than LwaCas13a (Cox et al. 2017). Cas13 has the ability to cleave the long noncoding RNAs (lncRNAs) in vivo and provide flexibility in manipulation. lncRNAs are present in plants which are 50–300-long nt, involved in many functions like guide protein translocation, regulate gene expression, modulate chromatin loop mechanics, and influence alternative splicing. However, functional characterization of lncRNA was hampered by the lack of mutants. Cas9 and Cas12 were used to create mutants but Cas13 is more versatile by having ability to directly cleave these lncRNAs. In addition, Cas13 is not limited to cytoplasmic transcripts but also targets the nuclear transcripts by fusing with NLS (Wang and Chekanova 2017; Liu et al. 2017b).

7.7.5 RNA Editing with CRISPR/Cas13

RNA editing system by Cas13 has multiple advantages over DNA editing systems. RNA editing can be used in non-dividing cells because it does not require HDR machinery. In addition, Cas13 is more flexible than Cas12 and Cas9 because it does not require PAM sequence for target cleavage. Some Cas13 effector nucleases show preference for PFS sequence but other orthologs do not require PFS such as LwaCas13a. RNA editing system based on Cas13 would be reversible, thus limiting off-targets in genome. In RNA editing system, two components ADAR and Cas13 are fused to convert adenine in RNA into inosine, treated like guanine. This technique is known as RNA editing for programmable A to I (G) replacement (REPAIR) system and used to treat the genetic diseases. Repetitive regions are targeted by ADAR1 and non-repetitive regions are targeted by ADAR2. ADAR1 and ADAR2 are the proteins which deaminase the adenosine to inosine and create a functional change in RNA. Cox et al. fused the dPspCas13b with ADAR deaminase domain and observed low level of RNA editing. They constructed hyperactive ADAR2 to increase the A > G editing. The PspCas13b with ADAR2 designated as REPAIRv1, displayed robust editing. This technique is potentially useful to treat Parkinson's and Duchene muscular dystrophy diseases. RNA editing is reversible but not stable. For example, all copies of a transcript would be effected by editing at DNA level, while in the case of RNA editing, non-effected and effected copies of transcript would exist at the same time suggesting dose control for effective manipulation (Cox et al. 2017).

7.7.6 Virus Interference

Various species of RNA viruses threaten the food security by damaging agricultural production (Romay and Bragard 2017). The most emerging diseases in the plants are due to viruses resulting in annual loss of approx. 30 billion USD. Plants develop immune system like immune reporter signaling, protein degradation, and RNAi-based gene silencing. RNAi-mediated gene silencing has been successfully used against plant viruses, but there are many drawbacks including incomplete knock-down, variability of target efficiency, and off-targeting effects (Sastry and Zitter 2014). The more practical, best, and safe strategy to control viral diseases in plants is using CRISPR/Cas system with strong antiviral nature. CRISPR/Cas9 is an efficient system against DNA viruses in plants and targets the coding and non-coding sequences of DNA. It showed strong interference with *Beet curly top* virus and *Tomato yellow leaf curl* virus leading to a delay in the accumulation of viruses (Rupaimoole and Slack 2017). In plants, the majority of viruses are RNA in nature; even many DNA viruses contain RNA intermediate in their life cycle (Nicaise 2014).

RNA viruses infect a wide range of commercially important plants, resulting in severe loses in quantity and quality of diverse key crops (Aman et al. 2018a). CRISPR/Cas13 is a system that is used efficiently against RNA viruses due to its robustness and simplicity and ability to cleave the RNA viruses (Fig. 7.7). The vitality of plant remains intact because collateral cleavage activity was not observed in plants in contrast to prokaryotes and in vitro (Gootenberg et al. 2017; Abudayyeh et al. 2016). Aman et al. used LshCas13 against turnip mosaic virus in *Nicotiana benthamiana*, and an estimated 50% reduction of viral GFP signals in 7 days was observed with no adverse effect on plant vitality. Cas13 can also target the newly evolved RNA viruses' variants because it can tolerate the single base mismatch. It can target the multiple RNA viruses simultaneously or single virus at multiple sites (Aman et al. 2018a). *Potyviridae* is a family of plant viruses which are RNA in nature and *Potyvirus* is the genus of *Potyviridae* containing 146 species. *Potyviruses* infect monocotyledon and dicotyledonous plants on wide range due to which significant losses of all over world occur. The genome of *Potyvirus* consists of linear ssRNA (~10 kb), with a long open reading frame that is translated as a polyprotein which cleaved into 10 functional proteins by protease. The recent study of Aman et al. demonstrated that Cas13 could be harnessed to confer resistance in plants against *Potyviruses* with the help of crRNA, designed to target the viral genome.

7.7.7 RNA Tracking and Imaging

RNA imaging is important for basic understanding of cellular mechanisms and studying diseases and tissue specificity. RNA imaging approaches are based on fusion of specific RNA binding domains to the florescent proteins, such as fusion of Pumilio homology domain with MS2 coat protein (MCP) (Yamada et al. 2011). Recently, Rcas9 system was used for RNA imaging purpose (Nelles et al. 2016). In addition, dCas13 was fused catalytically with fluorescent tags in the living cells and used to visualize the different types of RNA and trafficking of mRNA to specific cellular organelle (Fig. 7.7). However, this system requires negative feedback system due to the background noise of unbound protein. This system allowed the translocation of mRNA into stress granule in live cells. Fluorescent imaging is also established in plants and is used to visualize the dynamics of RNA (Abudayyeh et al. 2016).

7.7.8 Alternate Splicing

The alternative splicing of the pre-mRNA is a critical step in posttranscriptional gene regulation in eukaryotes responsible for diversity in proteome, and mis-splicing of RNA may lead to severe consequences in plants and humans (Scotti and Swanson 2016). Therefore, the capability of precisely alternative splicing of pre-mRNA would be important for many biotechnological and therapeutic applications. The splicing pattern of mRNA can be altering by targeting the splicing factor in pre-mRNA. To target the splicing factor, Graveley and Maniatis fused the RNA binding proteins like MS2 and PUF with splicing activator arginine/serine (RS)-rich domains or splicing suppressor glycine (Gly)-rich domain of hnRNP. In addition, using CRISPR-mediated base editing, cytidine deaminase was fused with Cas9 nickase or dCas9 for targeting the single nucleotide in DNA (Kim et al. 2017). For RNA editing such as cytosine-to-uracil (C-to-U) and adenosine-to-inosine (A-to-I), manipulation of mRNA could be achieved by dCas13, leading to modifications at RNA level without any permanent change at genomic level (Montiel-Gonzalez et al. 2013).

Cas13 can be reprogrammed to perform multiplex targeting, recruiting various splicing factors fused to dCas13 to intronic or exonic sequences of interest in many pre-mRNAs simultaneously, thus allowing correction of splicing defects. dCas13 can be modulated to support alternate splicing by fusing with splicing factor and promoting transcriptome plasticity which allows the plants to tolerate the particular stress conditions. Programing of Cas13 for alternate splicing can be used to eliminate the pathogenic splicing isoforms. Cas13d was fused with glycine-rich domain (hnRNP A1) to effect alternative splicing and promote exon exclusion (Wang et al. 2019a).

7.7.9 Regulation of Translation Through CRISPR/Cas13

Under stress conditions, the tRNA cleavage takes place in the anticodon loop by ribonucleases. In 2008, Thompson et al. revealed that *Arabidopsis* seedling subjected to the oxidative stress showed increase in cleavage activity of tRNA into fragments, thus inhibiting the translation process (Thompson and Parker 2009). dCas13 fused with translational repressors or enhancers may regulate the translation of specific mRNA. Therefore, dCas13 can be used to modify tRNA posttranscriptionally to influence the regulation of translation of mRNA into protein (Abudayyeh et al. 2016).

7.7.10 Programmed Cell Death and CRISPR/Cas13

The active catalytic site of HEPN domains is located on outer surface, making it possible for Cas13 to act on non-specific ssRNA targets and cleaving them in unspecific manner, leading to programmed cell death (PCD) in natural system. The variants of CRISPR/Cas13 can trigger the PCD in plants by targeting the multiple pathogens and degrade them in non-specific manner. Cas13 has promiscuous RNase activity, which has great importance for in vivo applications. The cytotoxicity of drugs in cancer therapy is reducing by targeting and killing the selective tumor cells without affecting the normal ones. Various cancerous cells are characterized by the expression of aberrant and unique biomarkers that distinguish these cancerous cells from healthy tissues. Many therapeutic strategies have been designed to target these selective biomarkers. Many pro-drugs such as protein toxins have been developed which target the cancer-specific biomarker and kill the cells by promoting the cell death (Bachran et al. 2014). The findings suggest that Cas13 can be used to target the aberrant transcript expressed in tumor cells using collateral cleavage activity, inducing PCD ultimately killing of selective cancerous cells.

7.7.11 Processing of mRNA for Stress Tolerance in Plants

Pre-crRNA processing ability of Cas13 may be used to target mRNAs in plants to cope with stresses and pathogens. Tissue- and organelle-specific silencing of RNA through Cas13 have great benefits to control diseases and also for studying metabolic pathways.

7.7.12 Gene Regulation with CRISPR/Cas

CRISPR/Cas13 is a flexible RNA-guided RNA targeting system which holds immense potential for robust, precise, and scalable posttranscriptional regulation of RNA transcripts. The importance of modulating the transcript level without effecting DNA has led to various approaches for posttranscriptional regulation (Brophy and Voigt 2014). Cas13 is the first tool of CRISPR system that is RNA specific and provides versatile RNA targeting. The efficient ssRNA targeting by Cas13 provides potential new strategies for RNA manipulations. The CRISPR/Cas13 is a system that is superior than other previously developed RNA targeting strategies with many unique and superior advantages. The system of CRISPR/Cas13 has two components, Cas13 and crRNA, which guide the Cas13 to the target RNA and facilitate its delivery in most of the organisms (Abudayyeh et al. 2016). The specificity of CRISPR/Cas13 system relies on the crRNA-spacer sequence, thus providing the scalability and versatility to this system. Cas13 has the ability to process its own crRNA from pre-crRNA transcript, which can be utilized for cytoplasmic RNA manipulations. For example, Cas13 can be programmed to study translational regulation, RNA localization, mRNA imaging, splicing isoforms, and RNA trafficking. In addition, the processing of crRNA will assist the expression and design of multiple gRNA for modulating and targeting the multiple transcripts (East-Seletsky et al. 2017).

7.7.13 Isolation of Specific RNA

Isolation of specific RNA and RNA-bound proteins is sometimes very important to characterize RNA and protein functions. Beads coated with dCas13 may facilitate isolation of particular RNA and also its associated proteins to characterize RNA-protein interactions, sequencing particular RNA and even studying protein-protein interaction between RNA binding proteins.

7.8 Future Prospect

The CRISPR/Cas system has already provided a remarkable set of tools with broad range of applications from cancer research to gene therapies. Recent applications of R-Cas9, type III, and especially type VI systems have opened new horizons with broader applications in RNA world beyond conventional DNA targeting CRISPR/Cas systems. While Cas13 has shown an immediate impact on disease diagnosis and posttranscriptional gene regulation in plant and animal biology. Other newly discovered effector proteins and applications, including CasX and EvolvR, are in the initial stage of characterization. EvolvR also offers an unprecedented control over rewriting genetic code leading to evolutions of traits beyond nature. The developments in RNA targeting tools with special focus on type VI CRISPR/Cas systems will definitely enhance knowledge and understanding about basic questions of RNA biology, tissue, and organism development and fundamental cellular pathways. In addition, CRISPR/Cas13 will help in exploring functions of non-coding RNAs, RNA processing, and identification of RNA binding proteins. With new developments, CRISPR/Cas13 has already been programmed as diagnostic and preventive tool for human viruses. However, with all these developments and promising tools like CRAVER, SHERLOCK, and PACMAN, suitable delivery methods for Cas13 and crRNA-spacer to the targeted tissues are some of the major challenges for therapeutic, translational, and future applications of CRISPR/Cas13.

Acknowledgments This work was partially supported by research grant from High Education Commission (HEC), Islamabad, Pakistan, to Dr. Aftab Ahmad under project # NRPU-6350. We are thankful to office of research, innovation, and commercialization (ORIC), UAF for their support in study and research.

Figures 7.2, 7.4 and 7.5 were reproduced from O'Connell (2019). All permissions obtained from Elsevier [License Number: 5197520784686].

References

Aagaard L, Rossi JJ (2007) RNAi therapeutics: principles, prospects and challenges. Adv Drug Deliv Rev 59(2–3):75–86

Abbott TR, Dhamdhere G, Liu Y, Lin X, Goudy L, Zeng L et al (2020) Development of CRISPR as an antiviral strategy to combat SARS-CoV-2 and influenza. Cell 181(4):865–876

Abudayyeh OO, Gootenberg JS, Konermann S, Joung J, Slaymaker IM, Cox DB, Shmakov S, Makarova KS, Semenova E, Minakhin L (2016) C2c2 is a single-component programmable RNA-guided RNA-targeting CRISPR effector. Science 353(6299):aaf5573

Abudayyeh OO, Gootenberg JS, Essletzbichler P, Han S, Joung J, Belanto JJ, Verdine V, Cox DB, Kellner MJ, Regev A (2017) RNA targeting with CRISPR–Cas13. Nature 550(7675):280–284

Adli M (2018) The CRISPR tool kit for genome editing and beyond. Nat Commun 9(1):1911

Ali Z, Mahas A, Mahfouz M (2018) CRISPR/Cas13 as a tool for RNA interference. Trends Plant Sci 23(5):374–378

Aman R, Ali Z, Butt H, Mahas A, Aljedaani F, Khan MZ, Ding S, Mahfouz M (2018a) RNA virus interference via CRISPR/Cas13a system in plants. Genome Biol 19(1):1

Aman R, Mahas A, Butt H, Ali Z, Aljedaani F, Mahfouz M (2018b) Engineering RNA virus interference via the CRISPR/Cas13 machinery in Arabidopsis. Viruses 10(12):732

Bachran C, Hasikova R, Leysath C, Sastalla I, Zhang Y, Fattah R, Liu S, Leppla S (2014) Cytolethal distending toxin B as a cell-killing component of tumor-targeted anthrax toxin fusion proteins. Cell Death Dis 5(1):e1003

Batra R, Nelles DA, Pirie E, Blue SM, Marina RJ, Wang H, Chaim IA, Thomas JD, Zhang N, Nguyen V (2017) Elimination of toxic microsatellite repeat expansion RNA by RNA-targeting Cas9. Cell 170(5):899–912.e810

Braun CJ, Bruno PM, Horlbeck MA, Gilbert LA, Weissman JS, Hemann MT (2016) Versatile in vivo regulation of tumor phenotypes by dCas9-mediated transcriptional perturbation. Proc Natl Acad Sci 113(27):E3892–E3900

Brophy JA, Voigt CA (2014) Principles of genetic circuit design. Nat Methods 11(5):508

Buchholz F, Kittler R, Slabicki M, Theis M (2006) Enzymatically prepared RNAi libraries. Nat Methods 3(9):696–700

Chaudhary K (2018) RNA base editing: programmable protein machine. Curr Sci 114(9):1808

Chen F, Ding X, Feng Y, Seebeck T, Jiang Y, Davis GD (2017) Targeted activation of diverse CRISPR-Cas systems for mammalian genome editing via proximal CRISPR targeting. Nat Commun 8(1):1–12

Chen S, Yu X, Guo D (2018) CRISPR-Cas targeting of host genes as an antiviral strategy. Viruses 10(1):40

Cox DB, Gootenberg JS, Abudayyeh OO, Franklin B, Kellner MJ, Joung J, Zhang F (2017) RNA editing with CRISPR-Cas13. Science 358(6366):1019–1027

Duan J, Lu G, Hong Y, Hu Q, Mai X, Guo J, Si X, Wang F, Zhang Y (2018) Live imaging and tracking of genome regions in CRISPR/dCas9 knock-in mice. Genome Biol 19(1):1–7

East-Seletsky A, O'Connell MR, Burstein D, Knott GJ, Doudna JA (2017) RNA targeting by functionally orthogonal type VI-A CRISPR-Cas enzymes. Mol Cell 66(3):373–383.e373

Fineran PC, Dy RL (2014) Gene regulation by engineered CRISPR-Cas systems. Curr Opin Microbiol 18:83–89

Freije CA, Myhrvold C, Boehm CK, Lin AE, Welch NL, Carter A, Metsky HC, Luo CY, Abudayyeh OO, Gootenberg JS (2019) Programmable inhibition and detection of RNA viruses using Cas13. Mol Cell 76(5):826–837.e811

Garcia-Doval C, Jinek M (2017) Molecular architectures and mechanisms of class 2 CRISPR-associated nucleases. Curr Opin Struct Biol 47:157–166

Gootenberg JS, Abudayyeh OO, Lee JW, Essletzbichler P, Dy AJ, Joung J, Verdine V, Donghia N, Daringer NM, Freije CA (2017) Nucleic acid detection with CRISPR-Cas13a/C2c2. Science 356(6336):438–442

Gootenberg JS, Abudayyeh OO, Kellner MJ, Joung J, Collins JJ, Zhang F (2018) Multiplexed and portable nucleic acid detection platform with Cas13, Cas12a, and Csm6. Science 360(6387):439–444

Granados-Riveron JT, Aquino-Jarquin G (2018) CRISPR–Cas13 precision transcriptome engineering in cancer. Cancer Res 78(15):4107–4113

Hale CR, Zhao P, Olson S, Duff MO, Graveley BR, Wells L, Terns RM, Terns MP (2009) RNA-guided RNA cleavage by a CRISPR RNA-Cas protein complex. Cell 139(5):945–956

Hutvagner G (2005) Small RNA asymmetry in RNAi: function in RISC assembly and gene regulation. FEBS Lett 579(26):5850–5857

Jia F, Li X, Zhang C, Tang X (2020) The expanded development and application of CRISPR system for sensitive nucleotide detection. Protein Cell 11:624–629

Jinek M, Chylinski K, Fonfara I, Hauer M, Doudna JA, Charpentier E (2012) A programmable dual-RNA–guided DNA endonuclease in adaptive bacterial immunity. Science 337(6096):816–821

Khambhati K, Bhattacharjee G, Singh V (2019) Current progress in CRISPR-based diagnostic platforms. J Cell Biochem 120(3):2721–2725

Kim D, Lim K, Kim S-T, S-h Y, Kim K, Ryu S-M, Kim J-S (2017) Genome-wide target specificities of CRISPR RNA-guided programmable deaminases. Nat Biotechnol 35(5):475

Konermann S, Lotfy P, Brideau NJ, Oki J, Shokhirev MN, Hsu PD (2018) Transcriptome engineering with RNA-targeting type VI-D CRISPR effectors. Cell 173(3):665–676.e614

Koonin EV, Makarova KS (2019) Origins and evolution of CRISPR-Cas systems. Philos Trans R Soc B 374(1772):20180087

Koonin EV, Makarova KS, Zhang F (2017) Diversity, classification and evolution of CRISPR-Cas systems. Curr Opin Microbiol 37:67–78

Liu L, Li X, Ma J, Li Z, You L, Wang J, Wang M, Zhang X, Wang Y (2017a) The molecular architecture for RNA-guided RNA cleavage by Cas13a. Cell 170(4):714–726.e710

Liu L, Li X, Wang J, Wang M, Chen P, Yin M, Li J, Sheng G, Wang Y (2017b) Two distant catalytic sites are responsible for C2c2 RNase activities. Cell 168(1–2):121–134.e112

Ma Y, Yu L, Zhang X, Xin C, Huang S, Bai L, Chen W, Gao R, Li J, Pan S (2018) Highly efficient and precise base editing by engineered dCas9-guide tRNA adenosine deaminase in rats. Cell Discov 4(1):1–4

Makarova KS, Zhang F, Koonin EV (2017) SnapShot: class 2 CRISPR-Cas systems. Cell 168(1):328–328.e321

Makarova KS, Wolf YI, Koonin EV (2018) Classification and nomenclature of CRISPR-Cas systems: where from here? CRISPR J 1(5):325–336

Marraffini LA (2016) The CRISPR-Cas system of streptococcus pyogenes: function and applications. In: Streptococcus pyogenes: basic biology to clinical manifestations [internet]. University of Oklahoma Health Sciences Center, Oklahoma

Martinez J, Patkaniowska A, Urlaub H, Lührmann R, Tuschl T (2002) Single-stranded antisense siRNAs guide target RNA cleavage in RNAi. Cell 110(5):563–574

Meeske AJ, Nakandakari-Higa S, Marraffini LA (2019) Cas13-induced cellular dormancy prevents the rise of CRISPR-resistant bacteriophage. Nature 570(7760):241–245

Mojica FJ, Díez-Villaseñor C, García-Martínez J, Almendros C (2009) Short motif sequences determine the targets of the prokaryotic CRISPR defence system. Microbiology 155(3):733–740

Molla KA, Karmakar S, Islam MT (2020) Wide horizons of CRISPR-Cas-derived Technologies for Basic Biology, agriculture, and medicine. In: CRISPR-Cas methods. Humana, New York, NY, pp 1–23

Montiel-Gonzalez MF, Vallecillo-Viejo I, Yudowski GA, Rosenthal JJ (2013) Correction of mutations within the cystic fibrosis transmembrane conductance regulator by site-directed RNA editing. Proc Natl Acad Sci 110(45):18285–18290

Nakatsuka T, K-i M, Abe Y, Kubota A, Kakizaki Y, Yamamura S, Nishihara M (2008) Flower color modification of gentian plants by RNAi-mediated gene silencing. Plant Biotechnol 25(1):61–68

Nelles DA, Fang MY, Aigner S, Yeo GW (2015) Applications of Cas9 as an RNA-programmed RNA-binding protein. Bioessays 37(7):732–739

Nelles DA, Fang MY, O'Connell MR, Xu JL, Markmiller SJ, Doudna JA, Yeo GW (2016) Programmable RNA tracking in live cells with CRISPR/Cas9. Cell 165(2):488–496

Nethery MA, Barrangou R (2019) Predicting and visualizing features of CRISPR–Cas systems. Methods Enzymol 616:1–25

Nicaise V (2014) Crop immunity against viruses: outcomes and future challenges. Front Plant Sci 5: 660

O'Connell MR (2019) Molecular mechanisms of RNA targeting by Cas13-containing type VI CRISPR-Cas systems. J Mol Biol 431(1):66–87. https://doi.org/10.1016/j.jmb.2018.06.029. Epub 2018 Jun 22. PMID: 29940185

O'Connell MR, Oakes BL, Sternberg SH, East-Seletsky A, Kaplan M, Doudna JA (2014) Programmable RNA recognition and cleavage by CRISPR/Cas9. Nature 516(7530):263–266

O'Geen H, Ren C, Nicolet CM, Perez AA, Halmai J, Le VM, Mackay JP, Farnham PJ, Segal DJ (2017) dCas9-based epigenome editing suggests acquisition of histone methylation is not sufficient for target gene repression. Nucleic Acids Res 45(17):9901–9916

Price AA, Sampson TR, Ratner HK, Grakoui A, Weiss DS (2015) Cas9-mediated targeting of viral RNA in eukaryotic cells. Proc Natl Acad Sci 112(19):6164–6169

Pushparaj P, Aarthi J, Manikandan J, Kumar S (2008) siRNA, miRNA, and shRNA: in vivo applications. J Dent Res 87(11):992–1003

Romay G, Bragard C (2017) Antiviral defenses in plants through genome editing. Front Microbiol 8:47

Rupaimoole R, Slack FJ (2017) MicroRNA therapeutics: towards a new era for the management of cancer and other diseases. Nat Rev Drug Discov 16(3):203

Sampson TR, Weiss DS (2014) Exploiting CRISPR/Cas systems for biotechnology. BioEssays: News Rev Mol Cell Dev Biol 36(1):34–38. https://doi.org/10.1002/bies.201300135

Sampson TR, Saroj SD, Llewellyn AC, Tzeng Y-L, Weiss DS (2013) A CRISPR/Cas system mediates bacterial innate immune evasion and virulence. Nature 497(7448):254–257

Sastry KS, Zitter TA (2014) Plant virus and viroid diseases in the tropics: volume 2: epidemiology and management. Springer Science & Business Media, Berlin

Scotti MM, Swanson MS (2016) RNA mis-splicing in disease. Nat Rev Genet 17(1):19

Shmakov S, Abudayyeh OO, Makarova KS, Wolf YI, Gootenberg JS, Semenova E, Minakhin L, Joung J, Konermann S, Severinov K (2015) Discovery and functional characterization of diverse class 2 CRISPR-Cas systems. Mol Cell 60(3):385–397

Shmakov S, Smargon A, Scott D, Cox D, Pyzocha N, Yan W, Abudayyeh OO, Gootenberg JS, Makarova KS, Wolf YI (2017) Diversity and evolution of class 2 CRISPR–Cas systems. Nat Rev Microbiol 15(3):169

Slaymaker IM, Mesa P, Kellner MJ, Kannan S, Brignole E, Koob J, Feliciano PR, Stella S, Abudayyeh OO, Gootenberg JS (2019) High-resolution structure of Cas13b and biochemical characterization of RNA targeting and cleavage. Cell Rep 26(13):3741–3751.e3745

Smargon AA, Cox DB, Pyzocha NK, Zheng K, Slaymaker IM, Gootenberg JS, Abudayyeh OA, Essletzbichler P, Shmakov S, Makarova KS (2017) Cas13b is a type VI-B CRISPR-associated RNA-guided RNase differentially regulated by accessory proteins Csx27 and Csx28. Mol Cell 65(4):618–630.e617

Staals RH, Agari Y, Maki-Yonekura S, Zhu Y, Taylor DW, Van Duijn E, Barendregt A, Vlot M, Koehorst JJ, Sakamoto K (2013) Structure and activity of the RNA-targeting type III-B CRISPR-Cas complex of Thermus thermophilus. Mol Cell 52(1):135–145

Sugimoto A (2004) High-throughput RNAi in Caenorhabditis elegans: genome-wide screens and functional genomics. Differentiation 72(2–3):81–91

Tambe A, East-Seletsky A, Knott GJ, Doudna JA, O'Connell MR (2018) RNA binding and HEPN-nuclease activation are decoupled in CRISPR-Cas13a. Cell Rep 24(4):1025–1036
Terns MP (2018) CRISPR-based technologies: impact of RNA-targeting systems. Mol Cell 72(3):404–412
Thompson DM, Parker R (2009) Stressing out over tRNA cleavage. Cell 138(2):215–219
Tijsterman M, Plasterk RH (2004) Dicers at RISC: the mechanism of RNAi. Cell 117(1):1–3
Wang H-LV, Chekanova JA (2017) Long noncoding RNAs in plants. In: Long non coding RNA biology. Springer, Berlin, pp 133–154
Wang KH-C, Tseng C-W, Lin H-Y, Chen I-T, Chen Y-H, Chen Y-M, Chen T-Y, Yang H-L (2010) RNAi knock-down of the Litopenaeus vannamei toll gene (LvToll) significantly increases mortality and reduces bacterial clearance after challenge with Vibrio harveyi. Dev Comp Immunol 34(1):49–58
Wang F, Wang L, Zou X, Duan S, Li Z, Deng Z, Luo J, Lee SY, Chen S (2019a) Advances in CRISPR-Cas systems for RNA targeting, tracking and editing. Biotechnol Adv 37(5):708–729
Wang L, Mo CY, Wasserman MR, Rostøl JT, Marraffini LA, Liu S (2019b) Dynamics of Cas10 govern discrimination between self and non-self in type III CRISPR-Cas immunity. Mol Cell 73(2):278–290.e274
Wright AV, Nuñez JK, Doudna JA (2016) Biology and applications of CRISPR systems: harnessing nature's toolbox for genome engineering. Cell 164(1–2):29–44
Xu C, Zhou Y, Xiao Q et al (2020) Novel miniature CRISPR–Cas13 systems from uncultivated microbes effective in degrading SARSCoV-2 sequences and influenza viruses. Res Square. https://doi.org/10.21203/rs.3.rs-30924/v1
Yamada T, Hayamizu Y, Yamamoto Y, Yomogida Y, Izadi-Najafabadi A, Futaba DN, Hata K (2011) A stretchable carbon nanotube strain sensor for human-motion detection. Nat Nanotechnol 6(5):296
Yan WX, Chong S, Zhang H, Makarova KS, Koonin EV, Cheng DR, Scott DA (2018) Cas13d is a compact RNA-targeting type VI CRISPR effector positively modulated by a WYL-domain-containing accessory protein. Mol Cell 70(2):327–339.e325
Yang L-Z, Wang Y, Li S-Q, Yao R-W, Luan P-F, Wu H, Carmichael GG, Chen L-L (2019) Dynamic imaging of RNA in living cells by CRISPR-Cas13 systems. Mol Cell 76(6):981–997.e987
Zetsche B, Gootenberg JS, Abudayyeh OO, Slaymaker IM, Makarova KS, Essletzbichler P, Volz SE, Joung J, Van Der Oost J, Regev A (2015) Cpf1 is a single RNA-guided endonuclease of a class 2 CRISPR-Cas system. Cell 163(3):759–771
Zhang J, Wang C, Ke N, Bliesath J, Chionis J, He QS, Q-x L, Chatterton JE, Wong-Staal F, Zhou D (2007) A more efficient RNAi inducible system for tight regulation of gene expression in mammalian cells and xenograft animals. RNA 13(8):1375–1383
Zhang B, Ye W, Ye Y, Zhou H, Saeed AF, Chen J, Lin J, Perčulija V, Chen Q, Chen C-J (2018) Structural insights into Cas13b-guided CRISPR RNA maturation and recognition. Cell Res 28(12):1198–1201
Zheng Y, Su T, Qi Q (2019) Microbial CRISPRi and CRISPRa systems for metabolic engineering. Biotechnol Bioprocess Eng 24:579–591

CRISPR for Rewriting Genetic Code

8

Zulqurnain Khan, Asim Razzaq, Ayesha Muneer, Maria Siddique, Umar Iqbal, Muhammad Asif Raza, and Tahmina Sattar

Contents

Z. Khan (✉) · A. Razzaq · A. Muneer · M. Siddique · T. Sattar
Institute of Plant Breeding and Biotechnology, Muhammad Nawaz Sharif University of Agriculture, Multan, Pakistan
e-mail: zulqurnain.khan@mnsuam.edu.pk

U. Iqbal
Better Cotton Initiatives (BCI), Lahore, Pakistan

M. A. Raza
Department of Veterinary and Animal Sciences, MNS University of Agriculture, Multan, Pakistan

A. Ahmad et al. (eds.), *The CRISPR/Cas Tool Kit for Genome Editing*,
https://doi.org/10.1007/978-981-16-6305-5_8

Abstract

Rewriting the genetic code has always been a big desire of researchers and scientists working in the field of life sciences. Manipulating DNA sequences can help scientists to study evolution, new phenotypes, and the screening of genetic libraries. The genome editing tools, especially CRISPR, enabled scientists to perform the rewriting of genetic code with an efficiency that was not possible before. The CRISPR system has also revolutionized the field of multiplex genome editing. Before Multiplex Automated Genome Engineering (MAGE), scientists were able to edit one gene at one time. MAGE is an efficient tool for the simultaneous manipulation of multiple genes at multiple loci and changing genetic codes to create genetic variants with different phenotypes. Another emerging tool named CAGE has also empowered the ability of MAGE to perform the rewriting of the genetic codes. Recently a CRISPR-based approach, EVOLVR, has also been introduced for efficient rewriting of genetic code. EVOLVR looks promising to overcome the shortcomings of the previously used techniques for genetic code rewriting. Conjugative Assembly Genome Engineering (CAGE) is important for transferring large chromosomal regions from donor to recipient DNA, subsequently, facilitating in writing large genomes. This chapter will introduce MAGE, CAGE, and EVOLVR and their applications in different fields.

Keywords

Genetic code · CRISPR · MAGE · CAGE · EVOLVR

Abbreviations

AID	Activation-induced deaminase
BET	Bromodomain and extraterminal domain
CAGE	Conjugative assembly genome engineering
CDA	Cytidine deaminase
CRE	*Cis*-regulatory elements
CRISPR	Clustered regularly interspaced short palindromic repeats
DSBs	Double-stranded DNA breaks
DXP	Deoxy xylulose phosphate
EMX	Empty spiracles homeobox
FDA	Food and Drug Administration
gRNA	Guided RNA
HIV	Human immunodeficiency virus
MACBETH	Multiplex automated *Corynebacterium glutamicum* base editing method
MAGE	Multiplex automated genome engineering
MP	Mutagenesis plasmids
PAM	Protospacer adjacent motif

ssDNA	Single-stranded DNA
TALEN	Transcription activator-like effector nucleases
TBD	Thioredoxin-binding domain
TRMR	Trackable multiplex recombineering
tRNA	Transfer RNA
YOGE	Yeast oligo-mediated genome engineering
ZFN	Zinc finger nucleases

8.1 Introduction

Genes, their variation, and their heredity indeed decide the life of an organism. The regulation of genes is a very complex system based on the type of cells. Eukaryotes have a far complex genetic system than prokaryotes (Vosseberg et al. 2021). The genes code for proteins and these proteins then regulate the whole metabolic processes running inside an organism. The expression of these genes is also controlled by other regulatory elements translated from other genes. This whole system codes the life of an organism.

The genetic system of all organisms is still not fully understood by scientists. Scientists always remained keen to somehow manipulate the genetic coding of life to harness its power for the greater good of the whole humanity, However, it was not until the discovery of synthetic biology which made this dream of scientists come true. It helped scientists in manipulating the biology of organisms for making new and relatively improved functions (Liang et al. 2011; Luo et al. 2013; Abil et al. 2015; Stano et al. 2013). This can be done either by the addition or deletion of the genetic material in or from the organism. This creates an impact on the phenotype of the organism, accordingly. The genetic engineering tools enabled the scientists to make the precise genetic modifications in the desired organism, at the desired place, for desired characteristics. This has led to the revolution in industry, agriculture, and medicine (Fiaz et al. 2021).

The rewriting of genetic code using genome engineering tools was being practiced by scientists for a long period (Boeke et al. 2016; Annaluru et al. 2015). However, these modifications were only possible to be made at only one location. Only one locus could have targeted using these genome engineering tools. It leads to the high cost of these protocols and also a reduction in the turnover of these procedures (Carr and Church 2009). They were, undoubtedly, good for the single gene-related functions and diseases or the functions related to single-nucleotide polymorphism. However, they were inefficient and costly for making a desirable functional change that was being controlled by multiple genes or the pleiotropic effect. The multiplex genome engineering provided a solution to this problem and enabled the scientists to perform the complex genome engineering tasks such as disease modeling and cell line development, which was previously impossible to be done (Sun et al. 2012; Alper and Wittmann 2013).

8.1.1 History of Rewriting Genetic Code

Biology and biotechnology have come a long way from having difficulty in reading the sequence of the DNA to manipulating it for the creation of novel and desirable functions. The journey from sequencing to synthetic biology took a long time but now it is evolving at a very fast speed (as shown in Fig. 8.1). The genome is also referred to be the blueprint of life and holds the most important position in the life of an organism. The rewriting of this genetic code having an enormous load of mysterious information has always remained a big goal of scientists. The tale of this dream is quite old and long. The genomes of the less complex organisms such as bacteria and other simple plats were first studied in detail, and the changes were made in them to check for the respective changes in the phenotypes of an organism. This modeling helped scientists to reach the level of synthetic biology where science stands today.

It all started in 1970, when research from the Khorana laboratory reported the synthesis of the first-ever synthetic gene, producing tRNA of alanine from artificial yeast. It was all encoded from deoxyribonucleotides. Another breakthrough after the creation of synthetic gene in 1970 was made in 2002 by the chemical synthesis of poliovirus in the laboratory by the assembly of oligonucleotides in 2002 (Cello et al. 2002). Yet another remarked footstep in the journey of genetic code rewriting was achieved by Chan and his team in 2005. They redesigned the T7 bacteriophage in a way that only its internal structure was manipulated while keeping the external system intact and as it was in the first place. This led to the separation of the genetic material responsible for performing the critical functions of bacteriophage from the genetic material which was not of critical importance for the functionality of the T7 phage (Chan et al. 2005).

In 2005, there was another big potential research performed in synthetic biology, when the scientists succeed in assembling a whole bacterial genome by the direct incorporation of the DNA fragments into the genome of *B. subtilis*. These DNA fragments were generated through PCR. The whole 3.57 Mb genome of

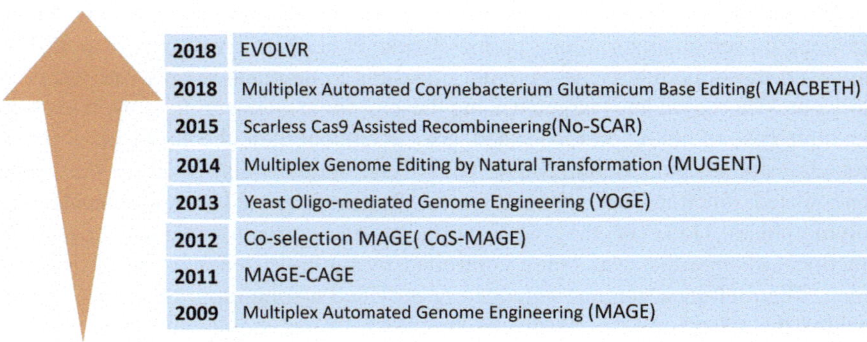

Fig. 8.1 The history of the invention of various multiplex genome rewriting techniques

Synechocystis PCC6803 was cloned in sets by the incorporation of the DNA fragments in a stepwise manner into the genome of *B. subtilis* (Itaya et al. 2005).

The assembling of the whole bacterial genome in 2005 led to another remarkable research that paved the way to achieve the goal of rewriting of genetic code of organisms in 2008. Scientists belonging to the J. Craig Venter Institute synthesized and assembled the whole genome of *M. genitalium* using the oligonucleotides. These oligonucleotides were too chemically synthesized. This complete genome was assembled in the *S. cerevisiae*. Then in 2010, the genome of *M. mycoides* was synthesized and incorporated into the recipient cell of *Mycoplasma*. It resulted in the creation of new *M. mycoides*. Their cells were wholly regulated by the artificially synthesized genome.

From the first synthetic biology experiment in 1970, science has witnessed a rapid evolution in the field of synthetic biology, and even the whole biochemical pathways are now being engineered, and to this date, the construction of a wholly synthetic organism is made possible by synthetic biology scientists. Yet the process of synthetic biology scientists is not fully achieved. They are now aiming at determining the least number of genes required for life. The experiment done in 2005 regarding the T7 bacteriophage has provided a deep insight into where the future of the rewiring of genetic code is destined. The experimentation in simpler organisms will lead to the information of knowing the essential part of the genome and then deleting the nonessential or nonfunctional part of the genome in a stepwise manner (Suzuki et al. 2015). It will be essential to perform it in a stepwise manner to validate that it does not have any drawback or to later troubleshoot the whole process quickly.

8.1.2 The Upper Hand of CRISPR on Other "Genetic Code Writing Techniques"

CRISPR has remained the favorite tool of scientists to perform the rewriting of the genetic code. The designing of the CRISPR construct is simpler than that of ZFN and TALEN. Moreover, it is also more efficient and less costly than other techniques of genetic code rewriting. However, the biggest advantage of CRISPR over the other genetic code rewriting techniques is its potential for performing multiplex genome editing (Zhang et al. 2021). The multiplex genome editing is a much-required tool for performing most of the crucial genetic code rewritings (Dalia et al. 2014), as most of the functions are controlled by multiple genes, which have to be edited via multiplex genome editing for not only saving time and money but also to achieve great efficiency. The easiness of CRISPR to be used as a multiplex genome rewriting tool was quite explicit during the earliest reports of it (Jinek et al. 2013; Wang et al. 2013; Li et al. 2013a, b), and it was further validated by the upcoming reports on gene activation or repression using CRISPR multiplex editing (Gilbert et al. 2013; Qi et al. 2013; Dominguez et al. 2015).

The ability of CRISPR to retarget its genetic modifications using the gRNA makes it a more favorable tool than its competitors to perform the genetic code rewriting (Gaj et al. 2013). Moreover, CRISPR has small-sized gRNA which makes the genome editing performed through it scalable and fast. It also facilitates the parallel screening of gRNA genome-wide libraries (Wang et al. 2014; Shalem et al. 2014; Adamson et al. 2016). Likewise, the small-sized gRNA also enables CRISPR to perform multiplex genome editing through the delivery of multiple gRNA into a cell. All these points validate the superiority of CRISPR over other genome code rewriting tools in terms of design, delivery, and efficiency.

Undoubtedly, multiplex genome editing has great potential for genetic code rewriting and the association of CRISPR further adds to its potential. But multiplex genome editing has some innate drawbacks associated with it, which can affect its potential to effectively achieve the intended results in genetic code rewriting, such as there is a limit to which a cell can tolerate the number of simultaneously made double-stranded breaks in its genome as the double-stranded breaks made by the genome editing tools activate the signaling of the apoptotic process within the cell and also the machinery of DNA damage repair. It can create limitations for the overall applications of multiplex genome editing involving a rigorous network of genes. The limit of DSB breaks made by the multiplex editing is already quite low as the previously published researches shows only four distinct targets in the case of orthologs of Cas9 while only seven targets for SpCas9, which too with an efficiency less than in the case of the single genome edit (Sakuma et al. 2015; Zetsche et al. 2017). However, taking comparatively, the multiplex genetic code rewriting performed through CRISPR is undoubtedly far efficient than that performed by ZFN or TALEN.

The variants of CRISPR Cas protein are described in another chapter of the book. However, this chapter is concentrated on exploring the relatively new avenues of genetic code rewriting through the chimera of CRISPR machinery with other molecules to harness the maximum potential of multiplex genetic code rewriting. The biggest hurdle in achieving the goal of harnessing an efficient genetic code rewriting in cells was the unavailability of automated and high throughput technologies. It costed the precision in not only designing the genetic code rewriting techniques but also in the whole process of genome manipulation. It was not early before the invention of high throughput and automated techniques of genetic code rewriting to overcome this problem. These high throughput and automated techniques include MAGE (Multiplex Automated Genome Engineering) and CAGE (Conjugative Assembly Genome Engineering).

MAGE was developed by the Church Lab in its series of efforts to develop efficient genetic engineering technologies. MAGE involves the oligonucleotides and single-stranded DNA to be introduced into the cell to make multiplex genome editing through invoking the DNA replication. MAGE enables scientists to perform the simultaneous recoding of hundreds of codons, while CAGE involves harnessing of the natural process of reproduction in bacterial conjugation. CAGE incorporates the different genomes from different strains of *E. coli* into a single genome. A detailed discussion on the principle, work, and applications of MAGE and CAGE will be provided in the upcoming sections.

The latest of all aforementioned techniques for CRISPR-based genetic code rewriting is EvolvR. It contains Cas9 nickase (nCas9) delivered along with an error-prone variant of DNA polymerase I. The gRNA in Cas protein finds the target location, which is then nicked by nCas9 and then replicated by the error-prone DNA polymerase and thus creating desired mutations at the desired place. Its detailed mechanism along with applications will be discussed in the section of EvolvR.

8.2 Multiplex Automated Genome Editing (MAGE)

MAGE is a multiplex genome rewriting technique that allows for the efficient and relatively rapid manipulation of the targeted genome. The previously present single gene rewriting techniques were quite inefficient in a way that they are required to perform again and again with a new construct to yield a compete for manipulation of a complex pathway or function (Gallagher et al. 2014). It inspired the scientists at Harvard University to develop a methodology which not only reduces the time and effort on genetic code rewriting but also is an automated process. The mutations generated by MAGE combine with the naturally occurring variations in the cells and collectively generate a big number of mutations within a cell.

MAGE involves a chemical combination of synthetically made single-stranded DNA (ssDNA) and oligonucleotides (Wang et al. 2009a, b). This combination is then introduced into the targeted cells at the targeted areas to create the desired rewriting of the genetic code at those sites. They are introduced into the desired cells using the electroporation method. Once being entered into the cells; they bind to the lagging strand of the replication fork at the targeted genetic code site and replicate with the rest of the process (as shown in Fig. 8.1). The MAGE runs in cycles, with each cycle approximately taking less than 3 h to complete and overall can be completed within some days generating a diverse genetic mutation in the cells (Gallagher et al. 2014).

The development of MAGE and then the continuous addition of innovations into the whole process have quite an interesting history. It started in the late twentieth century when scientists confirmed the ability of phage proteins to effectively substitute the DNA strands in *E. coli* through using the process of homologous recombination (Murphy 1998; Muyrers et al. 1999; Zhang et al. 1998). Moving on in 2001, the single-stranded oligonucleotides were successfully used to make recombination in the target genome, upon complementation with phage λ protein. This founded the base of the development of MAGE (Ellis et al. 2001). In 2009 the invention of MAGE provided an astounding way of high throughput recombination. It was performed in *E. coli* (Wang et al. 2009a, b).

In 2011 a major innovation was made into the whole process of MAGE by combining it with CAGE. MAGE was used to replace TAG codons with TAA codons in about 32 strains of *E. coli* (Isaacs 2011). Then using CAGE, the modified codons were integrated into genomes and eight changes were made. Thus, the different types of genome edit were merged into one organism. It yielded many benefits including the identification of associated phenotypes and the measurement of the frequencies of individual recombination.

MAGE requires an automated system that was once commercially unavailable. It was also relatively costly. This situation inspired the scientist to search for an alternate and it resulted in the discovery of co-selective MAGE which harnesses the procedure of co-selection. It performs the desired high throughput genetic code rewriting which too without requiring automated machinery (Wang et al. 2012). In this process, the combination of oligonucleotides and ssDNA is efficiently recombined using the λ-red protein Bet. The selectable antibiotic markers are introduced into the cells. The desired cells having the mutations can be identified by screening them on various selective media. Thus, the co-selective MAGE involves two-step genetic code rewriting of a genome. In the first step, the mutation is made while in the second step the desired cells are isolated from the rest of the cells.

Advancements in the field of science resulted in more innovations in the process of MAGE. After removing the need for automation through co-selection, the scientist in 2015 also removed the need for chromosomal markers to perform genome rewriting in *E. coli* using MAGE (Reisch and Prather 2015). The concept of recombineering in bacterial genome is quite old (Pines et al. 2003). This process was named "Scarless Cas9 Assisted Recombineering" (no-SCAR). It was named as no SCAR since it does not leave behind any scar site after the genetic code manipulation process. It has also used the Cas 9 with λ-red protein to yield a genetic code rewriting. λ-Red helps in the efficient integration of the DNA into the targeted site. The DSB made by Cas9 was used as the distinctive character to differentiate between the mutated and unmutated cells. Various types of mutations including deletions and insertions were reported to be made by this process.

8.2.1 Ways of Delivery

The way of delivery of a multiplex-based genome editing construct is of critical importance. The innovations being made into MAGE and other multiplex genome editing tools made it possible to increase the gene modification efficiency and decrease the editing toxicity of these tools. Now all of the success of these tools for genome code rewriting depends on the ability to deliver the required number of editing constructs into the targeted cell. Moreover, the delivery method should be capable of, in the case of multiplex editing, delivering the highly complex constructs effectively to the target cell. Besides, it is also natural for the different delivery methods to have their limitations for carrying several distinct genome editing components. Furthermore, the delivery methods will be judged and selected after being tested for various critical parameters including the amount of time, money, and toxicity associated with a particular method.

The simplest approach adopted by the scientists to deliver the multiplex genome editing construct was delivering ribonucleotide protein delivery of expressed Cas9 and gRNA. The delivery was made in two different ways. In the past, the delivery of ribonucleotide protein was done through electroporation (Kim et al. 2014). However, now ribonucleotide protein is delivered through lipid nanoparticles, capable of crossing the membrane to reach the inner surface of cells (Liang et al. 2015; Zuris et al. 2014). However, more recently tissue electroporation-based delivery of multiplex genome editing construct was made successfully (Maresch et al. 2016). This delivery method saves a lot of time and effort as it does not involve the exhausting procedures of g RNA cloning. Besides, due to less half-life of our delivery construct, the problems of toxicity and off-targets are also avoided by this method of multiplex construct delivery, and thus it can prove very useful in genetic code rewriting tasks especially gene knockout.

In 2017, besides Cas9 and gRNA delivery as DNA, the ribonucleoprotein delivery for Cas9-deaminase BE3 was made (Rees et al. 2017). It has a deaminase domain attached to it and thus does not create a double-stranded break at the target location; thus it further reduces the chances of toxicity associated with the genome editing procedures. Thus, the toxicity of the ribonucleotide protein-based delivery is much reduced, but it is still not negligible and especially in the scenario of rapidly performed simultaneous multiplex editing events.

Besides the simple ribonucleotide protein delivery through electroporation or lipoprotein transfection, scientists all over the world are utilizing quite interesting and innovative methods to perform multiplex genome editing. One of them is applying microfluid-based techniques to yield multiplex-based genetic code rewriting. These techniques have a setup with microfluid present in it, constrictions are made, and the diameters of these constrictions are made less than that of the cells which have to pass through them. The high-speed passing of those cells through these constrictions results in the disruption of their membranes and thus making the multiplex genome editing complex containing microfluid enter these cells (Sharei et al. 2013). However, the limitation of this method was that it was making only cytoplasmic delivery. In 2017, Ding and his team used an electric field to disintegrate the nuclear membrane too of the targeted cell, and thus the delivery of multiplex genome rewriting construct through microfluid was made possible not only to be delivered in the cytoplasm but also to the nucleus (Ding et al. 2017).

Nanoparticles are also gaining popularity to be used as a potent delivery source for the multiplex genome editing construct. One such example was discussed by Shalek, which was nanowire-based delivery of the multiplex genome rewriting complex (Shalek et al. 2010). From then an enormous amount of research has been performed in the field of nanotechnology and thus it also aided in the formation of relatively efficient ways of multiplex genome rewriting complex delivery into targeted cells. One such approach was using nanofiber to deliver the construct of ribonucleoprotein size (Saklayen et al. 2017). Undoubtedly, the efficiency of nanoparticle-based delivery is much better than other available methods; however, they have limitations for delivering the large size multiplex genome rewriting constructs.

The use of viral vectors is also an option to be used for the delivery of genetic code rewriting constructs. However, the possibility of unpredictable side effects of their usage and also their incapability to carry larger-sized constructs make them a relatively weak candidate to be used for delivery. Likewise, the nanoparticles also face the same issue of not able to deliver large-sized constructs. However, the continuously made research is expected to make some breakthroughs to not only enhance the capability of viral and nanoparticles but also to use some novel innovative approach towards delivering a multiplex genome editing construct. Nanotechnology is advancing with the speed of light; however, equal enthusiasm and focus are required to be made on the viral-based vectors to increase the potent available options for delivery of multiplex genome rewriting complex efficiently into the targeted place.

8.2.2 Applications

The applications of MAGE look promising to be used in various fields of science for the greater good of humanity. MAGE finds quite interesting applications in this modern era of genetic code rewriting. Previously the focus on the sequencing of organisms was made. But now enough protocols have been established for sequencing and quite remarkable data has been generated from it. Now the science is moving towards the manipulation of this genetic data. Although the manipulation of genomes was in practice for quite a long time, the invention of new techniques such as CRISPR has wholly evolved the way scientists manipulate the genome of organisms. It has provided an opportunity for scientists to perform the multiplex genetic code rewriting in a way to pursue applications that were not possible in the past.

The first-ever used microbe for MAGE was *E. coli* whose genome was manipulated by the scientists to rewrite the biosynthetic pathway of 1-deoxy-D-xylulose-5-phosphate (DXP) in *E. coli* by targeting 24 sites in its genome (Wang et al. 2009a, b). It started the era of automated multiplex genome engineering.

MAGE can be used for successful genetic code replacement with the sired sequence. Isaacs and his team were among the pioneering scientists to yield the power of MAGE to perform multiplex genome rewriting (Isaacs 2011). In 2011, they replaced all TAG codons with TAA codons in the *E. coli* strains and later measured the recombination frequency of the event. They were also the pioneering researchers to use CAGE along with MAGE to incorporate the MAGE-made mutations in a single chimeric genome. This single-codon replacement experiment along with the CAGE proved to be reference research for many upcoming multiplex editing attempts, aimed at rewriting the genetic code of target organisms.

The applications of MAGE can be diversified by using it in complementation with other techniques such as CAGE. The complementation of MAGE with CAGE yielded promising results, and it motivated scientists to make another complementation. This time MAGE was complemented with Trackable Multiplex Recombineering (TRMR). TRMR is a gene expression mapping tool that can predict the genes associated with the desired trait through high throughput screening (Sawitzke 2011). TRMR provided a map for gene expression change on the particular traits (acetate tolerance and cellulosic hydrolysate tolerance) and thus identifying the relevant genes and then MAGE was used to induce combinatorial diversity in these genes (Sandoval 2012).

The cost associated with automated genome multiplex made scientists adopt the altered way. In making such attempts, the multiplex genome engineering was performed in *E. coli* using the co-selection strategy and about eight sites in its genome were targeted successfully (Carr et al. 2012). A similar approach was applied by Wang and his team in 2012 to yield multiplex genome rewriting in *E. coli* by harnessing the co-selection strategy by the addition of phage λ protein with oligonucleotides (Wang et al. 2012).

E. coli was a widely targeted organism for making multiplex genome rewriting. However, in 2017, a great effort was made to achieve the multiplex in *S. cerevisiae*. The scientists also avoided the DSB-related genotoxicity issues by avoiding making them in the targeted genome (Barbieri et al. 2017). They used the synthetic oligonucleotides and annealed them to the lagging strand at the targeted location in the *S. cerevisiae* genome. Combinatorial genetic diversity was successfully observed. It produced a significant change in the beta carotene level in *S. cerevisiae* and thus paved a way for future multiplex genome rewriting attempts in eukaryotes.

In 2018, CRISPR-Cas12a-based multiplex genome editing was used to rewrite the genetic code of *E. coli*. They provided a system to rapidly and simultaneously integrate distinct genomic sequences at distinct places in the genome. Before this breakthrough, the targeting of a single site or targeting multiple sites only for point mutations (Shimatani et al. 2017) was in practice. Phage λ-red protein was used along with Cas12a to enhance the recombination process (Ao et al. 2018).

The CRISPR/Cas9-based multiplex genome editing is no doubt very effective and efficient; however, for some organisms such as bacteria, the double-stranded breaks in the genome bring great havoc in their genetic system and also cause them to die in certain extreme cases (Choi and Lee 2016). Multiplex genome editing is involved in making several DSB at distinct locations in the targeted genome, which can be in the case of bacteria, proved to be lethal. This problem inspired the scientists to bring innovation in the whole process of CRISPR-based multiplex genome editing by replacing the Cas9 with nCas9 and combining it with activation-induced cytidine deaminase (AID) (Wang et al. 2018a, b). The whole process was named as Multiplex Automated *Corynebacterium glutamicum* Base Editing Method (MACBETH). The production of glutamate was enhanced significantly that too without providing a template DNA. Moreover, the biggest achievement was making this whole system automatic by developing a robotic system and thus saving a lot of effort and money. This emphasizes the need and scope of cooperation among different disciplines of science.

The trend of using nCas9-based genome editing to avoid genotoxicity prevailed, and a year back, yet another microbe, *B. subtilis* was prone to nCas9-based multiplex genome editing to confer the modification in the riboflavin biosynthetic pathway by targeting three genes involved in this pathway (Liu et al. 2019). Various kinds of mutations involving point mutations, deletions, and insertions were made, and the remarkable efficiency of more than 80% was achieved while dealing with mutations smaller than 10 kb.

Most recently, *Streptomyces* was targeted to make CRISPR-based multiplex genome editing. A rather complex construct was made to achieve multiplex genome editing. The complex contained dCas9, cytidine deaminase (CDA), LVA protein degradation tag, and uracil glycosylase inhibitor (UGI) and was named as dCas9-CDA-ULstr (Zhao et al. 2020). It proved to be useful for *Streptomyces* for not having a homologous recombination repair. The three different kinds of point mutations were made (single, double, and triple) with the single having the highest while the triple point mutation having the least efficiency.

CRISPR-based multiplex genome editing has established itself for microbes. The model microbe such as *E. coli* has well-established protocols for performing MAGE and scientists are moving to other microbes. Similarly, the multiplex genome ending was performed in plants too. In plants too there remained some favorite crops for scientists such as rice and tomato. But due to having a much complex system than a microbe, the multiplex genome editing in plants and crops did not witness many diverse innovative approaches, as was seen in the case of microbes.

In 2013, Li and his team performed CRISPR-based multiplex genome rewriting in *A. thaliana* (Li et al. 2013a, b). The method of homologous recombination after the double-stranded made by Cas9 protein at two targets in the *A. thaliana* genome yielded multiplex genome editing. Rice remained one of the most favorite crops for scientists to target for CRISPR-based multiplex genome editing. It was due to its ease to perform multiplex genome editing successfully. In 2013, CRISPR type II-based CRISPR was used to rewrite the genetic code of rice. Gene knockout was performed to carry out the targeted mutations (Chen et al. 2013). Likewise, in 2013, another research performed biallelic mutation in rice using the CRISPR base multiplex genome editing (Shan et al. 2013). Moreover, the calculations for off-targets were also performed. In the comprehensive research, the wheat genome was also targeted using CRISPR-based multiplex genome editing. This research established the protocols for CRISPR-based multiplex genome editing of rice and wheat genomes.

A rather efficient toolkit for performing CRISPR-based multiplex genome editing in plant species was provided by Xing and his team (Xing et al. 2014). The favorability of scientists for rice continued and MPK genes in rice were targeted by multiplex genome editing to confer abiotic stress tolerance in it (Moustafa et al. 2014). Likewise, in 2016, CRISPR-based multiplex genome editing was applied to control cotton curl leaf disease (Iqbal et al. 2016). More recently, Shimatani and his group performed multiplex genome editing in rice and tomato by using a chimera of "activation activation-induced cytidine deaminase" with Cas9 (Shimatani et al. 2017). It was used to make point mutations in the targeted species. In tomato, gRNA was used to locate the target and then the substitutions in DNA were made. While in rice, point mutation intended to conferring herbicide resistance was made. More recently in 2018, CRISPR-based multiplex genome editing was used to target six loci in *Solanum pimpinellifolium* to rewrite the genetic code related to yield in it (Zsogon et al. 2018).

The *Arabidopsis*, tomato, and rice were targeted heavily by scientists to perform the CRISPR-based multiplex genome editing. It has much optimized the protocols for performing multiplex genome editing in plant species. However, combined with other high throughput techniques and with advancements in bioinformatics, the multiplex genome editing is hoped to be expanded to other plant species too.

Microbes were easy to perform multiplex genome editing; plants were a bit difficult, but the real difficulty and complication come while performing the multiplex editing in humans. Besides having complex genome and metabolic processes in mammals, the complexity associated with ethics and morality of performing the multiplex editing in humans makes it even harder to achieve. Stem cell-based editing in humans is being performed over a long period and protocols are being optimized and studied closely for their potential side effects.

In 2013, crRNA-based multiplex genome editing was performed in human cells for targeting EMX1 and PAVALB in the human genome (Cong et al. 2013a, b). The results indicated a 118 base pair excision in only 0.01% amplicons in the case of targeting EMX1. The results of this study indicate the amount of difficulty while performing multiplex genome editing in humans. The success rate is far away from even 1%. CRISPR-based multiplex genome engineering was performed in humans' cells by using the conventional Cas9 and gRNA chimera (Mali et al. 2013). Two gRNA were expressed simultaneously to achieve the targeting of the AAVS1 locus in the human genome.

The achievement of efficiency and cost reduction has remained an important goal of scientists performing multiplex genome engineering, as it was shown in the case of the development of co-selection-based MAGE instead of the costly automated MAGE. Likewise, the plasmids which can accommodate a relatively large number of gRNA constructs to carry out multiplex genome editing were in the focus of scientists, and "all-in-one expression vectors" were reported in 2014 to be used in humans to perform specific multiplex genome rewriting tasks (Sakuma et al. 2014).

Likewise, the golden gate assembly was used to accommodate four gRNA constructs in a single lentiviral plasmid (Kabadi et al. 2014). All four plasmids were provided their distinct promoter, under which they showed expression. Four distinct targets were targeted in the human genome and maximum efficiency of 33.3% was achieved. It makes it look more promising than the previously made attempt (Cong et al. 2013a, b). Rather more promising research was performed in the case of targeting Duchenne muscular dystrophy in humans. About 62% of the mutation related to Duchenne muscular dystrophy was rectified using CRISPR type II-based multiplex genetic code rewriting (Ousterout et al. 2015). This research added to the hope of scientists for combating diseases in the human genome using CRISPR-based multiplex genetic code rewriting. Likewise, another attempt was made in the medical field by targeting the hematopoietic stem in humans using CRISPR-based multiplex genome engineering to map the malignancies related to it to aid in future research in making cures of these malignancies (Tothova et al. 2017).

The extent of usage of CRISPR-based multiplex genome editing in humans is likely to expand in the future with the advancements in research. However, the legal, ethical, social, and moral obligations associated with performing CRISPR-based multiplex genetic engineering in humans indicate that we must wait for quite a long time before literally performing it in humans.

The multiplex genetic code rewriting finds application in the synthesis of the whole new synthetic organism, as it was reported in the case of the synthesis of fragments of *E. coli* (Richardson et al. 2017) and *Mycoplasma genitalium* (Gibson et al. 2008a, b, 2010). The scientists are also trying to synthetically make the whole *Saccharomyces cerevisiae*. All these efforts involve a huge application for multiplex genetic engineering.

Yet another interesting application of multiplex genetic engineering can be found in case of the efforts to revive extinct species. The portrays of mammoths have always fascinated human minds for a long period, and the urge to see them walking on the ground has been with *Homo sapiens* for quite a long time. However, the recent advances in multiplex genome engineering have made it possible to pursue this old dream of humans. The approach that will be adopted is the close study of the evolutionary history of these extinct species. The specific mutations and distinctions in their genome will be found that varied them from currently present animals, and then using multiplex genome engineering the genome of animals, it will be tried to revive those species. The extensive sequin data and evolutionary history can be really helpful in achieving this dream. As in the case of the revival of mammoths, some studies can be helpful such as the similarity of the genome of mammoths to an extent of 99.78% with the African elephants at the protein level, and just a difference of 0.6% at the sequence of DNA makes the scientists even more motivated for their aim (Miller et al. 2008). The phylogenetic studies even showed the similarity of mammoths more to the Asian elephants as compared to African elephants (Rohland et al. 2010). This will further reduce the amount of effort to put into chive the revival of mammoth species. However, the real de-extinction of these species is far to achieve and will require extensive data on transcriptomics, posttranslational, and posttranscriptional changes, and extensive use of multiplex genetic engineering at different levels will be required to achieve it.

Speed breeding is relatively a new tool of plant breeding and involves rapid breeding of plant species than the conventional methods (Watson et al. 2018). CRISPR-based multiplex genome engineering has shown the potential for making the breeding process fast and it looks promising in further optimizing the speed breeding process. Another application of CRISPR-based multiplex genetic engineering in plant breeding can be in double haploid production. Double haploid helps in the production of homozygous plants and thus is quite useful (Dwivedi et al. 2015). The multiplex genetic engineering combined with the double haploid production will yield promising results for reducing the time in the breeding process and development of the pure line.

Moreover, the CRISPR-based genome engineering applied to plants can have applications in solving the issues of gee redundancy in plants (Mao et al. 2013). Moreover, CRISPR-based multiplex genome engineering can have applications in combating the abiotic stress in plants. The redesigning of plants genome will help them to combat these stresses better such that the identification of cis-regulatory elements (CRE) involved in stress regulation can be performed through CRISPR-based multiplex genome engineering (Ha 2015). Likewise, multiplex genome engineering can be used to perform the domestication of wild species. An example of it can be the targeting of six loci in a wild species of tomato (*Solanum pimpinellifolium*), through multiplex genome engineering (Zsogon et al. 2018). It resulted in the production of the traits related to domestication such as productivity and yield close to the domesticated tomato species, *S. lycopersicum*.

The genetic code rewriting through multiplex can find many other applications in medical, agriculture, and other fields. The advances made in the automation of multiplex genome editing, in the future, will decide the extent to which multiplex genetic engineering will find its new applications.

8.2.3 Challenges

Undoubtedly, MAGE has enormous potential for performing multiplex genome editing and has a very diverse range of applications in various fields. However, it has some limitations which must be addressed to make it almost perfect. Currently, the biggest drawback of MAGE is that it is useful only for making genome editing in *E. coli*. The short oligonucleotides involved in these techniques are rather inefficient in doing genome rewiring in other species due to their inability to perform homologous recombination. Scientists tried to use a similar method with optimized oligonucleotides to perform genome editing in *Saccharomyces cerevisiae* and named this technique "yeast oligo-mediated genome engineering" (YOGE) (DiCarlo et al. 2013a, b). It provided a way for other scientists to use the basic technique of MAGE to perform genome editing in other species too. Likewise, yet another issue with multiplex genome editing is the decrease in efficacy with an increase in the size of the construct. This limitation severely affects the applications of MAGE. Likewise, MAGE can efficiently introduce only a very small modification, which is only 20 bp, which is too short for targeting the complex traits (Bao et al. 2015). Likewise, MAGE has also been found for performing undesired off-targets in the targeted species (Quintin et al. 2016). All of these limitations have to be addressed to utilize the true potential of MAGE.

8.3 Conjugative Assembly Genome Engineering (CAGE)

CAGE, as indicated from its name, is a genome assembly method that hierarchically assembles the genomes of various distinct strains of *E. coli* into one single genome, making it a chimeric genome. The assembly is made using the process of conjugation, which is the natural process of transfer of genetic material between bacterial species. CAGE thus facilitates the transfer of desired genetic material between bacterial strains on a large scale, which too by avoiding the limitations and drawbacks associated with in vitro manipulation of the genome. The CAGE involves two strains of bacteria, in which one acts a donor while the other acts as the recipient strain. The selectable markers are used in donor and recipient strains to perform the

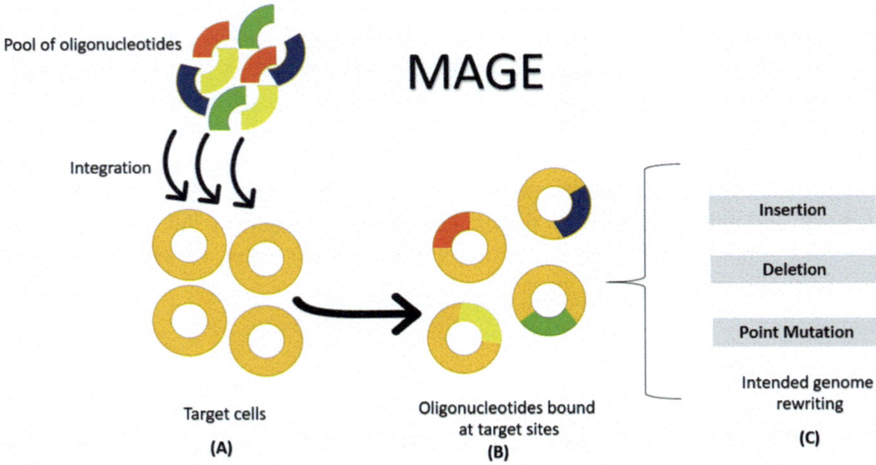

Fig. 8.2 The process of MAGE for genetic code rewriting: (**a**) the target cells with whom the pool of oligonucleotides must have interacted. (**b**) The pool of oligonucleotides gets integrated into targeted cells at the targeted sites. (**c**) The applications of MAGE have been shown

desired transfer of genomic material (As shown in Fig. 8.2). The CAGE is performed in rounds and one round can take up to 1 week, while four rounds, involving the assembly of 16 strains hierarchically, will take one complete month to be completed (Ma 2014).

CAGE is based on the process of conjugation. Conjugation is the significant process of horizontal gene transfer between bacteria while making cell-to-cell contact with other bacteria (Lorenz and Wackernagel 1994). The process is very well studied in the past and significant research has been performed (Ochman et al. 2000; Smith 1991; Lederberg 1946), which ultimately led to the manipulation of this natural process for the genome assembly. The basic process of conjugation is crucial to be studied to properly understand the underlying mechanism of genome manipulation through CAGE. A general description of CAGE workflow is given in Fig. 8.3.

The information for conjugation is found on the F plasmid in bacteria. The bacteria in which this plasmid is found uses conjugation as a method to distribute itself. The proteins involved in the recognition of the recipient strain are found on these plasmids. Moreover, the proteins involved in making cell-to-cell contact between the recipient and donor strain during conjugation and the proteins involved in the conjugative plasmid transfer are also found on a conjugative plasmid (Pansegrau 1994; Lanka and Wilkins 1995).

The donor strain having the conjugative plasmid recognizes the recipient strain lacking the conjugative plasmid through the specified portions, and then it uses its F pilus to attach to the recipient cell, thus making a cell-to-cell contact between both strains. The next step involves the transfer of DNA between both strains. A nick is made in conjugative plasmid at the origin of the transfer site and thus a single strand is formed, which is then transferred to the recipient bacteria through the *F pilus*

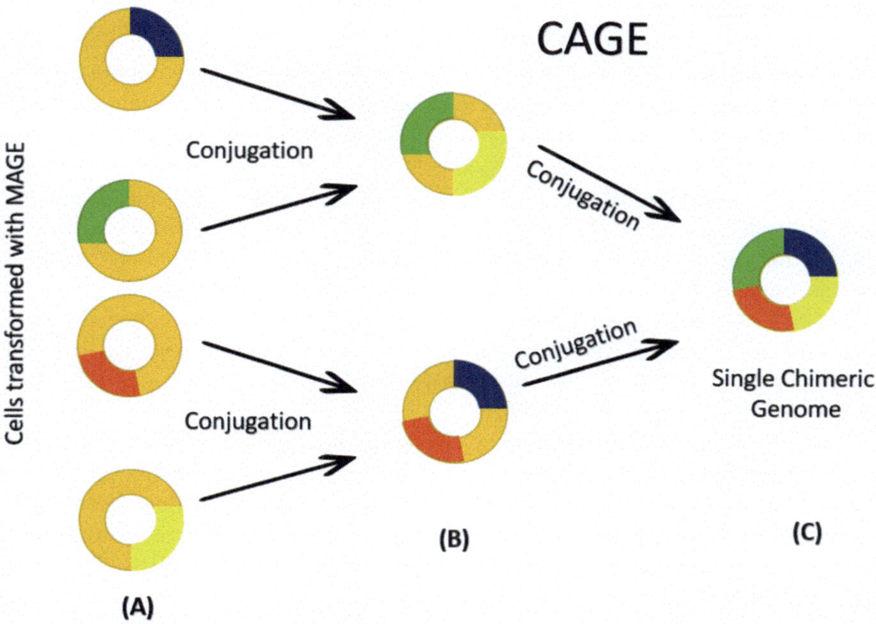

Fig. 8.3 The process of CAGE for genetic code rewriting: (**a**) the cell transformed with MAGE by integration of oligonucleotides in them. (**b**) The cells perform the conjugation, and the targeted genetic material is assembled into one cell. (**c**) The further conjugation of cells in the second cycle makes a single chimeric cell having all the targeted genetic material in it

(Furste and Pansegrau 1989; Guiney and Yakobson 1983). The single-stranded DNA from conjugative plasmid when reached inside the recipient cell replicates and forms its complementary strand. It thus creates a whole conjugative plasmid and itself becomes a donor strain, which will now donate the DN from conjugative plasmid to the other found recipient strain, lacking the conjugative plasmid (Pansegrau 1994; Curtiss 1969). The conjugative plasmids remain independent of the rest of the genome; however, sometimes the conjugative material incorporates into the rest of the genome through the recombination process (Smith 1991; Curtiss 1969).

The process of CAGE has some differences from the naturally occurring conjugation process. Firstly, it makes changes in the genetic information transferred between the bacteria. Secondly, it incorporates selective markers into recipient and donor bacteria to track the assembly of the genome into one chimeric genome. The phage red protein is utilized to isolate the desired portion of the chimeric genome having the targeted portion of the genome from donor and recipient cells (Court et al. 2002; Sharan et al. 2009). The selectable markers also play an important role in the whole process of CAGE because they can influence the fitness of the cells in which they are incorporated (Ma 2014).

8.3.1 Applications

CAGE has been used only in *E. coli* so far to make various kinds of genome editing. The MAGE was complemented with CAGE and the in vivo genetic code rewriting of the *E. coli* genome, not having a TAG codon was performed (Isaacs 2011; Lajoie 2013). In another research, all TAG codons in the *E. coli* genome were converted to TAA by first using MAGE, and then these genetically modified genomic fragments were hierarchically assembled in one genome using CAGE (Lajoie et al. 2012). Another insight from Lajoie research was the deletion of the potentially harmful spontaneous mutation in the E. coli genome using CAGE. Due to the in vivo nature of CAGE, it can be used to perform the testing of the genotypes before they assemble in the chimeric genome. Thus, it can be used to remove any potential harmful mutation present in the genome that can in future produce deleterious effects in the genome of the targeted organism.

CAGE finds many applications in diverse fields of science. For instance, it can be used in the recoding of the genome of an organism. It has some benefits such as the unwanted genome particles that can be removed before making the final assembly. It has proved its worth for genetic code rewriting and recoding in the *E. coli* genome through the appropriate placement of selection markers (Pal et al. 2014; Chari and Church 2017).

In 2014, another successful genome rewriting of *E. coli* was performed using CAGE (Ma 2014). In another research performed in 2016, it converted the TAG codons present in *E. coli* to TAA and a negligible phenotypic effect was observed after performing this genome editing using CAGE (Quintin et al. 2016). Release factor 1 in *E. coli* recognizes UAG and UAA stop codons, while release factor 2 recognizes the UAA and UGA stop codons. The TAG codon conversion to TAA would end the function of release factor 1 in *E. coli*.

The novel traits in the bacterial genome or another genome, in the future, can be produced using the CAGE. It has shown its usability for this purpose by repetitive experiments in *E. coli*. Likewise, it can be used in the biosynthetic pathways redesigning by the raged placement of the genome of the targeted species. The chimeric genome produced through CAGE can be made to have a completely redesigned and rather more desirable biosynthetic pathway for a component of commercial appeal.

8.3.2 Challenges

CAGE has been very promising in genetic code rewriting in *E. coli*; however, it has some drawbacks associated with it, which hinders its scope of application in various fields, such that it is currently limited to *E. coli* only and there are very less chances that it can be used successfully and efficiently in other genomes too. However, the innovative approaches, as they were done in the case of MAGE, can lead to the expansion of this technique to other species too. Moreover, it is not an automated process, unlike MAGE. It takes about a duration of 1 month for making an assembly of 16 *E. coli* genomes. Advances in the future are expected to overcome these drawbacks to ensure an increase in the scope of CAGE usage.

8.4 Targeted Mutagenesis with CRISPR-Guided DNA Polymerase (EvolvR)

EvolvR is yet another and the latest genome rewriting tool used by scientists to yield precise and efficient genome editing. It is CRISPR-based genome rewriting tool and can target the desired loci in the genome by making semi-random mutations in it (Halperin et al. 2018). The Cas9 variant, nCas9, is used in it, which has the advantage of not producing the double-stranded break and thus avoiding the genotoxicity in the cell inside which the genome rewriting process is being performed. The nCas9 forms a single-stranded nick at the targeted place in the target genome. It is coupled with an error-prone variant of DNA polymerase, DNA polymerase I (PolI3M). nCas9 creates the nick while error-prone polymerase carries out the mutation at the targeted place (as shown in Fig. 8.4). The mutation made in the Cas9 variant helps it to decrease its affinity for DNA, while the mutation in the PolI3M increases the error-making rate of this polymerase. Both together can efficiently perform the mutagenesis in the targeted genome at that targeted place. Targeted mutagenesis with advanced techniques has been performed early using advanced techniques such as using chimera of dCas9 with activation-induced deaminase (AID). But a limitation to it was that it was only converting/deaminating cytosine to uracil. It was quite limiting for making the large-scale changes in the desired way and desired genomes. However, EvolvR has overcome this limitation and can efficiently modify all four bases in the genome (Sadanand 2018).

A brief introduction to the working of EvolvR was discussed in the above paragraph. Going into some more details associated with the working of EvolvR, the CRISPR has a good reputation for being easy to program for making genome editing at the desired location in the genome. It commonly uses the gRNA which

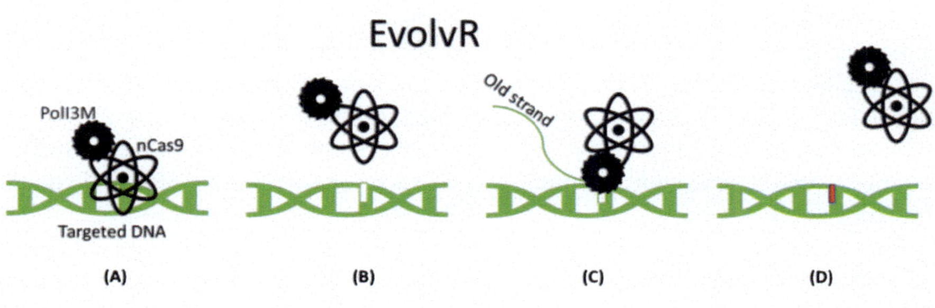

Fig. 8.4 EvolvR mechanism. (**a**) nCas9 binds to targeted location. (**b**) nCas9 generates a nick (in white color) at the targeted location and then dissociates from DNA. (**c**) PolI3M binds at the nicked place and generates a new DNA strand while dissociating the older nicked strand. (**d**) The old strand has been completely dissociated while PolI3M makes a single base mutation (red color) at the targeted place and the whole machinery of EvolvR dissociates from DNA

identifies the target sequence, as it is a complement to the target sequence. PolI3M was used in complementation to the nCas9 and gRNA complex. The PolI3M has three-point mutations in it and makes it more error-prone while incorporating the bases in the target location and also affects the proofreading ability of polymerase.

The process of EvolvR is provided in Fig. 8.4. As can be seen, it is comprised of some basic steps. Firstly, gRNA leads the whole complex of EvolvR to the targeted location in the genome. It binds to the targeted location. nCas9 produces nick in the targeted strand and then it disassociates from the targeted location. Now, the PolI3M binds at the location where nick was created and dissociates the nicked strand, while also making a new strand complementary to the template strand. The old strand completely detaches from the genome, while the new strand made by PolI3M has a mutation in it, which is created because of the point mutations in PolI3M itself. The EvolvR machinery dissociates from the strand and a nicked is left behind. The cell itself repairs that nick. EvolvR can retarget the same loci.

8.4.1 Applications

EvolvR was first used and invented by Halperin in 2018. Three mutations were made in the nCas9. These mutations increase the dissociation rate of nCas9 so that it can quickly dissociate after making nick so that PolI3M can associate with the target DNA. The processivity of PolI3M was increased by complementing it to the thioredoxin-binding domain (TBD), and it significantly improved the processivity of PolI3M. Moreover, two genes were targeted using EvolvR. One was ribosomal protein subunit E and the other was ribosomal protein subunit L. Their targeting through EvolvR made *E. coli* resistant to spectinomycin and streptomycin. More recently, Long and his team used EvolvR for screening ornithine aminotransferase for increasing the production of L-proline (Long et al. 2020).

EvolvR is the latest among all aforementioned genome rewriting techniques; it also seems to be the most promising among all of them for achieving remarkable applications in various science fields. It has better efficiency, precision, and flexibility as compared to other techniques. It represents the power of CRISPR for making genetic code rewriting. It can be readily combined with other technologies since it was seen in a recent case when EvolvR was combined with the Nobel Prize winner-directed evolution technique (Long et al. 2020). It shows that the potential of EvolvR is reengineering the biosynthetic pathway, as the modification of these pathways requires a very huge amount of optimization, which can be currently efficiently done through EvolvR.

EvolvR finds an interesting application in drug development for reducing the susceptibility of the microbes residing inside the human gut. These microbes can be used as a potential source of drugs within humans; however, currently, they face severe environments within the human gut, which makes it impossible to be used as a drug source (Riglar et al. 2019). However, EvolvR can make the required mutations for increasing the survival rate of microbes in the gut. Moreover, EvolvR can have other applications in the human medical field, such that the multiplex ability and processivity of EvolvR can be yielded for making the antibodies that are broadly neutralizing and can be then used against the viruses which produce the unusual type of mutations, such as those produced by HIV (Coleman et al. 2008).

EvolvR can also have application in the reconstruction of embryo development. The embryonic stage is the ideal stage to perform genome editing and EvolvR can be used to induce the semi-random mutations to completely reconstruct the embryo in the desired way. Likewise, the entire biosynthetic pathways can be alerted and reprogramed using simultaneous targeting by EvolvR.

8.4.2 Challenges

EvolvR has provided an exceptional genetic code rewriting technique. However, it has some limitations associated with it, such as it has potential for making the bias in base substitution, as it was seen that majority of the substitution made were of thymine and adenine (Halperin et al. 2018). Moreover, it has not been yet tested in mammals, and it is expected that the high mutation rate associated with EvolvR can cause genotoxicity in mammalian cells despite EvolvR not making the double-stranded breaks in their genome. Moreover, the DNA repair pathways and genetic composition of mammals are very complex (Findlay et al. 2014). They can create a hindrance in achieving successful genome rewriting. Moreover, as CRISPR-based targeting is always dependent on the presence of the Pam sequence, it can affect the applications of EvolvR in some cases. However, the variants of Cas protein may be used to increase the scale of PAM sequences (Timothy and Abbott 2018).

8.5 Comparison of EvolvR with Other Genome Editing Methods

There is no match to EvolvR when it comes to flexibility and accuracy in directed evolution and mutagenesis. It is not only better than MAGE and CAGE, but also other current technologies used for directed evolution and mutagenesis (Hess et al. 2016) including phage-assisted continuous evolution (Esvelt and Carlson 2011), orthogonal polymerase/plasmids (Camps et al. 2003), and mutagenesis plasmids (MP6) (Badran 2015). Moreover, EvolvR is better than CAGE and MAGE in terms of not requiring an extensive amount of exogenous material. A detailed difference and comparison of all techniques are provided in Table 8.1. Moreover, applications of all these technologies are provided in Fig. 8.5.

The size of the tunable window size in EvolvR is 350 bp (Halperin et al. 2018). It proves the usability of EvolvR for making a large number of random mutations at a targeted place Thus EvolvR will be more useful for making not only diverse but large-scale mutations in a targeted genome. It makes it even far useable than base editing (Wolter and Schindele 2019). Moreover, EvolvR is expected to produce comparatively large barcode diversity than other systems (Gaj and Perez 2018).

Table 8.1 Compassion of various genome rewriting tools

			CRISPR		
Features	ZFNs	TALENs	MAGE	CAGE	EVOLVR
Induce double-stranded break	Yes	Yes	No	No	No
Induce single-stranded break	No	No	No	No	Yes
Components	ZFP and Fok1	TALE and Fok1	Oligonucleotides	Plasmids for assembly	nCas9 and Poll3M
Scanning of target	Protein	Protein	DNA	DNA	RNA
Working principle	Blinding with DNA and DSB	Blinding with DNA and DSB	Blinding with DNA and DNA replication	Conjugation	Blinding with DNA and DNA replication
Target site	DNA	DNA	DNA	DNA	DNA
Year of invention	2000	2010	2009	2014	2018
Used in humans	Yes	Yes	Yes	No	No
Used in microbes	Yes	Yes	Yes	Yes	Yes
Diversity of the target	Moderate	High	High	High	High

8.6 Ethical, Legal, and Social Considerations in Rewriting Genetic Codes Using CRISPR

The potential of CRISPR is no doubt hindered and limited by some of its aforementioned limitations. However, the biggest limitation to the application of CRISPR in genome rewriting, especially in animals, plants, and humans, is based on legal and ethical issues surrounding the CRISPR-made genome rewriting. A detailed account of these issues will be provided in the last chapter. However, a brief introduction to these issues has been provided in the relevance of this chapter.

The funding to the CRISPR-based human genome editing has been badly sabotaged by the strict ethical and legal considerations attached to it. Although the experiment-based germline edition has been performed in animals (Campbell and Eichler 2013), in humans too, the embryo-based genome editing has been performed (Zhai and Lie 2016). Moreover, the controversial case of Chinese scientists using CRISPR for developing resistance against HIV was reported last year (Li et al. 2019). But it was not legal, and the scientist had to face the intensive legal

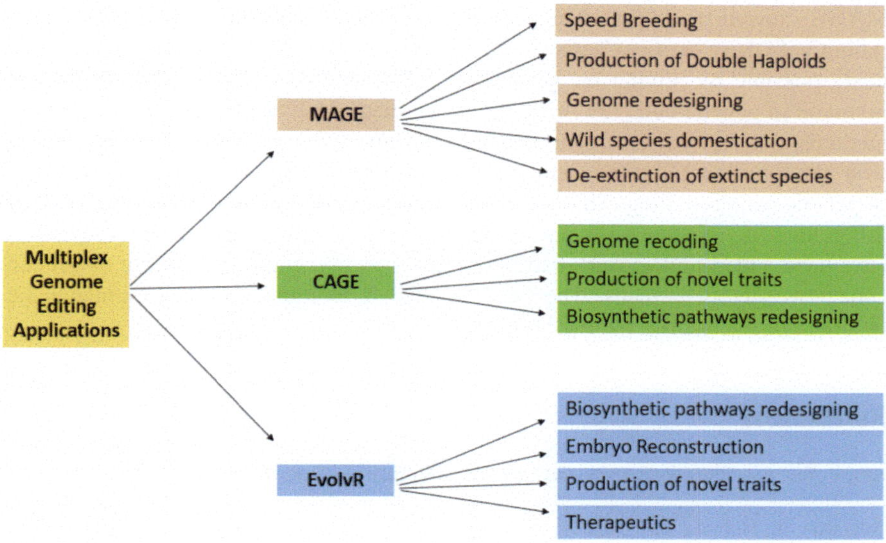

Fig. 8.5 The applications of various multiplex genetic code rewriting techniques

consequences for his actions. The germline-based CRISPR genome editing is not welcomed by the law. However, stem cell-based modifications, having no danger of including in human linage, are approved by the FDA. However, germline-based genome editing remains banned in Europe and the USA (Morrison and Saille 2019).

The genome editing in insects and plants, using CRISPR also faced heavy opposition from public and government sectors. The gene drive of CRISPR was feared to be unintentionally released in the environment by thug insects or the plants which are being genetically modified using CRISPR (Duensing et al. 2018). It will then keep on expanding and spreading in the environment, and in case it has some deleterious effects associated with it, it will lead to an ultimate catastrophe. It makes the base of the argument of anti-CRISPR people. Likewise, the application of CRISPR has some social issues too. A group of people strictly and religiously opposes the usage of CRISPR in making genetically modified organisms. They are afraid of the possible drawbacks of the whole technology. Some GM edible crops are available in the foreign markets; however, the success of the GMO using CRISPR depends on the acceptability of this technique in the eyes of the public, besides the government and regulatory authorities.

8.7 Conclusions

MAGE, CAGE, and EvolvR have undoubtedly a huge potential for achieving genetic code rewriting. However, some of the drawbacks associated with these technologies must be overcome. The large size multiplex genome editing through

MAGE showed fewer promising results. The efficiency of the technique was seen falling with the increase of targets. The large-scale genetic code rewriting cannot be made possible by relying on just factors. The changes in the number of factors involved in genetic code rewriting should be made such as changes in the delivery methods, delivery vehicles, and changes and improvement in donor DNA. Moreover, the significant improvements in viral vectors will also help to achieve the goals of genetic code rewriting. The complexity of the human genome has created a lot of difficulties for scientists to achieve genetic code rewriting in it. The overcautious laws have added unnecessary barriers to the already complex procedure of genetic code rewriting in humans. The need of time is to figure out and eliminate the pointless obstacles in scientific progress.

References

Abil Z, Xiong X, Zhao H (2015) Synthetic biology for therapeutic applications. Mol Pharm 12:322–331

Adamson B, Norman TM, Jost M, Weissman JS (2016) A multiplexed single-cell CRISPR screening platform enables systematic dissection of the unfolded protein response. Cell 167: 1867–1882

Alper HS, Wittmann C (2013) How multiplexed tools and approaches speed up the progress of metabolic engineering. Biotechnol J 8:506–507

Annaluru N, Ramalingam S, Chandrasegaran S (2015) Rewriting the blueprint of life by synthetic genomics and genome engineering. Genome Biol 16:125

Ao X, Yao Y, Li T, Yang TT, Dong X, Zheng ZT, Guo Y (2018) A multiplex genome editing method for *Escherichia coli* based on CRISPR-Cas12a. Front Microbiol 9(2307):1–13

Badran DR (2015) Development of potent in vivo mutagenesis plasmids with broad mutational spectra. Nat Commun 6:8425

Bao Z, Cobb RE, Zhao H (2015) Accelerated genome engineering through multiplexing. Wiley Interdiscip Rev Syst Biol Med 8(1):5–21

Barbieri EM, Muir P, Akhuetie BO, Yellman CM, Isaacs FJ (2017) Precise editing at DNA replication forks enables multiplex genome engineering in eukaryotes. Cell 171(6):1453–1467

Boeke JD, Church G, Hessel A, Kelley NJ, Arkin A, Cai Y, Carlson R (2016) Genome engineering. The genome project-write. Science 353:126–127

Campbell CD, Eichler EE (2013) Properties and rates of germline mutations in humans. TIG 29(10):575–584

Camps M, Naukkarinen J, Johnson BP, Loeb LA (2003) Targeted gene evolution in Escherichia coli using a highly error-prone DNA polymerase I. Proc Natl Acad Sci 100(17):9727–9732

Carr PA, Church GM (2009) Genome engineering. Nat Biotechnol 27(12):1151–1162

Carr PA, Wang HH, Sterling B (2012) Enhanced multiplex genome engineering through co-operative oligonucleotide co-selection. Nucleic Acids Res 40(17):e132

Cello J, Paul AV, Wimmer E (2002) Chemical synthesis of poliovirus cDNA: generation of infectious virus in the absence of natural template. Science 297:1016–1018

Chan LY, Kosuri S, Endy D (2005) Refactoring bacteriophage T7. Mol Syst Biol 1(1):2005–0018

Chari R, Church GM (2017) Beyond editing to writing large genomes. Nat Rev Genet 18(12):749–760

Chen B, Gilbert LA, Cimini BA, Schnitzbauer J, Zhang W, Li G-W, Park J, Blackburn EH, Weissman JS, Qi LS (2013) Dynamic imaging of genomic loci in living human cells by an optimized CRISPR/Cas system. Cell 155:1479–1491

Choi KR, Lee SY (2016) CRISPR technologies for bacterial systems: current achievements and future directions. Biotechnol Adv 34:1180–1209

Coleman JR, Papamichail D, Skiena S, Futcher B, Wimmer E, Mueller S (2008) Virus attenuation by genome-scale changes in codon pair bias. Science 320(5884):1784–1787
Cong L, Ran FA, Cox D, Lin S, Barretto R, Habib N, Hsu PD, Wu X (2013a) Multiplex genome engineering using CRISPR/Cas systems. Science 339:819–823
Cong L, Ran FA, Cox D, Lin SL, Barretto R, Habib N, Hsu PD, Wu XB, Jiang WY, Marraffini LA (2013b) Multiplex genome engineering using CRISPR/ Cas systems. Science 339:819–823
Court DL, Sawitzke JA, Thomason LC (2002) Genetic engineering using homologous recombination. Annu Rev Genet 36:361–388
Curtiss R (1969) Bacterial conjugation. Annu Rev Microbiol 23:69–136
Dalia AB, McDonough E, Camilli A (2014) Multiplex genome editing by natural transformation. Proc Natl Acad Sci 111:8937–8942
DiCarlo JE, Conley AJ, Penttila M, Jantti J, Wang HH, Church GM (2013a) Yeast oligo-mediated genome engineering (YOGE). ACS Synth Biol 2:741–749
DiCarlo JE, Norville JE, Mali P, Rios X, Aach J, Church GM (2013b) Genome engineering in *Saccharomyces cerevisiae* using CRISPR-Cas systems. Nucleic Acids Res 41:4336–4343
Ding X, Stewart MP, Sharei A, Weaver JC, Langer RS, Jensen KF (2017) High-throughput nuclear delivery and rapid expression of DNA via mechanical and electrical cell-membrane disruption. Nat Biomed Eng 1:00039
Dominguez AA, Lim WA, Qi LS (2015) Beyond editing: repurposing CRISPR-Cas9 for precision genome regulation and interrogation. Nat Rev Mol Cell Biol 17:5–15
Duensing N, Sprink T, Parrott WA, Fedorova M, Lema MA, Wolt JD, Bartsch D (2018) Novel features and considerations for ERA and regulation of crops produced by genome editing. Front Bioneng Biotechnol 6:79
Dwivedi SL, Britt AB, Tripathi L, Sharma S, Upadhyaya HD, Ortiz R (2015) Haploids: constraints and opportunities in plant breeding. Biotechnol Adv 33:812
Ellis HM, Yu D, DiTizio T, Court DL (2001) High-efficiency mutagenesis, repair, and engineering of chromosomal DNA using single-stranded oligonucleotides. Proc Natl Acad Sci 98:6742–6746
Esvelt JC, Carlson DR (2011) A system for the continuous directed evolution of biomolecules. Nature 472:499–503
Fiaz S, Khan SA, Ali Noor M, Younas A, Ali H, Ali K, Anis GB (2021) Genome engineering for food security. In: Genome engineering for crop improvement, pp 380–390
Findlay GM, Boyle EA, Hause RJ, Klein JC, Shendure J (2014) Saturation editing of genomic regions by multiplex homology-directed repair. Nature 513:120–123
Furste J, Pansegrau W (1989) Conjugative transfer of promiscuous IncP plasmids: interaction of plasmid encoded products with the transfer origin. Proc Natl Acad Sci U S A 86:1771–1775
Gaj T, Perez P (2018) The continuously evolving CRISPR barcoding toolbox. Genome Biol 19:143
Gaj T, Gersbach CA, Barbas CF (2013) ZFN, TALEN, and CRISPR/Cas-based methods for genome engineering. Trends Biotechnol 31:397–405
Gallagher RR, Li Z, Lewis AO, Isaacs FJ (2014) Rapid editing and evolution of bacterial genomes using libraries of synthetic DNA. Nat Protoc 9(10):2301–2316
Gibson DG, Benders GA, Andrews C, Denisova EA, Baden H, Zaveri J, Stockwell TB (2008a) Complete chemical synthesis, assembly, and cloning of a Mycoplasma genitalium genome. Science 319:1215–1220
Gibson DG, Benders GA, Andrews C, Denisova EA, Baden-Tillson H, Zaveri J (2008b) Complete chemical synthesis, assembly, and cloning of a Mycoplasma genitalium genome. Science 319: 1215–1220
Gibson DG, Glass JI, Lartigue C, Noskov VN, Chuang RY, Algire MA (2010) Creation of a bacterial cell controlled by a chemically synthesized genome. Science 329:52–56
Gilbert LA, Larson MH, Morsu L, Liu Z, Brar GA, Torres SE (2013) CRISPR-mediated modular RNA-guided regulation of transcription in eukaryotes. Cell 154:442–451
Guiney D, Yakobson E (1983) Location and nucleotide sequence of the transfer origin of the broad host range plasmid RK2. Proc Natl Acad Sci 80:3595–3598

Ha CD (2015) The role of cis-regulatory elements in abiotic-stress responses in plants. Tap Chi Sinh Hoc 37(3):370

Halperin SO, Tou CJ, Wong EB, Modavi C, Schaffer DV, Dueber JE (2018) CRISPR-guided DNA polymerases enable diversification of all nucleotides in a tunable window. Nature 560:248

Hess GT, Fresard L, Han K, Lee CH, Li A, Cimprich KA (2016) Directed evolution using dCas9-targeted somatic hypermutation in mammalian cells. Nat Methods 13:1036–1042

Iqbal Z, Sattar MN, Shafiq M (2016) CRISPR/Cas9: a tool to circumscribe cotton leaf curl disease. Front Plant Sci 7:475

Isaacs FJ (2011) Precise manipulation of chromosomes in vivo enables genome-wide codon replacement. Science 333:348–353

Itaya M, Tsuge K, Koizumi M, Fujita K (2005) Combining two genomes in one cell: stable cloning of the Synechocystis PCC6803 genome in the Bacillus subtilis 168 genomes. Proc Natl Acad Sci 102:15971–15976

Jinek M, East A, Cheng A, Lin S, Ma E, Doudna J (2013) RNA-programmed genome editing in human cells. eLife 2:e00471

Kabadi AM, Ousterout DG, Hilton IB, Gersbach CA (2014) Multiplex CRISPR/Cas9-based genome engineering from a single lentiviral vector. Nucleic Acids Res 42:e147

Kim S, Kim D, Cho SW, Kim J (2014) Highly efficient RNA-guided genome editing in human cells via delivery of purified Cas9 ribonucleoproteins. Genome Res 24:1012–1019

Lajoie MJ (2013) Gnomically recoded organisms expand biological functions. Science 342:357–360

Lajoie MJ, Gregg CJ, Mosberg JA, Washington GC, Church GM (2012) Manipulating replisome dynamics to enhance lambda red-mediated multiplex genome engineering. Nucleic Acids Res 42:e170

Lanka E, Wilkins BM (1995) DNA processing reactions in bacterial conjugation. Annu Rev Biochem 64:141–169

Lederberg J (1946) Gene recombination in *Escherichia coli*. Nature 53:673–684

Li JF, Aach J, Norville JE, McCormack M, Zhang D, Bush J, Church GM, Sheen J (2013a) Multiplex and homologous recombination-mediated plant genome editing via guide RNA/Cas9. Nat Biotechnol 31:688–691

Li JF, Norville JE, Aach J, McCormack M, Zhang D, Bush J, Church GM, Sheen J (2013b) Multiplex and homologous recombination-mediated genome editing in Arabidopsis and Nicotiana benthamiana using guide RNA and Cas9. Nat Biotechnol 31:688–691

Li J, Walker S, Zhang X (2019) Experiments that led to the first gene-edited babies: the ethical failings and the urgent need for better governance. J Zhejiang Univ Sci B 20(1):32–38

Liang J, Luo Y, Zhao H (2011) Synthetic biology: putting synthesis into biology. WIREs Syst Biol Med 3:7–20

Liang X, Potter J, Kumar S, Zou Y, Quintanilla R (2015) Rapid and highly efficient mammalian cell engineering via Cas9 protein transfection. J Biotechnol 208:44–53

Liu D, Huang C, Guo J (2019) Development and characterization of a CRISPR/Cas9n-based multiplex genome editing system for Bacillus subtilis. Biotechnol Biofuels 12:197

Long M, Xu M, Qiao Z, Ma Z, Osire T, Yang T, Rao Z (2020) Directed evolution of *ornithine Cyclodeaminase* using an EvolvR-based growth-coupling strategy for efficient biosynthesis of l-proline. ACS Synth Biol 9:1855

Lorenz MG, Wackernagel W (1994) Bacterial gene transfer by natural genetic transformation in the environment. Microbiol Rev 58:563–602

Luo Y, Lee JK, Zhao H (2013) Challenges and opportunities in synthetic biology for chemical engineers. Chem Eng Sci 103:115–119

Ma J (2014) Precise manipulation of bacterial chromosomes by conjugative assembly genome engineering. Nature 9:2285–2300

Mali P, Yang L, Esvelt KM, Aach J, Guell M, DiCarlo JE, Norville JE, Church GM (2013) RNA-guided human genome engineering via Cas9. Science 339:823–826

Mao Y, Zhang H, Xu N, Zhang B, Gou F, Zhu JK (2013) Application of the CRISPR–Cas system for efficient genome engineering in plants. Mol Plant 6(6):2008–2201

Maresch R, Varela I, Unger K, Rad R (2016) Multiplexed pancreatic genome engineering and cancer induction by transfection-based CRISPR/Cas9 delivery in mice. Nat Commun 7:10770

Miller W, Drautz DI, Ratan A, Schuster SC (2008) Sequencing the nuclear genome of the extinct woolly mammoth. Nature 456:387–390

Morrison M, Saille S (2019) CRISPR in context: towards a socially responsible debate on embryo editing. Palgrave Commun 5(1):1–9

Moustafa K, AbuQamar S, Jarrar M, Al-Rajab AJ, Guiller J (2014) MAPK cascades and major abiotic stresses. Plant Cell Rep 33(8):1217–1225

Murphy KC (1998) Use of bacteriophage lambda recombination functions to promote gene replacement in Escherichia coli. J Bacteriol 180:2063

Muyrers JP, Zhang Y, Testa G, Stewart AF (1999) Rapid modification of bacterial artificial chromosomes by ET-recombination. Nucleic Acids Res 27:1555–1557

Ochman H, Lawrence JG, Groisman EA (2000) Lateral gene transfer and the nature of bacterial innovation. Nature 405:299–304

Ousterout DG, Kabadi AM, Thakore PI, Majoros WH, Reddy TE, Gersbach CA (2015) Multiplex CRISPR/Cas9- based genome editing for correction of dystrophin mutations that cause Duchenne muscular dystrophy. Nat Commun 6:6244

Pal C, Papp B, Posfai G (2014) The dawn of evolutionary genome engineering. Nat Rev Genet 15(7):504–512

Pansegrau W (1994) Complete nucleotide sequence of Birmingham IncPα plasmids: compilation and comparative analysis. J Mol Biol 239:623–663

Pines G, Freed EF, Winkler JD, Gill RT (2003) Bacterial recombineering—genome engineering via phage-based homologous recombination. Proc Natl Acad Sci 100:9727–9732

Qi LS, Larson MH, Gilbert LA, Doudna JA, Weissman JS, Arkin AP, Lim WA (2013) Repurposing CRISPR as an RNA-guided platform for sequence-specific control of gene expression. Cell 152: 1173–1183

Quintin M, Ma NJ, Ahmed S, Bhatia S, Lewis A, Isaacs FJ, Densmore D (2016) Merlin: computer-aided oligonucleotide design for large scale genome engineering with MAGE. ACS Synthetic Biol 5(6):452–458

Rees HA, Komor AC, Yeh WH, Caetano J, Warman M, Edge A, Liu DR (2017) Improving the DNA specificity and applicability of base editing through protein engineering and protein delivery. Nat Commun 8:15790

Reisch C, Prather K (2015) The no-SCAR (Scarless Cas9 assisted recombineering) system for genome editing in *Escherichia coli*. Sci Rep 5:15096

Richardson SM, Mitchell LA, Stracquadanio G, Yang K, Dymond JS, DiCarlo JE, Lee D, Bader JS (2017) Design of a synthetic yeast genome. Science 355:1040–1044

Riglar DT, Richmond DL, Trottier L (2019) Bacterial variability in the mammalian gut captured by a single-cell synthetic oscillator. Nat Commun 10:4665

Rohland N, Reich D, Mallick S, Meyer M, Hofreite M (2010) Genomic DNA sequences from mastodon and woolly mammoth reveal deep speciation of forest and savanna elephants. PLoS Biol 8:e1000564

Sadanand S (2018) EvolvR-ing to targeted mutagenesis. Nat Biotechnol 36:819

Saklayen N, Huber M, Madrid M, Nuzzo V, Mazur E (2017) Intracellular delivery using nanosecond-laser excitation of large-area Plasmonic substrates. ACS Nano 11:3671–3680

Sakuma T, Nishikawa A, Kume S, Chayama K, Yamamoto T (2014) Multiplex genome engineering in human cells using all-in-one CRISPR/Cas9 vector system. Sci Rep 4:5400

Sakuma T, Takenaga M, Kawabe Y, Nakamura T, Kamihira M, Yamamoto T (2015) Homologous recombination-independent large gene cassette knock-in in CHO cells using TALEN and MMEJ-directed donor plasmids. Int J Mol Sci 16(10):23849–23866

Sandoval NR (2012) Strategy for directing combinatorial genome engineering in *Escherichia coli*. Proc Natl Acad Sci U S A 109:10540–10545

Sawitzke JA (2011) Probing cellular processes with oligo-mediated recombination and using the knowledge gained to optimize recombineering. J Mol Biol 407:45–59
Shalek AK, Robinson JT, Karp ES, Lee JS, Ahn DR, Yoon MH, Park H (2010) Vertical silicon nanowires as a universal platform for delivering biomolecules into living cells. Proc Natl Acad Sci 107:1870–1875
Shalem O, Sanjana NE, Hartenian E, Shi X, Scott ZF (2014) Genome-scale CRISPR-Cas9 knockout screening in human cells. Science 343:84–87
Shan Q, Wang Y, Li J, Zhang Y, Chen K, Liang Z, Zhang K, Liu J, Xi JJ, Qiu JL (2013) Targeted genome modification of crop plants using a CRISPR-Cas system. Nat Biotechnol 31:686–688
Sharan SK, Thomason LC, Kuznetsov SG, Court DL (2009) Recombineering: a homologous recombination-based method of genetic engineering. Nat Protoc 4:206–223
Sharei A, Zoldan J, Adamo A, Sim WY, Cho N, Jackson E, Mao S (2013) A vector-free microfluidic platform for intracellular delivery. Proc Natl Acad Sci 110:2082–2087
Shimatani Z, Kashojiya S, Takayama M (2017) Targeted base editing in rice and tomato using a CRISPR–Cas9 cytidine deaminase fusion. Nat Biotechnol 35:441–443
Smith GR (1991) Conjugational recombination in *E. coli*: myths and mechanisms. Cell 64:19–27
Stano P, Souza TP, Kuruma Y, Carrara P, Luisi PL (2013) Semi-synthetic minimal cells: biochemical, physical and technological aspects. Elsevier, Amsterdam, pp 261–276
Sun N, Liang J, Abil Z, Zhao H (2012) Optimized TAL effector nucleases (TALENs) for use in treatment of sickle cell disease. Mol BioSyst 8:1255–1263
Suzuki Y, Assad N, Kostylev M, Noskov VN, Wise KS, Karas BJ (2015) Bacterial genome reduction using the progressive clustering of deletions via yeast sexual cycling. Genome Res 25:435–444
Timothy R, Abbott S (2018) Evolution at the cutting edge: CRISPR-mediated directed evolution. Mol Cell 72:402–403
Tothova Z, Krill-Burger JM, Popova KD, Landers CC, Sievers QL, Yudovich D, Ebert BL (2017) Multiplex CRISPR/Cas9-based genome editing in human hematopoietic stem cells models clonal hematopoiesis and myeloid neoplasia. Cell Stem Cell 21(4):547–555
Vosseberg J, Hooff JJ, Houben M, Vlimmeren A, Wijk LM, Gabaldon T (2021) Timing the origin of eukaryotic cellular complexity with ancient duplications. Nat Ecol Evol 5(1):92–100
Wang HH, Isaacs FJ, Carr PA, Sun ZZ, Xu G, Forest CR (2009a) Programming cells by multiplex genome engineering and accelerated evolution. Nature 460:894–898
Wang HH, Farren J, Peter A, George M (2009b) Programming cells by multiplex genome engineering and accelerated evolution. Nature 460:894–898
Wang HH, Kim H, Cong L, Jeong J, Bang D, Church GM (2012) Genome-scale promoter engineering by coselection MAGE. Nat Methods 9:591–593
Wang H, Yang H, Shivalila CS, Dawlaty MM, Cheng AW, Zhang F (2013) One-step generation of mice carrying mutations in multiple genes by CRISPR/Cas-mediated genome engineering. Cell 153:910–918
Wang T, Wei JJ, Sabatini DM, Lander ES (2014) Genetic screens in human cells using the CRISPR-Cas9 system. Science 343:80–84
Wang P, Su L, Gao H, Jiang X, Wu X, Li Y, Zhang Q, Wang Y, Ren F (2018a) Genome-wide characterization of bHLH genes in grape and analysis of their potential relevance to abiotic stress tolerance and secondary metabolite biosynthesis. Front Sci Plant 9:64
Wang Y, Liu Y, Liu J, Guo Y, Fan L (2018b) MACBETH: multiplex automated Corynebacterium glutamicum base editing method. Metab Eng 47:200–210
Watson A, Ghosh S, Williams MJ, Cuddy WS, Simmonds J, Rey M-D, Hatta MAM, Hinchliffe Steed A, Reynolds D (2018) Speed breeding is a powerful tool to accelerate crop research and breeding. Nat Plants 4(1):23
Wolter F, Schindele PH (2019) Plant breeding at the speed of light: the power of CRISPR/Cas to generate directed genetic diversity at multiple sites. BMC Plant Biol 19:176
Xing HL, Dong L, Wang ZP, Zhang HY, Han CY, Liu B, Wang XC (2014) A CRISPR/Cas9 toolkit for multiplex genome editing in plants. BMC Plant Biol 14:327

Zetsche B, Heidenreich M, Mohanraju P, Fedorova I, Kneppers J, DeGennaro EM, Zhang F (2017) Multiplex gene editing by CRISPR–Cpf1 using a single crRNA array. Nat Biotechnol 35(1):31–34

Zhai X, Lie R (2016) No ethical divide between China and the west in human embryo research. Dev World Bioeth 16(2):116–120

Zhang Y, Buchholz F, Muyrers JP, Stewart AF (1998) A new logic for DNA engineering using recombination in Escherichia coli. Nat Genet 20:123–128

Zhang Y, Held MA, Kaur D, Showalter AM (2021) CRISPR-Cas9 multiplex genome editing of the hydroxyproline-O-galactosyltransferase gene family alters arabinogalactan-protein glycosylation and function in Arabidopsis. BMC Plant Biol 21(1):1–16

Zhao Y, Tian J, Zheng G (2020) Multiplex genome editing using a dCas9-cytidine deaminase fusion in Streptomyces. Sci China Life Sci 63:1053–1062

Zsogon A, Cermak T, Naves ER, Notini MM, Edel KH, Weinl S, Peres LEP (2018) De novo domestication of wild tomato using genome editing. Nat Biotechnol 36(12):1211–1216

Zuris JA, Thompson DB, Shu Y, Guilinger JP (2014) Cationic lipid-mediated delivery of proteins enables efficient protein-based genome editing in vitro and in vivo. Nat Biotechnol 33:73–80

Applications of CRISPR/Cas System in Plants

9

Sultan Habibullah Khan, Hooran Tariq, Iqra Farooq, Hira Tasleeem, Muhammad Zubair Ghouri, Muhammad Salman Mubarik, and Zulqurnain Khan

Contents

Abstract

Over decades, success of genetic engineering has been proved by provision of several solutions to the problems related to biotic or abiotic stress, growth, yield, nutrition, and quality of the plants. Value addition and aesthetic improvement

S. H. Khan
Center of Agricultural Biochemistry and Biotechnology (CABB)/US-Pakistan Center for Advanced Studies in Agriculture and Food Security, University of Agriculture, Faisalabad, Pakistan
e-mail: sultan@uaf.edu.pk

H. Tariq · I. Farooq · H. Tasleeem · Z. Khan (✉)
Institute of Plant Breeding and Biotechnology, Muhammad Nawaz Sharif University of Agriculture, Multan, Pakistan

M. Z. Ghouri · M. S. Mubarik
Center for Advanced Studies in Agriculture and Food Security (CAS-AFS), University of Agriculture, Faisalabad, Pakistan

Centre of Agricultural Biochemistry and Biotechnology (CABB), University of Agriculture, Faisalabad, Pakistan

A. Ahmad et al. (eds.), *The CRISPR/Cas Tool Kit for Genome Editing*,
https://doi.org/10.1007/978-981-16-6305-5_9

have also been addressed by genetic engineering techniques. Last decades witness a steeply development and progress in genetic engineering of plants using genome editing tools with more precision and accuracy. CRISPR/Cas, being an advanced genome editing tool, has been used more frequently for this purpose owing to its simplicity, designing, ease in cloning, and high modularity with low cost and high adaptability. Several plant parameters have been modified and improved using CRISPR technology. CRISPR toolbox has a variety of tools for gene editing/modification such as Cas9, Cas12, Cas13, base editors, prime editors, etc. Along with gene modifications, CRISPR/Cas has also been successfully deployed for gene regulation using CRISPRi and CRISPRa and epigenetic modifications using epigenetic modifiers such as LSD, TET1, etc. Moreover, multiplexing feature of CRISPR/Cas9 has given this technology an advantage over all other contemporary gene editing technologies to effectively target genomes. This chapter encompasses applications of CRISPR technology in several plant species for genetic improvement of multiple traits. We describe potential applications of CRISPR system in model as well as horticultural, legumes, and tree species. Finally, we discussed biosafety rules, regulations, and prospects of CRISPR technology in plant genome engineering.

Keywords

CRISPR/Cas · Gene editing · Base editing · Prime editing · Multiplexing · Applications

Abbreviations

AFLP	Amplified fragment length polymorphism
AGPase	ADP-glucose pyrophosphorylase
BAC	Bacterial artificial chromosome
cDNA	Complementary DNA
CL	Coumarate ligase
Cpf1	CRISPR from *Prevotella* and *Francisella*
cr RNA	CRISPR RNA
CRISPR	Clustered regularly interspaced short palindromic repeats
CRISPRa	CRISPR activation
CRISPRi	CRISPR interference
dCas9	Deactivated nuclease
DNA	Deoxyribonucleic acid
DREB	Dehydration-responsive element-binding protein
ERF	Ethylene response factor
GBSS	Granule-bound starch synthase
GenEd	Genome editing
GEOs	Genome edited organisms
GFP	Green fluorescent protein
GM	Genetically modified

GMOs	Genetically modified organisms
gRNA	Guided RNA
HDR	Homology directed repair
KO	Knockout
KRAB	Kruppel-associated box
LSD	Lysergic acid diethylamide
MDH	Malate dehydrogenase
nGM	Novel genetic modification techniques
NHEJ	Non-homologous end joining
PAM	Protospacer adjacent motif
PDS	Phytoene desaturase
PEPC	Phosphoenolpyruvate carboxylase
RGEN	RNA-guided endonuclease
RNA	Ribonucleic acid
RNAi	RNA interference
RNP	Ribonucleoprotein
SgRNA	Single guided RNA
SSNs	Sequence-specific nucleases
TALEN	Transcription activator-like effector nucleases
TCA	Tricarboxylic acid cycle
TILLING	Targeting induced local lesions in genomes
tracrRNA	*Trans*-activating CRISPR RNA
USDA	United States Department of Agriculture
WDV	Wheat dwarf virus
ZFN	Zinc finger nucleases

9.1 Introduction

Genome engineering via designer nucleases has dramatically changed our way of editing the genomes of organisms in a precise and highly well-organized manner. CRISPR/Cas9 system, an RNA-guided endonucleases (RGENs) which are discovered in bacteria, is used to destroy invading viruses. In comparison to ZNFs and TALENs, the CRISPR technology is exceedingly easy to develop and clone. In artificial CRISPR/Cas9 system, the Cas9 protein is drove to the target sequence by a single guided RNA (sgRNA). This single guided RNA is quite easy to design according to the target sequence. This gRNA is generally composed of 20 nt, complementary to the mark DNA sequence following base pairing of Watson-Crick (Cong et al. 2013; Wang et al. 2013). With its high throughput, gene editing with CRISPR and TALENs techniques have also changed agricultural biotechnology. The natural TAL effector proteins have a DNA binding domain and an effector domain. This binding domain can be edited by binding it with a specific sequence of DNA in the targeted genome. In addition, the engineered binding domain can be linked to custom effector domains like nuclease, repressor, or activator to achieve precise DNA manipulation. One prominent feature of CRISPR technology is that it brings DNA, RNA, and protein close together in a programmable and predictable pattern meaning that CRISPR can be used to recruit enhancers and repressors to a specific DNA sequence. This property of CRISPR can be utilized to control expression of gene either through CRISPR-based activation (CRISPRa) or repression (CRISPRi) (interference). We are using different online platforms such as CHOPCHOP, CRISPR-P, MultiTargetor, etc. and reagents provided by Addgene, Vectorbuilder, GeneCopoeia, Home—Nootropics Frontline, etc.

CRISPR has emerged as a useful tool for studying gene function or constructing regulatory networks on a genome-wide scale, while it also provides sequence-specific control of gene expression. CRISPR has recently arisen as the convenient and useful GenEd technique for targeting any DNA sequence in any desired and targeted genome (Cong et al. 2013; Wang et al. 2013). Microbial natural commutable immune system recognizes conquering plasmids and bacteriophages based on RNAs: trancrRNA and crRNA. The CRISPR system optimized for gene targeting consists of a single guided RNA designed to target the genome at the specific DNA sequences (Jinek et al. 2012). Cas9 protein is made of two endonucleases disciplines, RuvC and HNH. The prime requirement of Cas9 is availability of protospacer adjacent motif (PAM) region, and there are different PAM requirements for different types of Cas proteins but generally used Cas9 required 5′NGG3′ along with 20 nt target sequence (sgRNA). Although, Cas9 has also been engineered with altered PAM specificities (Kleinstiver et al. 2015). Apart from Cas9 with nuclease activity for creation of DSBs, another nuclease-deactivated Cas9 version (dCas9) has also been developed and used in bacterial and human cell for targeted DNA binding (Qi et al. 2013). dCas9 can also be utilized for successful transcriptional blockage to suppress gene expression. Moreover, different protein effectors, repressors or activators, can be used to adjust gene expression (Gilbert et al. 2014). CRISPR activator is denoted as CRISPRa while CRISPR repressor is denoted as CRISPRi. Kruppel-associated box (KRAB) repressor domain is fused with dCas9 to result efficient transcriptional suppression/interference (Gilbert et al. 2014; Kearns et al. 2014). Furthermore, multiplexed CRISPRi was employed to modulate endogenous gene expression with no off-target consequences (Gilbert et al. 2013).

In a variety of plants, CRISPR was proven as an ideal way for targeted genome alterations with high ploidy level. Recently, transgenic cotton lines expressing GFP gene were embattled with CRISPR/Cas9 system to confirm targeting efficiency of Cas9 in cotton (Janga et al. 2017). The facility of multiplexing in the case of CRISPR system has evolved into a leading trait of this GenEd tool. Gene targeting with multiple gRNAs has been reported in *Nicotiana benthamiana* for monoclonal antibody production and glyco-engineering (Li et al. 2009). Interventions in the gene editing tools like prime and base editing facilitate the genome editing to overcome the barriers of unclean editing, low efficiency, off-targets mutations, and inefficiency of (HDR) (Zhou et al. 2018). Prime editing holds an eminent potential to improve traits in plants by transitions, transversions, and small indels without double-stranded break requirements. In the mechanism of prime editing, reverse transcriptase fuses with Cas9 nickase to form a complex and allows it to enter in genome along with prime editing guide RNA (pegRNA) having desired edited bases to target the specific sites. Similarly, base editing is another promising implement in which base editors like cytidine deaminase is utilized to transmute purine and pyrimidines with each other (Lu and Zhu 2017).

The progressive gene editing method has allowed plants to improve specific traits by CRISPR-Cas. A novel technique of effective genome editing, containing DNA-free editing in plants, CRISPR from bacteria *Prevotella* and *Francisella* has recently emerged as a new technology with improved productivity, particularly and possibly broader applications than CRISPR. This was named as Cpf1. CRISPR/Cpf1 has been successfully used for genome editing in plants (Kim et al. 2017). Targeted gene engineering is need of the hour for the scientists and public as well. With the increasing world population, there is a need to explore cutting edge and state-of-the-art technologies to increase crop production and protection. Several abiotic stresses have been evolved due to the climate change. To increase crop production with sustainability, genetic improvement in the plants is of prime importance. Precise genetic mutation for crop improvement has been appreciated by researchers and regulatory authorities as well. The USDA has announced that the organism having precise deletion in their genomes created by using GenEd tool will not be regulated as GMOs. So, the future genome editing technology is very bright.

CRISPR system has been applied to edit plant genomes, for instance, *Arabidopsis thaliana,* tomato, rice, wheat, tobacco, maize, potato, soybean, and cotton, for important traits including drought tolerance, cold tolerance, yield increase, virus disease resistance, fungal disease resistance, herbicide resistance, quality improvement, and yield of crops (Fig. 9.1). Additionally, non-GM crops may also be developed utilizing the potential of CRISPR (Kanchiswamy et al. 2015). In the USA, a number of crops have been approved for profitable use (Waltz 2016). Cas9 proteins of the type II system are broadly applied for gene editing in various plants due to its easiness and high modularity (Jinek et al. 2012). Cpf1 is another tool in CRISPR kit which may be used for genome editing in another fashion (Qi 2018; Zaidi et al. 2017; Kleinstiver et al. 2016). Cpf1 has been used to modify the genomes of tobacco and rice (Endo et al. 2016; Begemann et al. 2017; Yin et al. 2017; Hu et al. 2017). However, it was found that with more improvements in Cpf1 like using transcriptional enhancers, codon optimization, strong promoters, and terminators may be proved more beneficial for targeted modification in plant genome (Zaidi et al. 2017). That's how we use CRISPR for the development of living organisms. It is an indispensable need of the society to reduce the disease or to save the crops by using different techniques of CRISPR that we listed above.

9.2 Applications of CRISPR/Cas in Model Plants

Model plants are plant species that have been extensively studied for the ease with which they can be used to investigate specific biological processes or for their value in biotechnology or agronomy. Throughout their lives, plants are subjected to several environmental challenges. Plants have evolved complicated protective responses to a variety of abiotic challenges because of their sessile nature and

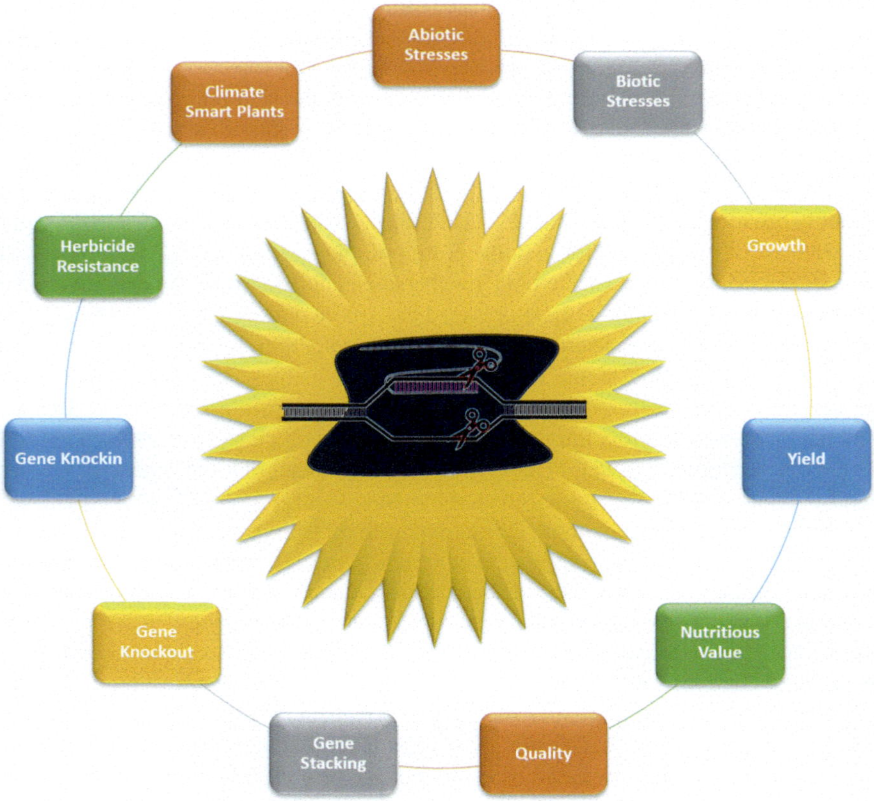

Fig. 9.1 Applications of CRISPR technology for genetic improvement of plants

inability to escape, as explained by Takashi Hirayama and Kazuo Shinozaki. Extensive investigation of *Arabidopsis* responses to infections also yielded essential insights into what Marc Nishimura and Jeff Dangl refer to as the "plant immune system" in their timeline of the field's development. It is beyond a doubt that advances in biotic and abiotic stress research in *Arabidopsis* will result in increased crop plant productivity in the near future (McCourt and Benning 2010).

To knock out the targeted gene single guided RNA, CRISPR-mediated genome editing modifications in tobacco model plants were carried out. *Agrobacterium tumefaciens* was used in gene transformation. The mutated gene targeted Cas9-cleaved 5′ sites in coding areas (Jiang et al. 2017). To observe crossover inverted sequence on homologous chromosomes in *Arabidopsis*, CRISPR/Cas-mediated chromosome engineering was utilised. The rearranged chromosome 4 by crossing Col O harboring with Ler-1 meiotic crossover can be restored into a region with no detectably genetic exchange previously (Schmidt et al. 2020). Some important model plants targeted via CRISPR technology have been summarized in Table 9.1.

Table 9.1 Possible applications of CRISPR system in model plants

Model plant	Technique	Trait/gene	References
Tomato	CRISPR-mediated cis engineering	SlWUS expression	van der Knaap et al. (2014)
Tobacco	CRISPR-Cas/sgRNA via mediated gene editing	Mutated gene	Jiang et al. (2017)
Arabidopsis	CRISPR mediated via chromosome engineering	Crossing over	Schmidt et al. (2020)

9.3 Application of CRISPR Crops

Several applications of genetic engineering techniques have been reported in crops from growth and yield improvement to quality enhancement. Over the past two decades, genetic engineering techniques have been evolving and new tools have emerged which are more sophisticated and predictable. Genome editing has played a vital role in improving crops to feed the masses. Genetic engineering approaches are need of the day to produce crops to meet global demands for food security (Wang et al. 2019a, b). Advance level expertise is now being introduced in this field and several advancements have been made successfully in CRISPR/Cas9, such that Cpf1 could make it easier to employ this technique in wheat development. Genome editing techniques could be utilized in wheat development in the future to address abiotic or biotic stress tolerance, production, condition, and nutritional value, as well as other agronomically intricate variables involving several genes (Kumar et al. 2019).

Recently in a CRISPR with SgRNA-based study, Wang et al. (2016) insert indels near the OsERF922 translation initiation codon in the Japanese rice variety Kuiku13. Six transgene-free homozygous T2 mutant lines with distinct change indels resulted in improved pathogen response toward isolates of *M. oryzae* 06-47-6. Studies have also been conducted to examine other agronomic parameters on edited lines such as seed weight, seed setting rate, flag leaf length, flag leaf width, number of panicles, and plant height. None of the reported characteristics were significantly different from native plants, indicating that regulating OsERF922 can improve plant immune responses to abiotic or biotic stressors without compromising plant breed and cultivate (Wang et al. 2016; Langner et al. 2018). Cas9 was coupled to the NLS and guided RNAs were run by the pol III type promoter of U3 snRNA to reassemble the cas9 ribonucleoprotein complex in the nucleus. For rice expression, the Cas9 coding sequence was codon optimized (Hu et al. 2017).

Wheat is grown as a main food staple crop worldwide (Wang et al. 2014). The *TaMLO* genome (Locus O; Mildew resistance) was effectively targeted using the CRISPR/Cas9-mediated gene modulation method in wheat protoplast. Increased resistance to Powdery mildew resulted by CRISPR-based TaMLO knockdown which is caused by Blumeria graminis f. sp. tritici. T7 endonuclease I restriction analysis was performed on 72 T0 knockout wheat MLO homologs (TaMLO-A) recombinant lines (T7E1) (Wang et al. 2014). The number of transgenic lines acquired can be improved or increased using efficient construct delivery strategies. SSNs and gRNA are typically delivered using T-DNA-based delivery methods. The use of wheat dwarf virus (WDV)-based DNA replicons for transitory and endogenous expression of CRISPR cassette improved gene targeting efficiencies by several orders of magnitude. High-efficiency genome engineering with WDV-based DNA replicons is a new future consideration for high-frequency gene targeting of complicated genomes.

Lee et al. (2018) have reported that wheat protoplasts were edited using CRISPR gene editing method for two abiotic stress-related genes, wheat ethylene responsive factor 3 (WER3) and wheat dehydration responsive element binding protein 2 (TaDREB2). The changed genes' expression was confirmed over 70% of protoplasts which were effectually transfected by revealing T7 endonuclease assay. Off-target mutations and transgene integration are major challenges with the use of CMGE in crops. To address these concerns, researchers devised a successful genome editing approach created on the biolistic delivery of CRISPR ribonucleoproteins (RNPs). In general, CRISPR DNA will be combined with the host genome and expressed in a steady way, whereas the biolistic method of delivering RNPs will give transient expression and be destroyed quickly, decreasing off-targets significantly (Liang et al. 2017). In bread wheat, the CRISPR RNP complex was utilized to edit two separate genes (TaGASR7 and TaGW2) in two different divergent environments. Off-target effects are greatly reduced when this complex is degraded in vivo, and no off-targets were observed in the transformed bread wheat population. A new RNP transport methodology has been made accessible by Ling et al. (2018). This DNA-free editing methodology prevents laborious transgene removal procedures, for example, backcross breeding, and allows the creation of transgenic-free seedlings at T0. However, because the expression is transitory, this technology has constraints, involving lower proficiency levels than CRISPR DNA binary delivery systems, and it also demands time-consuming altered examination with no marker selection during development. The RNP approach will be a feasible option for CRISPR gene editing in agricultural species if these disadvantages can be overcome. Multiplexed genome editing using CRISPR/Cas9 has been shown for crops to alter numerous essential agronomic qualities at the same time. Wang et al. (2018a, b) conducted a multiplexed genome engineering and reported the resulted heritability and mutation frequency in hexaploidy wheat. They studied three wheat genes: TaGW2 (acts as a negative regulator of grain traits), TaLpx-1 (lipoxygenase to confer fighting against *Fusarium graminearum*), and TaMLO (acts as a negative regulator of grain traits) (its knock leads to increased resistance against powdery mildew). They targeted the abovementioned three wheat genes with three gRNAs in combination with tRNA spaced polycistronic cassette under the transcriptional control of a single promoter TaU3. The editing efficiency of wheat protoplasts was assessed by using next-generation sequencing, and also the DNA was analyzed for modulation before mutant screening and *Agrobacterium*-mediated transformation. Then the statistical and phenotypical examination in three sequential generations (T0, T1, T2, and T3) and editing proficiencies of these three

Table 9.2 CRISPR system-based genome engineering in crop plants

Crop	Technique	Trait	References
Rice	Targeted base editing of CRISPR/Cas9 system	NRT1	Lu and Zhu (2017)
	CRISPR cpf1	Multiplex gene editing	Wang et al. (2017)
	CRISPR-via Cas9 system	Knockout of OsNarmp5	Tang et al. (2017)
Wheat	CRISPR/Cas9 system via targeted mutagenesis	Exogenous DsRed gene	Bhowmik et al. (2018)
	Chimeric guided RNAs CRISPR/Cas system	(Inox and pds) genes	Upadhyay et al. (2013)
	CRISPR/Cas ribonucleoprotein (RNP) method	(TaGW2 and TaGASR7)	Ling et al. (2018)
Cotton	CRISPR induced genome editing	Editing mutation in GhCLA1 gene	Gao et al. (2017)
	CRISPR-Cas9 via mediated system	Distinct sites of cotton vacuolar H+ pyrophosphatase (GhVP) genes and cloroplastos alterados 1 (GhCLA1)	Chen et al. (2017)
	Tissue-specific CRISPR ca9 system	sRNA targeting to CLA1, ERA1 and GGB gene	Lei et al. (2021)
Maize	CRISPR/Cas9 system	Pre-assembled cas9-gRNA ribonucleoproteins generated edited alleles	Svitashev et al. (2016)
	PEG-delivered CRISPR ribonucleoprotein technique	The inositol phosphate kinase gene targeting by gRNAs sequences	Sant'Ana et al. (2020)

homologous copies were detected. Transgenerational gene editing activity was found to be a way to bring new genetic recombinations in the progeny of the CRISPR expressing plant involved in this study. This method of multiplex genome engineering in polyploid crops will be effective (Jaganathan et al. 2018). GeneCopoeia is a commercial tool for creating genome-wide sgRNA clones. HDR donor cloning vectors and bespoke HDR donor construction in cas 9 stable lines were produced (Noman et al. 2016). Uses of CRISPR system in crop plant species have been mentioned in Table 9.2.

9.4 Applications in Horticultural Crops

CRISPR/Cas system is utilized to increase vegetable quality. The value of potato thickener is critical for food or also some several technical applications. Using CRSIPR/Cas9 to alter the GBSS gene which produces only amylopectin-containing starch, the "waxy genotype" was generated in hexaploidy potato (Klösgen et al. 1986). Indicating that all four alleles of the GBSS gene had been knocked out, the starch analysis revealed that one of the genetic engineered lines had no amylose and only generated amylopectin. The application of CRISPR/Cas9 in polyploid crops for efficient multi-allelic mutagenesis is established in this study. In potato, the ACETOLACTATE SYNTHASE1 (StALS1) gene was altered to produce multi-allelic mutations similar to those seen in humans (Butler et al. 2016). It has also been used to identify the function of the StMYB44 gene in potatoes (Karkute et al. 2017). Targeted genome editing of sweet orange using Cas9/sgRNA has been demonstrated succefully (Jia and Wang 2015). CRISPR has been successfully deployed for gene

Table 9.3 Findings of GenEd through CRISPR system in horticultural crops

Horticultural crops	Technique	Trait/gene	References
Potato	CRISPR/Cas9	Mutating granule-bound starch synthase gene	Andersson et al. (2018)
	CRISPR/Cas9	Mutating ACETOLACTATE SYNTHASE1 (StaLS1) gene	Butler et al. (2016)
Tomato	CRISPR/Cas9 deletion strategy	Reduced expression of SlOFP20	Ye et al. (2017)
Apple	cDNA-AFLP-related approach	Designated mal-DDNA (DQ417661) for fruit acidity	Yao et al. (2009)
	CRISPR/Cas9 based	Identification of two aluminum-activated malate transporter (ALMT)-like genes	Bai et al. (2012)
Capsicum	CRISPR via mediated point mutation	Base editor changes the motif TTGGC to W-box (TTGAC) resulting in increased expression of MYB31	Li et al. (2020)
Cabbage	Genome editing via CRISPR/Cas9	*BolC.GA4.a*	Lawrenson et al. (2015)
Carrot	Genome editing via CRISPR/Cas9	Flavanone 3-hydroxylase (DcF3H) blockage of the anthocyanin biosynthesis in purple-colored carrot	Klimek Chodacka et al. (2018)
Cucumber	CRISPR/Cas9-mediated genome editing	*elF4E*	Chandrasekaran et al. (2016)
Strawberry	Genome editing via CRISPR/Cas9	*FveTAA1, FvARF8*	Zhou et al. (2018)

editing in horticultural plants ranging from vegetable to fruit plants for several purposes (Zhang et al. 2017). Table 9.3 represents genome engineering findings of CRISPR system in horticultural crops.

9.5 CRISPR/Cas9-Mediated Genome Editing in Legumes

CRISPR technology is undoubtedly one of the most powerful and accurate gene editing techniques ever discovered. In the face of a rapidly growing human population, researchers, breeders, and policymakers must preserve food security. Crop development through genetic recombination or random mutagenesis, on the other side, takes time and can't keep up with the ever-rising food demand. However, CRISPR has opened novel avenues to edit any sequence in genome more efficiently with any targeted desired gene. Furthermore, CRISPR results in the creation of, technically, nongenetically modified plants with desirable characteristics, which can help boost agricultural output in abiotic stress situations (Nadeem et al. 2019).

In a variety of animals, several genetic engineering technologies based on modified nucleases have been created, and they have effectively mutated certain variety of loci in recent years (Shan et al. 2013a, b; Meng et al. 2017). Soybean genome is extensively replicated, posing a significant barrier for traditional genetic methods to use for gene function analysis. Another difficulty is that the effectiveness of *Agrobacterium*-mediated transformation in soybeans is relatively low, and this is modified by tissue and cultivar. As a result, the quality of gRNAs prior to whole-plant transformation is assessed by the broad application of *Agrobacterium rhizogenes*-mediated hairy root transformation, because transgenic hairy roots can be generated in as little as a few weeks. Soybean has a significant importance when it comes to crop for oil and protein. It is the first species from legumes that have been targeted for genetic engineering by the CRISPR system. The CRISPR technique has been used to effectively modify the genome of soybean (Jacobs et al. 2015). CRISPR/Cas was used, by testing the 11 GmU6 promoters to see their potential in achieving gRNA expression in hairy soybean roots, for gene knockout in soybean by disrupting regulatory or noncoding regions resulting in substantial genomic deletions (ranging from 1 to 4.5 kb) induced by CRISPR genes on the same chromosome (Cai et al. 2018). CRISPR has also been used to change seed oil, plant architecture, and flowering timing, among other agronomic features. Resistance to soybean mosaic virus and generation of variants of seed storage protein genes that are beneficial for breeding food type has been achieved by CRISPR technology (Do et al. 2019). Such community efforts to gather mutants of all soybean genes will be crucial for soybean genetics and biotechnology. These research findings, which add to the scientific community's collection of all mutant soybean genes, will be important for soybean genetics and biotechnology.

In 2015, using the CRISPR/Cas9 technology, several organizations claimed effective gene editing function in soybeans (Bai et al. 2020; Guo et al. 2020; Sankar et al. 2015). Following the results, the soybean-scientific community has continued to work to enhance gene editing technologies. GmU6-8 and GmU6-10 promoters have been demonstrated to improve editing efficiency with better activity (20.3% and 20.6%, individually) for gRNA expression in soybean hairy root (Tan et al. 2019). The large deletions in genome could be attained in soybeans by the utilization of dual guided RNAs for cleaving two neighboring loci on the same chromosome found by dual inquiries (Cai et al. 2020c). Vast deletion technique will help researchers understand the role of regulatory elements and noncoding genes as well as ensure that target genes are completely knocked out (Zhao et al. 2018). Egg cell-specific promoters were recently studied, and one of them was shown in stable transgenic lines of soybean to have a high gene editing efficiency of about 26.8% (Zheng et al. 2020).

Agronomic parameters, for example, storage proteins and seed oil, plant architecture, and blooming time, were also studied by scientists to better understand the functioning of gene (Kanazashi et al. 2018). They altered two homologous genes, GmPPD2 and GmPPD1, which encode *Arabidopsis* PEAPOD orthologs with a single gRNA. The subsequent twin mutants highly demonstrated pod features and irregular leaf (Cai et al. 2018). In FLOWERING LOCUS T2a (GmFT2a) generated by CRISPR-mediated mutagenesis, delayed flowering in soybean was discovered in homozygous mutants. The researchers were able by crossing the ft2a mutant with GmFT5a to create ft2aft5a double mutants. Under short-day conditions, the double mutants blossomed 31 days later and generated seeds and more pods than the plants of wild type (Wang et al. 2020). Scientists recently carried out the modification of four GmSPL9 genes and used a CRISPR gene editing technique yielding a variety of soybean mutants by altering branch number and node number on the main stem with different combinations of mutated loci (Bao et al. 2019). CRISPR technology has enhanced seed-related features, for instance, bitter beany taste of soybean seed product (Bai et al. 2020; Cai et al. 2020a, b), seed oil profile (Hall et al. 2019), soybean mosaic virus resistance (Liu et al. 2020a, b), and isoflavone content. Furthermore, the utilization of gRNAs to generate seed storage protein genes mutants will be beneficial in the breeding of food-type soybeans as examined by Wang et al. (2019a, b).

It's worth noting that the 100 candidate genes have been created by the first soybean CRISPR library. Using better techniques, a library of mutant soybean lines was also created (Liu et al. 2020a, b). Such community-wide initiatives to gather mutants of all soybean genes will be critical for soybean biotechnology and genetics (Bhowmik et al. 2021). Because of its nutritional value and nitrogen-fixing symbiosis, one of the most important dietary legumes is chickpea (*C. arietinum L.*). After the common bean, chickpea is the second major food legume. In more than 50 countries, chickpeas are farmed, with an annual report production of 14.24 million tons. In 2013, despite the publishing of the Draught Sequence of the desi-type chickpea genome, efforts to generate new chickpea varieties using genomics and genetic engineering methods remain constrained due to a lack of viable and repeatable plant regeneration mechanisms (Singh and Jain 2015). In the regeneration of transgenic plants containing a chimeric gene expressing the *Bacillus thuringiensis* pesticide crystal protein Cry1Aabc for pod borer resistance, chickpea has mostly been used (Das et al. 2017). Scientists recently developed a stable transformation

Table 9.4 Applications of CRISPR system in legumes

Legume	Technique	Trait	References
Soybean	CRISPR/Cas9	*Tested multiple GmU6 promoters*	Wang et al. (2020)
	CRISPR/Cas9-mediated mutagenesis	*Homologous mutants in flowering LOCUST2a (GmFT2a)*	Cai et al. (2018)
	CRISP-cas9	*Edit two homologous genes by single gRNA (GmPPDI or GmPPD2)*	Kanazashi et al. (2018)
Chickpea	CRISPR-mediated gene editing	*Knockouts of reveille 7 (RVE7) genes and 4-coumarate ligase (4CL)*	Badhan et al. (2021)
	CRISPR genome editing	*For resistance the* $Ala_{251}Thr$ *substitution in psbA chloroplast gene being responsible*	Bhowmik et al. (2021)
Lotus japonicus	CRISPR genome editing	LjSYMRK, LjLbs	Wang et al. (2016)

system for generating stress tolerance (AtBAG4 and TIBAG) which was successfully used to generate stress tolerance (AtBAG4 and TIBAG) and transgenic lines expressing GUS (uidA) (Bhowmik et al. 2019). Gene editing technologies can be utilized to define gene functions and improve agricultural qualities in chickpea due to the accessibility of transcriptome sequences and reference genome alongside transformation processes. The first research using CRISPR-mediated engineering of chickpea protoplasts was newly printed, documenting the creation of knockouts of Reveille 7 (RVE7) genes and the 4-coumarate ligase (4CL), both of which are associated with chickpea drought tolerance (Panchamoorthy and Kannan 2021). Researches have been established for providing a scientific basis for future quality innovation and development in the viability of gene editing in chickpea.

The focus of future gene editing efforts in chickpea would be herbicide resistance and increased carotenoids content. Many post-emerging herbicides are used to suppress the broadleaf weeds in lentil and peas are harmful to chickpea. In chickpea and other legumes, the detection of the Ala251Thr mutation in the psbA chloroplast gene as the reason of other legumes and herbicide resistance in chickpea using genome engineering has created a new path for producing herbicide resistance (McMurray et al. 2019). CRISPR is a capable technique for crop improvement and functional genomics. The CRISPR/Cas technology was utilized to investigate the role of NFR genes in peanut hairy roots (Shu et al. 2020). To date, CRISPR applications have been represented in Table 9.4.

9.6 CRISPR-Based Editing in Tree Plants

Owing to the CRISPR system's simplicity, versatility, and species independence, researchers can now achieve previously unthinkable levels of control and precision over genetic alterations. The ability of CRISPR technology to induce useless mutations in the earliest production has been established in fruit, nut trees and forest (Fernandez i Marti and Dodd 2018; Parsons and Mackay 2018). Traditional breeding takes time, especially in woody plants, and preexisting genetic variability limits the development of novel plant traits. Genome editing (GE) using CRISPR/Cas technology allows quick, easy, and targeted alterations in the genomes of plants, while it has also been found equally effective for the introduction of new genes. The findings point to attractive potential for CRISPR-induced mutations and related traits persisting beyond multiple clonal generations for the lucrative production of high-valued trees reliant on the vegetative mode of reproduction. The crossing of trees with high genome heterozygosity is an opportunity and a challenge for genome editing because CRISPR editing on end location can translate sequence polymorphisms unproductive, but also a challenge because CRISPR's potential and specificity can be used to edit alone.

Forest trees are a vital commodity, providing the global population with fiber, energy, materials, and climate buffers and CRISPR can further enhance this key feature (Cai et al. 2018; Soyk et al. 2017). The current revolution of genome editing is being driven by CRISPR's sensitivity and efficacy for targeted DNA modifications, as well as its ease of adoption in practically any species. Making the system even more versatile for the moment, CRISPR's expanding admiration is lashing the invention and classification of new CRISPR-related endonucleases with advance properties (Murovec et al. 2017; Yubing et al. 2019). CRISPR has been successfully employed in targeting genes in tropical trees and grape (Ren et al. 2016; Lin et al. 2016) and in *Parasponia andersonii* for putative development and biosynthetic pathway (Van Zeijl et al. 2018). The latest CRISPR elements were produced to increase the genome editing characteristics. The utilization of other CRISPR/Cas systems, for example, SaCas9, could enhance the number of possible target guideline RNA (gRNA), especially in areas with a wealth of ATs, which may facilitate the editing of the promoter. In the evaluation of CRISPR in new research systems, PDS (phytoene desaturase) has been a prominent indicator. It interferes with the production of chlorophyll, enabling a visual knockout evaluation. Cassava (Odipio et al. 2017), apple (Nishitani et al. 2016), grape (Nakajima et al. 2017), coffee, and kiwifruit (Sanchez-Vega et al. 2018; Elorriaga et al. 2018) have been genetically modified using CRISPR technology. CRISPR's first stably altered tree, poplar, has achieved the most progress in woody species to date (Schwarz et al. 2015). Bioinformatics resources based on the poplar system are available to help genome editing in heterozygous species (Xue and Tsai 2015; Zhou et al. 2015). Critical flowering genes may be successfully altered in both female and male poplar genotypes, according to a recent study (Bewg et al. 2018). The researchers assembled a huge alteration dataset (Elorriaga et al. 2018) drawn from over 500 transgenic events, which should help researchers better understand CRISPR editing patterns which might aid in understanding CRISPR/Cas editing patterns. Several forest tree species have now effectively used genetic transformation systems to enhance features, for example, salt tolerance (Nishitani et al. 2016), architecture of tree (Busov et al. 2003), lignin content (Chaw et al. 2019), and abiotic or biotic stress response (Bewg et al. 2018).

Like findings from other plants and mammals, small alterations like insertions and deletions are the most ordinary mending consequence of gRNA-mediated cleavage in the genome of trees via Cas9, with 1 bp additions (+1), mainly +T or +A, prevailing in many instances (Bewg et al. 2018). According to reported mutation patterns, many repair processes are implicated differently in *Populus* and other tree species, with cNHEJ apparently contributing to minor (1–4 bp) deletions (Fan et al. 2015). The fluctuating reliance of these paths on sequence perspectives found in diverse studies is most likely to blame for CRISPR/non-random Cas9's nature (Bewg et al. 2018). Smaller size frameshift insertions and deletions are the most prevalent repair consequence of single guided RNA-guided Cas9 cleavage in trees, with 1 bp insertions (+1) predominating in most incidents, related to research in other plants and mammals (Bewg et al. 2018). According to published mutation patterns, many repair processes are engaged differently in *Populus* and other tree species, with cNHEJ apparently contributing to +1, +2, and MMEJ (and SSA) to bigger deletions, minor (1–4 bp) deletions, and TMEJ to complicated indels. The fluctuating dependence of these paths on sequence perspectives is likely to explain the non-accidental kind of CRISPR mending results observed in several experiments which included trees (Bewg et al. 2018). Because of their extended generation cycles and strict restrictions on flowering transgenic plants, cross-generational screening for transgenic trees is difficult (Dalla Costa et al. 2017). The regeneration of protoplasts in other tree species, on the other hand, remains a difficulty.

CRISPR is a revolutionary technique that has been used to modify features in grains, vegetables, and even fruit trees. There is a tremendous motivation to defeat the hurdle because lowering the negative impressions and imprint foreign DNA will boost the prospects of merging CRISPR technology with profitable positioning of stylish trees. Although editing is commonly followed by the transfer of the bacterial transfer DNA into the host gemini, *Agrobacterium tumefaciens*-mediated gene transfer remains the most prevalent method for delivering CRISPR/Cas9 components into dicotyledonous plants. This research has revealed that while creating effective binary vectors to reduce the amount of foreign DNA in CRISPR fruit plants, certain aspects must be addressed. CRISPR/Cas GE has now been able to offer a new variation in woody and leafy plants (Cao et al. 2005). Some of the examples of CRISPR/Cas have been summarized in Table 9.5.

9.7 Biosafety and CRISPR-Edited Plants

Genome editing, genome engineering, or gene modifications are terms used to describe changes (deletions, substitutions, and insertions) made to a living organism's genome. CRISPR and its related Cas9 protein are a widely utilized genome editing approach nowadays (CRISPR-Cas9). The CRISPR/Cas system defends cells from DNA viral infection in prokaryotes. CRISPR has an advanced tool that can be used in health, agriculture, and basic gene function research. CRISPR was used to introduce or improve tolerance to abiotic and biotic stresses, also to make better yield, quality, and nutritional value in an increasing number of

Table 9.5 Genome engineering in tree species via CRISPR/Cas9 system

Tree plants	Technique	Trait	References
Forest trees	CRISPR-associated Cas endonucleases	Gene modification	Yubing et al. (2019)
Tropical tress	SaCas9 system	(gRNA) target sites, particularly in AT rich areas	Van Zeijl et al. (2018)
Rubber trees	A Cas9/sgRNA system	Targeted mutations by Cas9 proteins in vitro	Strauss et al. (2015); Fan et al. (2020)
Woody trees	On/off targeting CRISPR editing	Direct transport of pre-assembled Cas9-gRNA ribonucleoproteins	Dalla Costa et al. (2020)

dicot and monocot species of plant, among other things. While there are still concerns about biosafety, gene editing is a gifted skill that could help with food supply for the world's growing population (El-Mounadi et al. 2020).

Plant genomes could be edited while circumventing national biosafety restrictions, thanks to a novel variation on a pioneering gene editing technology. Plant scientists have jumped on board with the well-known CRISPR technology, which uses Cas9 to cut exact segments of DNA in a genome while being guided by two RNA strands. Researchers hope to deactivate specific genes in wheat and rice to generate disease-resistant variants (Cyranoski 2015). The prospect that novel genetic modification techniques in plant development would have unavoidable detrimental effects on environment and on the human health is critical in the argument over their regulation (Lemgo et al. 2013).

For most nGMs, existing expertise to cope with this issue is insufficient, especially for newly discovered nGMs like genome editing and its variants, such as base editing. As a result, there are concerns about the risk and safety of plants created utilizing various nGMs, such as genome editing and also other nGMs including trans-grafting. Some genome editing technologies' limited level of genomic alteration and increased efficiency of targeting, i.e., precision, is not considered a guarantee of safety. It is especially true when it comes to the unique traits created because of such modifications. All the nGMs we've looked at so far have the potential to induce unintended changes of various kinds or frequencies. On the other hand, the quick advancement of nGM plants may jeopardize the discovery and removal of undesirable consequences (Eckerstorfer et al. 2019a, b).

nGMs and genome editing, which differ from traditional breeding procedures and genetic engineering techniques, are applied to create features of plants and a mixture of characteristics ideally used in agricultural (Eckerstorfer et al. 2019a, b). For directed genetic or random alterations at specific genomic zones, CRISPR technologies and other approaches based on site-directed nucleases have been developed (Fauser et al. 2014). To repair the break in double strand, the cellular DNA repair mechanism involves NHEJ or HDR. During the technique DNA recombination substitutions, deletions, and insertion may happen (Puchta et al. 1996). Three types of programmable endonucleases are utilized in plant genetic engineering: ZFN, CRISPR-Cas9, and TALENs (Malzahn et al. 2017).

9.7.1 Biosafety Concerns

Methodological, biosafety, and sociological issues still abound when it comes to using genome editing in plants. Most of them are focused on targeting the selection of gene site, designing of guide RNA, delivery method, and effects of off-targeting. The main worry is the risk of off-target alterations generating unexpected genetic variations in plants (Gómez-Pineda and Gómez-Pineda 2019). CRISPR fragments could be further fragmented into filler DNA and can be then inserted into expected or unexpected positions in genome during the process of DNA repair (Gorbunova and Levy 1997). Transgene integration and mutations via off-targeting can be avoided by in vitro delivering pre-assembled ribonucleoproteins of CRISPR (Malnoy et al. 2016). Despite the fact that this procedure used in a range of crop species, it even has several application issues, for example, high pricing, limited stability, and high requirements for a technical staff, all of which must be solved (Murovec et al. 2018).

Enhancing ribonucleoprotein, RNA guide-design techniques, engineering of protein, spatiotemporally regulating gRNAs and Cas9 through a variety of environmental or chemical inducers, or employing artificial genetic circuits that can adjust CRISPR action according to established logic have all been tried to reduce Cas9's off-targeting (Svitashev et al. 2016). More concerns concerning CRISPR stem from the protein Cas9 itself, which has been shown to cause an immunogenic response in mice; it has ringed alarm bells for potential immunogenic side effects associated with CRISPPR (Chew et al. 2016). The lack of understanding of the principles and applications of genome editing adds to societal concerns. A critical feature here is the distinction between transgenic plants, genome edited plants, and genetically modified plants (Eckerstorfer et al. 2019a, b). Plants using genome editing could be transgenic or not. The transgenic harboring the CRISPR cassette could be deleted if genes are segregated. If this suggestion is being acted upon, a genome-edited plant could be then considered as a non-transgenic entity. Misconceptions concerning genome editing principles can be corrected and avoided if the public understands them (Eckerstorfer et al. 2019a, b).

9.8 Future Prospects of CRISPR/Cas in Plant Genome Engineering

A novel breeding technique called genome editing allows for concentrated or directional breeding. Crop attributes such as nutritional value, stress tolerance, yield, and insect and herbicide resistance have all been improved using the CRISPR/Cas9 system. Multiplex genome editing will have a significant impression on agricultural plants' ability to improve complex agronomic features cheaply. Genome editing, with its low cost, precision, and speed, being used on an always-increasing variety of plant species, also offers an unrivalled potential for plant breeding. The success of CRISPR editing and control the effect of CRISPR editing are predicted by the future models (Hess et al. 2017; Sakuma and Yamamoto 2017). CRISPR-induced mutations are now allowing researchers to make precise edits without using knock-ins (Wilson et al. 2018). Bioinformatic methods for guided RNA strategy can be used to improve productivity and eliminate off-target effects. More CRISPR datasets are needed to produce new design tools because they rely on activity prediction models and off-target identification techniques. The correct delivery or packaging of the CRISPR composite to the targeted plant cells is a major barrier to CRISPR technology adoption in agriculture. In order to attain high-proficiency genetic engineering in plants, new delivery systems must be developed, as a result, developing smaller Cas9 proteins or progress in falling the size of current Cas proteins (Afzal et al. 2020).

Regarding regulation and acceptance of genome-edited organisms (GEOs), the USDA has announced no regulation for EN-based precise deletions in the genome. This development has been found very encouraging for the scientists which are working in this field. Many crop plants and animals have been targeted with ENs or artificial DNA binding proteins and promising results have been found. The variety is increasing in the GenEd toolbox which further broadens the scope and applications of genome editing. Suppression of gene at DNA level by creating deletions/insertions in the target DNA has been proved more fascinating than previously developed technologies such as RNAi. Mutations created by GenEd tools are more precise, specific, and efficient with predictable results, while other techniques, e.g., RNAi, TILLING, and use of other mutagens do not have these features. Moreover, tunable and remote control regulation of gene expression has become possible using GenEd tools. Expression of indigenous genes can be regulated efficiently using TALEs, ZFs, and dCas alone or fused with effector domains. Researchers have also found that after transformation of GenEd reagents, further generations of the transgenic plants can be produced free of these proteins through segregation. So, these tools can also be used for production of transgene free plants and for clean gene technology as well. Researchers and scientists working in the field of genome editing are very enthusiastic or optimistic about the bright feature of this field. All fields of biology are now using these tools to produce desirable genetic improvements in plants and animals.

References

Afzal S, Sirohi P, Singh NK (2020) A review of CRISPR associated genome engineering: application, advances and future prospects of genome targeting tool for crop improvement. Biotechnol Lett 42(9):1611–1632. https://doi.org/10.1007/s10529-020-02950-w

Andersson M, Turesson H, Olsson N, Fält A-S, Ohlsson P, Gonzalez MN, Samuelsson M, Hofvander P (2018) Genome editing in potato via CRISPR-Cas9 ribonucleoprotein delivery. Physiol Plant 164(4):378–384. https://doi.org/10.1111/ppl.12731

Badhan S, Ball AS, Mantri N (2021) First report of CRISPR/Cas9 mediated DNA-free editing of 4CL and RVE7 genes in chickpea protoplasts. Int J Mol Sci 22(1):396

Bai Y, Dougherty L, Li M, Fazio G, Cheng L, Xu K (2012) A natural mutation-led truncation in one of the two aluminum-activated malate transporter-like genes at the Ma locus is associated with low fruit acidity in apple. Mol Genet Genomics 287(8):663–678. https://doi.org/10.1007/s00438-012-0707-7

Bai M, Yuan J, Kuang H, Gong P, Li S, Zhang Z, Liu B, Sun J, Yang M, Yang L, Wang D, Song S, Guan Y (2020) Generation of a multiplex mutagenesis population via pooled CRISPR-Cas9 in soya bean. Plant Biotechnol J 18(3):721–731. https://doi.org/10.1111/pbi.13239

Bao A, Chen H, Chen L, Chen S, Hao Q, Guo W, Qiu D, Shan Z, Yang Z, Yuan S (2019) CRISPR/Cas9-mediated targeted mutagenesis of GmSPL9 genes alters plant architecture in soybean. BMC Plant Biol 19(1):1–12

Begemann MB, Gray BN, January E, Gordon GC, He Y, Liu H, Wu X, Brutnell TP, Mockler TC, Oufattole M (2017) Precise insertion and guided editing of higher plant genomes using Cpf1 CRISPR nucleases. Sci Rep 7(1):1–6

Bewg WP, Ci D, Tsai CJ (2018) Genome editing in trees: from multiple repair pathways to long-term stability. Front Plant Sci 9:1732

Bhowmik P, Ellison E, Polley B, Bollina V, Kulkarni M, Ghanbarnia K, Song H, Gao C, Voytas DF, Kagale S (2018) Targeted mutagenesis in wheat microspores using CRISPR/Cas9. Sci Rep 8(1):1–0

Bhowmik SS, Cheng AY, Long H, Tan GZH, Hoang TML, Karbaschi MR, Williams B, Higgins TJV, Mundree SG (2019) Robust genetic transformation system to obtain non-chimeric transgenic chickpea. Front Plant Sci 10:524

Bhowmik P, Konkin D, Polowick P, Hodgins CL, Subedi M, Xiang D, Yu B, Patterson N, Rajagopalan N, Babic V, Ro DK (2021) CRISPR/Cas9 gene editing in legume crops: opportunities and challenges. Legum Sci 3:e96

Busov VB, Meilan R, Pearce DW, Ma C, Rood SB, Strauss SH (2003) Activation tagging of a dominant gibberellin catabolism gene (GA 2-oxidase) from poplar that regulates tree stature. Plant Physiol 132(3):1283–1291

Butler NM, Baltes NJ, Voytas DF, Douches DS (2016) Geminivirus-mediated genome editing in potato (Solanum tuberosum L) using sequence-specific nucleases. Front Plant Sci 7:1045. https://doi.org/10.3389/fpls.2016.01045

Cai Y, Chen L, Sun S, Wu C, Yao W, Jiang B, Han T, Hou W (2018) CRISPR/Cas9-mediated deletion of large genomic fragments in soybean. Int J Mol Sci 19(12):3835

Cai L, Chen C, Wang W, Gao X, Kuang X, Jiang Y, Li L, Wu G (2020a) Acid-free epoxidation of soybean oil with hydrogen peroxide to epoxidized soybean oil over titanium silicalite-1 zeolite supported cadmium catalysts. J Ind Eng Chem 91:191–200

Cai Y, Chen L, Zhang Y, Yuan S, Su Q, Sun S, Wu C, Yao W, Han T, Hou W (2020b) Target base editing in soybean using a modified CRISPR/Cas9 system. Plant Biotechnol J 18(10):1996–1998. https://doi.org/10.1111/pbi.13386

Cai Y, Wang L, Chen L, Wu T, Liu L, Sun S, Wu C, Yao W, Jiang B, Yuan S, Han T, Hou W (2020c) Mutagenesis of GmFT2a and GmFT5a mediated by CRISPR/Cas9 contributes for expanding the regional adaptability of soybean. Plant Biotechnol J 18(1):298–309. https://doi.org/10.1111/pbi.13199

Cao Y, Wu Y, Zheng Z, Song F (2005) Overexpression of the rice EREBP-like gene OsBIERF3 enhances disease resistance and salt tolerance in transgenic tobacco. Physiol Mol Plant Pathol 67(3–5):202–211. https://doi.org/10.1016/j.pmpp.2006.01.004

Chandrasekaran J, Brumin M, Wolf D, Leibman D, Klap C, Pearlsman M, Sherman A, Arazi T, Gal-On A (2016) Development of broad virus resistance in non-transgenic cucumber using CRISPR/Cas9 technology. Mol Plant Pathol 17(7):1140–1153

Chaw S-M, Liu Y-C, Wu Y-W, Wang H-Y, Lin C-YI, Wu C-S, Ke H-M, Chang L-Y, Hsu C-Y, Yang H-T (2019) Stout camphor tree genome fills gaps in understanding of flowering plant genome evolution. Nat Plants 5(1):63–73

Chen X, Lu X, Shu N, Wang S, Wang J, Wang D, Guo L, Ye W (2017) Targeted mutagenesis in cotton (Gossypium hirsutum L.) using the CRISPR/Cas9 system. Sci Rep 7(1):1–7

Chew WL, Tabebordbar M, Cheng JK, Mali P, Wu EY, Ng AH, Zhu K, Wagers AJ, Church GM (2016) A multifunctional AAV–CRISPR–Cas9 and its host response. Nat Methods 13(10):868–874

Cong L, Ran FA, Cox D, Lin S, Barretto R, Habib N, Hsu PD, Wu X, Jiang W, Marraffini LA, Zhang F (2013) Multiplex genome engineering using CRISPR/Cas systems. Science 339(6121):819–823

Cyranoski D (2015) CRISPR tweak may help gene-edited crops bypass biosafety regulation. Nature News

Dalla Costa L, Malnoy M, Gribaudo I (2017) Breeding next generation tree fruits: technical and legal challenges. Hortic Res 4(1):1–11

Dalla Costa L, Piazza S, Pompili V, Salvagnin U, Cestaro A, Moffa L et al (2020) Strategies to produce T-DNA free CRISPRed fruit trees via Agrobacterium tumefaciens stable gene transfer. Sci Rep 10(1):1–14

Das A, Rushton PJ, Rohila JS (2017) Metabolomic profiling of soybeans (Glycine max L.) reveals the importance of sugar and nitrogen metabolism under drought and heat stress. Plants 6(2):21

Do PT, Nguyen CX, Bui HT, Tran LT, Stacey G, Gillman JD, Zhang ZJ, Stacey MG (2019) Demonstration of highly efficient dual gRNA CRISPR/Cas9 editing of the homeologous GmFAD2–1A and GmFAD2–1B genes to yield a high oleic, low linoleic and α-linolenic acid phenotype in soybean. BMC Plant Biol 19(1):1–4

Eckerstorfer MF, Dolezel M, Heissenberger A, Miklau M, Reichenbecher W, Steinbrecher RA, Waßmann F (2019a) An EU perspective on biosafety considerations for plants developed by genome editing and other new genetic modification techniques (nGMs). Front Bioeng Biotechnol 7:31

Eckerstorfer MF, Engelhard M, Heissenberger A, Simon S, Teichmann H (2019b) Plants developed by new genetic modification techniques—comparison of existing regulatory frameworks in the EU and non-EU countries. Front Bioeng Biotechnol 7:26

El-Mounadi K, Morales-Floriano ML, Garcia-Ruiz H (2020) Principles, applications, and biosafety of plant genome editing using CRISPR-Cas9. Front Plant Sci 11:56

Elorriaga E, Klocko AL, Ma C, Strauss SH (2018) Variation in mutation spectra among CRISPR/Cas9 mutagenized poplars. Front Plant Sci 9:594

Endo A, Masafumi M, Kaya H, Toki S (2016) Efficient targeted mutagenesis of rice and tobacco genomes using Cpf1 from Francisella novicida. Sci Rep 6(1):1–9

Fan D, Liu T, Li C, Jiao B, Li S, Hou Y, Luo K (2015) Efficient CRISPR/Cas9-mediated targeted mutagenesis in Populus in the first generation. Sci Rep 5(1):1–7

Fan Y, Xin S, Dai X, Yang X, Huang H, Hua Y (2020) Efficient genome editing of rubber tree (Hevea brasiliensis) protoplasts using CRISPR/Cas9 ribonucleoproteins. Ind Crop Prod 146: 112146

Fauser F, Schiml S, Puchta H (2014) Both CRISPR/C as-based nucleases and nickases can be used efficiently for genome engineering in A rabidopsis thaliana. Plant J 79(2):348–359

Fernandez i Marti A, Dodd RS (2018) Using CRISPR as a gene editing tool for validating adaptive gene function in tree landscape genomics. Front Ecol Evol 6:76

Gao W, Long L, Tian X, Xu F, Liu J, Singh PK, Botella JR, Song C (2017) Genome editing in cotton with the CRISPR/Cas9 system. Front Plant Sci 8:1364

Gilbert LA, Larson MH, Morsut L, Liu Z, Brar GA, Torres SE, Stern-Ginossar N, Brandman O, Whitehead EH, Doudna JA, Lim WA (2013) CRISPR-mediated modular RNA-guided regulation of transcription in eukaryotes. Cell 154(2):442–451

Gilbert LA, Horlbeck MA, Adamson B, Villalta JE, Chen Y, Whitehead EH, Guimaraes C, Panning B, Ploegh HL, Bassik MC, Qi LS (2014) Genome-scale CRISPR-mediated control of gene repression and activation. Cell 159(3):647–661

Gómez-Pineda JG, Gómez-Pineda JG (2019) The natural interest rate in Latin America. Borradores de Economía, No. 1067

Gorbunova V, Levy AA (1997) Non-homologous DNA end joining in plant cells is associated with deletions and filler DNA insertions. Nucleic Acids Res 25(22):4650–4657

Guo W, Chen L, Chen H, Yang H, You Q, Bao A, Chen S, Hao Q, Huang Y, Qiu D, Shan Z, Yang Z, Yuan S, Zhang C, Zhang X, Jiao Y, Tran L-SP, Zhou X, Cao D (2020) Overexpression of GmWRI1b in soybean stably improves plant architecture and associated yield parameters,

and increases total seed oil production under field conditions. Plant Biotechnol J 18(8):1639–1641. https://doi.org/10.1111/pbi.13324
Hall SJ, Russell AE, A'lece RM (2019) Do corn-soybean rotations enhance decomposition of soil organic matter? Plant and Soil 444(1):427–442
Hess GT, Tycko J, Yao D, Bassik MC (2017) Methods and applications of CRISPR-mediated base editing in eukaryotic genomes. Mol Cell 68(1):26–43
Hu X, Wang C, Liu Q, Fu Y, Wang K (2017) Targeted mutagenesis in rice using CRISPR-Cpf1 system. J Genet Genomics 44(1):71–73
Jacobs TB, LaFayette PR, Schmitz RJ, Parrott WA (2015) Targeted genome modifications in soybean with CRISPR/Cas9. BMC Biotechnol 15(1):1–10
Jaganathan D, Ramasamy K, Sellamuthu G, Jayabalan S, Venkataraman G (2018) CRISPR for crop improvement: an update review. Front Plant Sci 9:985. https://doi.org/10.3389/fpls.2018.00985
Janga MR, Campbell LM, Rathore KS (2017) CRISPR/Cas9-mediated targeted mutagenesis in upland cotton (Gossypium hirsutum L.). Plant Mol Biol 94(4):349–360
Jia H, Wang N (2015) Targeted genome editing of sweet orange using Cas9/sgRNA. PLoS One 9(4):e93806
Jiang Y, Guo L, Ma X, Zhao X, Jiao B, Li C, Luo K (2017) The WRKY transcription factors PtrWRKY18 and PtrWRKY35 promote Melampsora resistance in Populus. Tree Physiol 37(5):665–675
Jinek M, Chylinski K, Fonfara I, Hauer M, Doudna JA, Charpentier E (2012) A programmable dual-RNA-guided DNA endonuclease in adaptive bacterial immunity. Science 337(6096):816–821. https://doi.org/10.1126/science.1225829
Kanazashi Y, Hirose A, Takahashi I, Mikami M, Endo M, Hirose S, Toki S, Kaga A, Naito K, Ishimoto M (2018) Simultaneous site-directed mutagenesis of duplicated loci in soybean using a single guide RNA. Plant Cell Rep 37(3):553–563
Kanchiswamy CN, Malnoy M, Velasco R, Kim J-S, Viola R (2015) Non-GMO genetically edited crop plants. Trends Biotechnol 33(9):489–491
Karkute SG, Singh AK, Gupta OP, Singh PM, Singh B (2017) CRISPR/Cas9 mediated genome engineering for improvement of horticultural crops. Front Plant Sci 8:1635. https://doi.org/10.3389/fpls.2017.01635
Kearns NA, Genga RM, Enuameh MS, Garber M, Wolfe SA, Maehr R (2014) Cas9 effector-mediated regulation of transcription and differentiation in human pluripotent stem cells. Development 141(1):219–223
Kim H, Kim S-T, Ryu J, Kang B-C, Kim J-S, Kim S-G (2017) CRISPR/Cpf1-mediated DNA-free plant genome editing. Nat Commun 8(1):1–7
Kleinstiver BP, Prew MS, Tsai SQ, Topkar VV, Nguyen NT, Zheng Z, Gonzales APW, Li Z, Peterson RT, Yeh J-RJ, Aryee MJ, Joung JK (2015) Engineered CRISPR-Cas9 nucleases with altered PAM specificities. Nature 523(7561):481–485. https://doi.org/10.1038/nature14592
Kleinstiver BP, Tsai SQ, Prew MS, Nguyen NT, Welch MM, Lopez JM, McCaw ZR, Aryee MJ, Joung JK (2016) Genome-wide specificities of CRISPR-Cas Cpf1 nucleases in human cells. Nat Biotechnol 34(8):869–874
Klimek-Chodacka M, Oleszkiewicz T, Lowder LG, Qi Y, Baranski R (2018) Efficient CRISPR/Cas9-based genome editing in carrot cells. Plant Cell Rep 37(4):575–586
Klösgen RB, Gierl A, Schwarz-Sommer Z, Saedler H (1986) Molecular analysis of the waxy locus of Zea mays. Mol Gen Genet 203(2):237–244. https://doi.org/10.1007/BF00333960
Kumar R, Kaur A, Pandey A, Mamrutha HM, Singh GP (2019) CRISPR-based genome editing in wheat: a comprehensive review and future prospects. Mol Biol Rep 46(3):3557–3569. https://doi.org/10.1007/s11033-019-04761-3
Langner T, Kamoun S, Belhaj K (2018) CRISPR crops: plant genome editing toward disease resistance. Annu Rev Phytopathol 56(1):479–512. https://doi.org/10.1146/annurev-phyto-080417-050158

Lawrenson T, Shorinola O, Stacey N, Li C, Østergaard L, Patron N, Uauy C, Harwood W (2015) Induction of targeted, heritable mutations in barley and Brassica oleracea using RNA-guided Cas9 nuclease. Genome Biol 16(1):1–3

Lee JK, Jeong E, Lee J, Jung M, Shin E, Kim Y, Lee K, Jung I, Kim D, Kim S, Kim J-S (2018) Directed evolution of CRISPR-Cas9 to increase its specificity. Nat Commun 9(1):3048. https://doi.org/10.1038/s41467-018-05477-x

Lei J, Dai P, Li J, Yang M, Li X, Zhang W, Zhou G, Liu X (2021) Tissue-specific CRISPR/Cas9 system of cotton pollen with GhPLIMP2b and GhMYB24 promoters. J Plant Biol 64(1):13–21

Lemgo GNY, Sabbadini S, Pandolfini T, Mezzetti B (2013) Biosafety considerations of RNAi-mediated virus resistance in fruit-tree cultivars and in rootstock. Transgenic Res 22(6):1073–1088

Li Y, Su X, Zhang B, Huang Q, Zhang X, Huang R (2009) Expression of jasmonic ethylene responsive factor gene in transgenic poplar tree leads to increased salt tolerance. Tree Physiol 29(2):273–279

Li G, Zhou Z, Liang L, Song Z, Hu Y, Cui J, Chen W, Hu K, Cheng J (2020) Genome-wide identification and analysis of highly specific CRISPR/Cas9 editing sites in pepper (Capsicum annuum L.). PLoS One 15(12):e0244515

Liang P, Ding C, Sun H, Xie X, Xu Y, Zhang X, Sun Y, Xiong Y, Ma W, Liu Y, Wang Y, Fang J, Liu D, Songyang Z, Zhou C, Huang J (2017) Correction of β-thalassemia mutant by base editor in human embryos. Protein Cell 8(11):811–822. https://doi.org/10.1007/s13238-017-0475-6

Lin D, Anderson-Teixeira KJ, Lai J, Mi X, Ren H, Ma K (2016) Traits of dominant tree species predict local scale variation in forest aboveground and topsoil carbon stocks. Plant and Soil 409(1):435–446

Ling H-Q, Ma B, Shi X, Liu H, Dong L, Sun H, Cao Y, Gao Q, Zheng S, Li Y, Yu Y, Du H, Qi M, Li Y, Lu H, Yu H, Cui Y, Wang N, Chen C et al (2018) Genome sequence of the progenitor of wheat A subgenome Triticum urartu. Nature 557(7705):424–428. https://doi.org/10.1038/s41586-018-0108-0

Liu S, Zhang M, Feng F, Tian Z (2020a) Toward a "Green Revolution" for Soybean. Mol Plant 13(5):688–697

Liu Y, Du H, Li P, Shen Y, Peng H, Liu S, Zhou G-A, Zhang H, Liu Z, Shi M (2020b) Pan-genome of wild and cultivated soybeans. Cell 182(1):162–176

Lu Y, Zhu JK (2017) Precise editing of a target base in the rice genome using a modified CRISPR/Cas9 system. Mol Plant 10(3):523–525

Malnoy M, Viola R, Jung M-H, Koo O-J, Kim S, Kim J-S, Velasco R, Nagamangala Kanchiswamy C (2016) DNA-free genetically edited grapevine and apple protoplast using CRISPR/Cas9 ribonucleoproteins. Front Plant Sci 7:1904

Malzahn A, Lowder L, Qi Y (2017) Plant genome editing with TALEN and CRISPR. Cell Biosci 7(1):1–18

McCourt P, Benning C (2010) Arabidopsis: a rich harvest 10 years after completion of the genome sequence. Plant J 61(6):905–908. https://doi.org/10.1111/j.1365-313X.2010.04176.x

McMurray LS, Preston C, Vandenberg A, Mao D, Oldach KH, Meier KS, Paull JG (2019) Development of high levels of metribuzin tolerance in lentil. Weed Sci 67(1):83–90

Meng Y, Hou Y, Wang H, Ji R, Liu B, Wen J, Niu L, Lin H (2017) Targeted mutagenesis by CRISPR/Cas9 system in the model legume Medicago truncatula. Plant Cell Rep 36(2):371–374. https://doi.org/10.1007/s00299-016-2069-9

Murovec J, Pirc Ž, Yang B (2017) New variants of CRISPR RNA-guided genome editing enzymes. Plant Biotechnol J 15(8):917–926

Murovec J, Guček K, Bohanec B, Avbelj M, Jerala R (2018) DNA-free genome editing of Brassica oleracea and B. rapa protoplasts using CRISPR-Cas9 ribonucleoprotein complexes. Front Plant Sci 9:1594

Nadeem M, Li J, Yahya M, Sher A, Ma C, Wang X, Qiu L (2019) Research progress and perspective on drought stress in legumes: a review. Int J Mol Sci 20(10):2541. https://doi.org/10.3390/ijms20102541

Nakajima I, Ban Y, Azuma A, Onoue N, Moriguchi T, Yamamoto T et al (2017) CRISPR/Cas9-mediated targeted mutagenesis in grape. PLoS One 12(5):e0177966
Nishitani C, Hirai N, Komori S, Wada M, Okada K, Osakabe K, Yamamoto T, Osakabe Y (2016) Efficient genome editing in apple using a CRISPR/Cas9 system. Sci Rep 6(1):1–8
Odipio J, Alicai T, Ingelbrecht I, Nusinow DA, Bart R, Taylor NJ (2017) Efficient CRISPR/Cas9 genome editing of phytoene desaturase in cassava. Front Plant Sci 8:1780
Panchamoorthy PS, Kannan RR (2021) Enhanced saccharfication of pomegranate peel by Trichoderma Reesei and statiscal optimization of medium composition by response surface methodology. https://doi.org/10.21203/rs.3.rs-226555/v1
Parsons J, MacKay J (2018) The theory and applications of CRISPR in plant and tree improvement. CAB Rev 13(07)
Puchta H, Dujon B, Hohn B (1996) Two different but related mechanisms are used in plants for the repair of genomic double-strand breaks by homologous recombination. Proc Natl Acad Sci 93(10):5055–5060
Qi Y (2018) A CRISPR–Cpf1 system for efficient genome editing and transcriptional repression in plants. In: Plant and animal genome XXVI conference, 13–17 Jan 2018
Qi LS, Larson MH, Gilbert LA, Doudna JA, Weissman JS, Arkin AP, Lim WA (2013) Repurposing CRISPR as an RNA-guided platform for sequence-specific control of gene expression. Cell 152(5):1173–1183
Ren C, Liu X, Zhang Z, Wang Y, Duan W, Li S, Liang Z (2016) CRISPR/Cas9-mediated efficient targeted mutagenesis in Chardonnay (Vitis vinifera L.). Sci Rep 6(1):1–9
Sakuma T, Yamamoto T (2017) Magic wands of CRISPR—lots of choices for gene knock-in. Cell Biol Toxicol 33(6):501–505
Sanchez-Vega F, Mina M, Armenia J, Chatila WK, Luna A, La KC, Dimitriadoy S, Liu DL, Kantheti HS, Saghafinia S (2018) Oncogenic signaling pathways in the cancer genome atlas. Cell 173(2):321–337
Sankar P, Zachariah B, Vickneshwaran V, Jacob SE, Sridhar MG (2015) Amelioration of oxidative stress and insulin resistance by soy isoflavones (from Glycine max) in ovariectomized Wistar rats fed with high fat diet: the molecular mechanisms. Exp Gerontol 63:67–75. https://doi.org/10.1016/j.exger.2015.02.001
Schmidt C, Fransz P, Rönspies M, Dreissig S, Fuchs J, Heckmann S, Houben A, Puchta H (2020) Changing local recombination patterns in Arabidopsis by CRISPR/Cas mediated chromosome engineering. Nat Commun 11(1):1–8
Sant'Ana RR, Caprestano CA, Nodari RO, Agapito-Tenfen SZ (2020) PEG-delivered CRISPR-Cas9 ribonucleoproteins system for gene-editing screening of maize protoplasts. Gene 11(9):1029
Schwarz K, Fragkias M, Boone CG, Zhou W, McHale M, Grove JM, O'Neil-Dunne J, McFadden JP, Buckley GL, Childers D (2015) Trees grow on money: urban tree canopy cover and environmental justice. PLoS One 10(4):e0122051
Shan Q, Wang Y, Chen K, Liang Z, Li J, Zhang Y, Zhang K, Liu J, Voytas DF, Zheng X, Zhang Y, Gao C (2013a) Rapid and efficient gene modification in rice and Brachypodium using TALENs. Mol Plant 6(4):1365–1368. https://doi.org/10.1093/mp/sss162
Shan Q, Wang Y, Li J, Zhang Y, Chen K, Liang Z, Zhang K, Liu J, Xi J, Qiu J-L, Gao C (2013b) Targeted genome modification of crop plants using a CRISPR-Cas system. Nat Biotechnol 31: 686–688. https://doi.org/10.1038/nbt.2650
Shu H, Luo Z, Peng Z, Wang J (2020) The application of CRISPR/Cas9 in hairy roots to explore the functions of AhNFR1 and AhNFR5 genes during peanut nodulation. BMC Plant Biol 20(1):1–5
Singh VK, Jain M (2015) Genome-wide survey and comprehensive expression profiling of Aux/IAA gene family in chickpea and soybean. Front Plant Sci 6:918
Soyk S, Müller NA, Park SJ, Schmalenbach I, Jiang K, Hayama R, Zhang L, Van Eck J, Jiménez-Gómez JM, Lippman ZB (2017) Variation in the flowering gene SELF PRUNING 5G promotes day-neutrality and early yield in tomato. Nat Genet 49(1):162–168

Strauss SH, Costanza A, Séguin A (2015) Genetically engineered trees: paralysis from good intentions. Science 349(6250):794–795
Svitashev S, Schwartz C, Lenderts B, Young JK, Cigan AM (2016) Genome editing in maize directed by CRISPR–Cas9 ribonucleoprotein complexes. Nat Commun 7(1):1–7
Tan C, Zhou H, Wang X, Mai K, He G (2019) Resveratrol attenuates oxidative stress and inflammatory response in turbot fed with soybean meal based diet. Fish Shellfish Immunol 91:130–135. https://doi.org/10.1016/j.fsi.2019.05.030
Tang L, Mao B, Li Y, Lv Q, Zhang L, Chen C, He H, Wang W, Zeng X, Shao Y, Pan Y (2017) Knockout of OsNramp5 using the CRISPR/Cas9 system produces low Cd-accumulating indica rice without compromising yield. Sci Rep 7(1):1–2
Tsai C-J, Xue L-J (2015) CRISPRing into the woods. GM Crops Food 6(4):206–215
Upadhyay SK, Kumar J, Alok A, Tuli R (2013) RNA-guided genome editing for target gene mutations in wheat. G3: genes, genomes. Genetics 3(12):2233–2238
van der Knaap E, Chakrabarti M, Chu YH, Clevenger JP, Illa-Berenguer E, Huang Z, Keyhaninejad N, Mu Q, Sun L, Wang Y, Wu S (2014) What lies beyond the eye: the molecular mechanisms regulating tomato fruit weight and shape. Front Plant Sci 5:227
Van Zeijl A, Wardhani TA, Seifi Kalhor M, Rutten L, Bu F, Hartog M, Linders S, Fedorova EE, Bisseling T, Kohlen W (2018) CRISPR/Cas9-mediated mutagenesis of four putative symbiosis genes of the tropical tree Parasponia andersonii reveals novel phenotypes. Front Plant Sci 9:284
Waltz E (2016) CRISPR-edited crops free to enter market, skip regulation. Nature Publishing Group, London
Wang Y, Li Z, Xu J, Zeng B, Ling L, You L, Chen Y, Huang Y, Tan A (2013) The CRISPR/Cas system mediates efficient genome engineering in Bombyx mori. Cell Res 23(12):1414–1416
Wang Y, Cheng X, Shan Q, Zhang Y, Liu J, Gao C, Qiu J-L (2014) Simultaneous editing of three homoeoalleles in hexaploid bread wheat confers heritable resistance to powdery mildew. Nat Biotechnol 32(9):947–951. https://doi.org/10.1038/nbt.2969
Wang F, Wang C, Liu P, Lei C, Hao W, Gao Y, Liu Y-G, Zhao K (2016) Enhanced rice blast resistance by CRISPR/Cas9-targeted mutagenesis of the ERF transcription factor gene OsERF922. PLoS One 11(4):e0154027. https://doi.org/10.1371/journal.pone.0154027
Wang M, Mao Y, Lu Y, Tao X, Zhu JK (2017) Multiplex gene editing in rice using the CRISPR-Cpf1 system. Mol Plant 10(7):1011–1013
Wang W, Simmonds J, Pan Q, Davidson D, He F, Battal A, Akhunova A, Trick HN, Uauy C, Akhunov E (2018a) Gene editing and mutagenesis reveal inter-cultivar differences and additivity in the contribution of TaGW2 homoeologues to grain size and weight in wheat. Theor Appl Genet 131(11):2463–2475. https://doi.org/10.1007/s00122-018-3166-7
Wang Z, Wang S, Li D, Zhang Q, Li L, Zhong C et al (2018b) Optimized paired-sgRNA/Cas9 cloning, and expression cassette triggers high-efficiency multiplex genome editing in kiwifruit. Plant Biotechnol J 16(8):1424–1433
Wang T, Zhang H, Zhu H (2019a) CRISPR technology is revolutionizing the improvement of tomato and other fruit crops. Hortic Res 6(1):1–13. https://doi.org/10.1038/s41438-019-0159-x
Wang X, Yong H, Gao L, Li L, Jin M, Liu J (2019b) Preparation and characterization of antioxidant and pH-sensitive films based on chitosan and black soybean seed coat extract. Food Hydrocoll 89:56–66
Wang L, Sun S, Wu T, Liu L, Sun X, Cai Y, Li J, Jia H, Yuan S, Chen L (2020) Natural variation and CRISPR/Cas9-mediated mutation in GmPRR37 affect photoperiodic flowering and contribute to regional adaptation of soybean. Plant Biotechnol J 18(9):1869
Wilson LOW, O'Brien AR, Bauer DC (2018) The current state and future of CRISPR-Cas9 gRNA design tools. Front Pharmacol 9:749. https://doi.org/10.3389/fphar.2018.00749
Xue LJ, Tsai CJ (2015) AGEseq: analysis of genome editing by sequencing. Mol Plant 8(9):1428–1430
Yao Y-X, Li M, Liu Z, You C-X, Wang D-M, Zhai H, Hao Y-J (2009) Molecular cloning of three malic acid related genes MdPEPC, MdVHA-A, MdcyME and their expression analysis in apple fruits. Sci Hortic 122(3):404–408. https://doi.org/10.1016/j.scienta.2009.05.033

Ye J, Wang X, Hu T, Zhang F, Wang B, Li C, Yang T, Li H, Lu Y, Giovannoni JJ, Zhang Y, Ye Z (2017) An InDel in the promoter of Al-ACTIVATED MALATE TRANSPORTER9 selected during tomato domestication determines fruit malate contents and aluminum tolerance. Plant Cell 29(9):2249–2268. https://doi.org/10.1105/tpc.17.00211

Yin X, Biswal AK, Dionora J, Perdigon KM, Balahadia CP, Mazumdar S, Chater C, Lin H-C, Coe RA, Kretzschmar T (2017) CRISPR-Cas9 and CRISPR-Cpf1 mediated targeting of a stomatal developmental gene EPFL9 in rice. Plant Cell Rep 36(5):745–757

Yubing HE, Min ZHU, Lihao W, Junhua WU, Qiaoyan W, Rongchen W, Yunde Z (2019) Improvements of TKC technology accelerate isolation of transgene-free CRISPR/Cas9-edited rice plants. Rice Sci 26(2):109–117

Zaidi SS-A, Mahfouz MM, Mansoor S (2017) CRISPR-Cpf1: a new tool for plant genome editing. Trends Plant Sci 22(7):550–553. https://doi.org/10.1016/j.tplants.2017.05.001

Zhang F, Leblanc C, Irish VF, Jacob Y (2017) Rapid and efficient CRISPR/Cas9 gene editing in Citrus using the YAO promoter. Plant Cell Rep 36:1883–1887. https://doi.org/10.1007/s00299-017-2202-4

Zhao S, Wang Z, Kang H, Li J, Zhang S, Han C, Huang A (2018) Fully bio-based soybean adhesive in situ cross-linked by interactive network skeleton from plant oil-anchored fiber. Ind Crop Prod 122:366–374

Zheng N, Li T, Dittman JD, Su J, Li R, Gassmann W, Peng D, Whitham SA, Liu S, Yang B (2020) CRISPR/Cas9-based gene editing using egg cell-specific promoters in Arabidopsis and soybean. Front Plant Sci 11:800

Zhou X, Jacobs TB, Xue LJ, Harding SA, Tsai CJ (2015) Exploiting SNP s for biallelic CRISPR mutations in the outcrossing woody perennial Populus reveals 4-coumarate: CoA ligase specificity and redundancy. New Phytol 208(2):298–301

Zhou J, Wang G, Liu Z (2018) Efficient genome editing of wild strawberry genes, vector development and validation. Plant Biotechnol J 16(11):1868–1877

Challenges and Future Prospects of CRISPR Technology

10

Zulqurnain Khan, Sultan Habibullah Khan, and Aftab Ahmad

Contents

Abstract

Genome editing (GE) tools have been revolutionizing life sciences by various marvelous applications for targeted gene modifications in organisms. GenEd tools, for example, zinc finger nucleases (ZFN), transcription activator-like

Z. Khan (✉)
Institute of Plant Breeding and Biotechnology, Muhammad Nawaz Sharif University of Agriculture, Multan, Pakistan
e-mail: zulqurnain.khan@mnsuam.edu.pk

S. H. Khan
Center of Agricultural Biochemistry and Biotechnology (CABB)/US-Pakistan Center for Advanced Studies in Agriculture and Food Security, University of Agriculture, Faisalabad, Pakistan
e-mail: sultan@uaf.edu.pk

A. Ahmad
Department of Biochemistry/US-Pakistan Center for Advanced Studies in Agriculture and Food Security, University of Agriculture, Faisalabad, Pakistan
e-mail: aftab.ahmad@uaf.edu.pk

A. Ahmad et al. (eds.), *The CRISPR/Cas Tool Kit for Genome Editing*,
https://doi.org/10.1007/978-981-16-6305-5_10

effector nucleases (TALEN), and CRISPR (clustered regularly interspaced short palindromic repeats)/Cas (CRISPR-associated) systems have been widely used in the last decade for genetic manipulation of plants, animals, microbes, and other organisms. The utility of the CRISPR/Cas tool is widespread compared to other contemporary tools due to its simplicity, efficiency, cost-effectiveness, and accuracy. Moreover, there are a number of variants in the CRISPR/Cas toolkit which increased its usefulness. Along with a number of benefits of the CRISPR/Cas including unique feature of multiplexing, the system has also some critical limitations and concerns. Off-targeting is one of the biggest limitations of the system which hinders its extensive use. There is a need to address these concerns associated with CRISPR/Cas system to take more benefits from this system along with increasing biosafety and public acceptance. Efficient bioinformatics tools and in silico analysis should be used for designing CRISPR/Cas-gRNA to a specific target and predicting possible off-targets. The future CRISPR/Cas system application in life sciences particularly human therapeutics and animal genome editing may be increased by mitigating the off-targets and other limitations of the system. The chapter outlines application of CRISPR system along with addressing its concerns specifically off-targeting.

Keywords

Genome editing · ZFNs · TALENs · CRISPR · Limitations · Off-targeting

10.1 CRISPR/Cas and off-Targeting

The success of the CRISPR/Cas9 may be measured by the fact that CRISPR system was applied in a diversity of living creatures for efficient gene editing, including important model plants and commodities, animal models and human cell lines, microorganisms, insects, etc. Nonetheless, GE systems; ZFNs, TALENs, and CRISPR, may also create unexpected and unpredicted off-target mutations. Cas9 is reported to cause off-target effects by binding to the unintended/undesirable regions resulting in cleavage (Alkan et al. 2018). It has been found that the off-targets are homologous to the gRNA on-target regions of the target sequence. Many reports discovered that CRISPR system is more vulnerable to the off-targeting compared to ZFNs and TALENs owing to its monomeric nature. ZFNs and TALENs are basically dimeric in nature facilitating the modularity in identification of shorter to longer target DNA sequences. Off-target regions in CRISPR/Cas9-gRNA may contain up to six mismatches compared to on-target regions (Martin et al. 2016). Fewer mismatches in the off-targets may lead to an inclination for more prominent binding and cleavage. These undesirable off-targets raise many concerns on the specificity and precision of the CRISPR. Although for the identification of off-target regions for selected sequence of gRNA, a number of tools have been developed which reduces mismatches. First study of Cas9-associated off-target impacts was identified in human cancer cell lines (Fu et al. 2013). A study indicated that the occurrence of these effects was remarkably very high, due to inappropriate working of DNA repair mechanism in tumor cells (Zischewski et al. 2017). While, in case of base editing approach, off-target effects can be observed due to gRNA independent or gRNA dependent editing events (Rees and Liu 2018; Kim et al. 2019). For the reduction of off-target effects associated with gRNA-dependent base editing, several strategies have been used (Kim et al. 2019; Jin et al. 2019). DNA specificity of Cas9 components of base editors have been enhanced by modifying the sgRNA through addition of 5′-guanosine nucleotides or delivery of BEs as a complex of ribonucleoprotein (RNP) (Kim et al. 2019; Rees et al. 2017; Yeh et al. 2018). The off-targeting associated with independent gRNA in base editing is owing to Cas9's irregular attachment to the deaminase domain of base editor to C or A bases (Zuo et al. 2019).

The fundamental concern with CRISPR-based gene editing is the off-target effect, and researchers have attempted many times to mitigate these impacts. For overcoming the off-target effects, various approaches have now been developed such as functional screening and transcriptome analysis after treatment with dCas9, optimization and specificity of sgRNA design, and utilization of effective variants of Cas9 (Konermann et al. 2015). Moreover, the off-target effects have been minimized by increasing specificity of cleavage site or reducing the time interval of nuclease expression (Tsai and Joung 2016). Other approaches may include delivery directly in the form of ribonucleoproteins or using paired Cas9 nickases, truncated sgRNAs, and tunable systems (Kim et al. 2017; Ran et al. 2013a; Dow et al. 2015). Tunable or inducible system is useful in reducing the unpredictable DNA cleavage into the genome which is considered as potential concern in off-target mutations. For instance, two AcrIIA2 and AcrIIA4 (anti-CRISPR) proteins can be utilized to block the cleavage activity of Cas9 after Cas9 cuts at specific target site (Rauch et al. 2017). Several Cas9 variants with high fidelity were developed in recent years through optimization and delineation of structure of Cas9, for example, SpCas9, SpCas9-HF1, HypaCas9, evoCas9, xCas9, and SpCas9-NG (Kleinstiver et al. 2016; Slaymaker et al. 2016; Casini et al. 2018; Nishimasu et al. 2018).

Basically, off-targets have been characterized to be majorly of three types. The first type includes region at other PAMs such as 5′-NGG-3′, containing substitutions/mismatches (Tsai et al. 2015), while the second type includes PAMs regions which contain insertions/deletions comparable to target DNA or gRNA spacer (Lin et al. 2014). Moreover, DNA/RNA forms a small hairpin structure with remaining nucleotides for accurate annealing thereby facilitating Cas9 activity, though off-targets identified at these locations are more than on-target activities (Lin et al. 2014). The third type of off-targets corresponds to the sequence cleavage with the distinct sites of PAM (5′-NAG-3′) (Tsai et al. 2015). However, it was reported that genome editing based on CRISPR has two types of off-targets: the first type off-targets with sequence similarity to the target in the genomic regions, while the second type is not relevant to the target sequence and may be considered as unexpected off-target.

The off-target effects lead to inconsistencies, ambiguities into scientific breakthroughs which resulted in misperceiving potential in beneficial agricultural and medical applications of CRISPR/Cas9 and other GenEd tools. Off-targeting can be checked using a number of methods and procedures, such as mathematical/computer-aided prediction, validating, Cas9-sgRNA delivery modification, off-target cleavage, high fidelity SpCas9 engineering, and gRNA engineering (Li et al. 2019).

The technology of CRISPR has upgraded genome editing field because of its swiftness, simplicity, high efficiency, and affordability than customary strategies. In a brief timeframe, its applications have gotten exceptionally broad; however it can cause undesirable mutations at off-target sites subsequently after targeting on-target sequence (Kim et al. 2015). Recently, this innovation has been applied in embryo of human to address pathogenic mutations and has re-lighted the moral discussion of germline cell editing (Tang et al. 2017). Other than the moral contemplations, the chance of making off-target mutation with obscure outcomes is a worry. There are numerous reports of high-recurrence off-targeting changes in human and mouse cell lines (Fu et al. 2013). Moreover, undesirable mutations in mammalian embryo may cause irregularity that seems to be a major concern (Iyer et al. 2015; Hay et al. 2017; Iyer et al. 2018).

The nonspecificity of the cleavage and binding of particular nucleotides in the guide RNA and the target 20-nucleotide DNA raise the issue of off-target. The insertions and deletions between gRNA and the targeted DNA likewise comprise legitimate off-targeting effects (Lin et al. 2014) which is considered as in silico off-targeting indicators (Bae et al. 2014). A few methodologies have been received to minimize off-target events. By increasing the length of binding sequence, the chance of off-target effects has been considerably reduced. Complete detection of off-target sites is still a key challenge in editing a gene (Gabriel et al. 2015). Some of them are discussed in Table 10.1.

10.1.1 Off-Targeting Minimization Strategies

A number of methods have been reported to reduce off-targeting (Kim et al. 2015). The sgRNA sequence can be changed or 3′ end truncation of sgRNA (got from domain of tracrRNA), and shortening of the region corresponding to the targeted location at the sgRNA 5′ end by upwards of 3 nt and two G nucleotides addition to the sgRNA 5′ end (before 20-nt) improve target specificity and minimize off-target events by 5000-folds (Cho et al. 2014).

Controlling the Cas9-sgRNA complex concentration by titrating the amount of sgRNA and Cas9 is another possible method for reducing off targeting. In some cases, specificity is expanded by diminishing the measure of transfected DNA; additionally it prompts a decrease in on-target cleavage. The advancement of both sgRNA and Cas9 configuration may improve specificity of Cas9 without sacrificing cleavage efficiency (Kuscu et al. 2014).

Mutated nickase (D10) Cas9 variant can replace wild-type Cas9 nuclease combined with two sgRNAs that each cleaves just one strand. These mutant variants of Cas9 considerably reduce the activity of off-target by 50 to 1500-fold. This strategy

Table 10.1 Methods of off-target detection

Methods	Benefits	Disadvantages	References
IDLV	Delicate (1%), programmable	Many genuine off-target locations are impossible to capture	Wang et al. (2015)
T7E1 assay	Easy and simple	Very costly and have low sensitivity	Kim et al. (2009)
HTGTS	Can identify translocations	There are false negatives, which are incomplete because of chromatin convenience	Frock et al. (2015)
Digenome-seq	Sensitive, impartial, and cost-effective	Not very commonly used	Kim et al. (2015)
ChIP-seq	Unbiased in detection of binding sites of Cas9	Not all off-target DNA-binding sites recognized by dCas9 are being cleaved	Heigwer et al. (2014)
GUIDE-seq	Balanced and can identify hotspots of cutoff point	No removal of false negatives, chromatin accessibility is limited	Tsai et al. (2015)
FISH	Quick	Less precise	Paulis et al. (2015)
In silico prediction	Prediction of off-target mutation sites	Miscarries bona fide sites of off-target	Heigwer et al. (2014)
Deep sequencing	Highly précised	It can miss off-target sites in genome somewhere	Cho et al. (2014), Cong et al. (2013)

enhanced zygotes gene knockout of mouse without compromising on the on-targeting efficiency cleavage (Ran et al. 2013b). This flexible procedure empowers a wider range of gene altering applications that usually need higher specificity. Additionally, the cleavage specificity of DNA is enhanced by combining FokI nuclease domain (fCas9) with chemically inactive Cas9 which alters targeted DNA with greater than 140-folds, with relatively higher specificity than Cas9 of wild type (Tsai et al. 2014). Moreover, this study provided the establishment to the further improvement and enhancement of specificity of Cas9.

10.1.2 CRISPR Tools for Minimizing off-Targets

Modified Cas9 is used to regulate off-targeting in various ways such as to produce DSB at the target location, with deactivated nuclease (dCas9), a pair of Cas9 was employed in conjunction with a dimer of domains of FokI nuclease. Off-targeting can be controlled with these methods, enhancing the chances of CRISPR/Cas technology being used safely.

Based on the CRISPR system, "base editor," the base editing technology is an advanced and upcoming genome modification method. This method is more effective, straightforward, and well developed by scientists (Komor et al. 2016). The replacement of specific nucleotides does not require DNA templates or double-stranded DNA breaks. The HNH and RuvC domain inactivation in Cas9 nuclease results in the formation of nCas9 and dCas9, which cleave DNA's double strands. The nCas9 retains its capacity to be programmed with single guided RNA and then cuts single-stranded (ssDNA) sections by targeting certain DNA sequences. After that, the cytidine (C) in a ssDNA is changed to uracil (U), which is subsequently replaced with thymine during DNA replication or repair via cytidine deaminase. BE3 is currently among the most widely used systems, and it is equipped with a UGI, which suppresses the endogenous action of base excision. As a result, the ensuing base editor converts a cytidine to a thymine on nontarget sequences and completes a C-G to a T-A conversion (Standage-Beier et al. 2015).

Prime editing is a precise and versatile GE method that utilizes an engineered reverse transcriptase coupled with catalytically impaired Cas9 endonuclease and programmed with pegRNA that not only specifies the site of target but also encodes the required edit to write genetic information directly into a specified site in DNA. More than 175 changes have been done in human cells, such as targeted deletions, insertions, and all 12 types of point mutations, all without the use of templates of DSBs or donor DNA (Anzalone et al. 2019). Various CRISPR tools may be used to mitigate off-targets. A number of tools are given in Fig. 10.1 which may be used to decrease off-targets according to the user's objectives.

10.2 Specificity Problem with CRISPR/Cas

The specificity of Cas9 is controlled by the 20 nt guide sequences of sgRNA and PAM region; adjacent to the target sequence. It was observed that off-target cleavage could still occur with even 3–5 base pair mismatches in the PAM-distal part of the sgRNA sequence (Cong et al. 2013; Mali et al. 2013; Hsu et al. 2013; Pattanayak et al. 2013; Fu et al. 2013). Previously, it has been reported that on-target and off-target cleavage can be affected by different gRNA structures (Hsu et al. 2013). Some important studies suggest that along with PAM site which is essential for the initiation of Cas9 binding, seed sequence (adjacent to PAM) corresponding to 3′ end of the crRNA complementary sequence is also critically important for subsequent Cas9 binding, loop formation, and nuclease activity in Cas9 (Nishimasu et al. 2014; Jinek et al. 2014).

CRISPR/Cas technology has been updated and advanced to work on moderately small genomes in microscopic organisms; however, modification in larger genomes requires a higher specificity (Tsai and Joung 2016; Zischewski et al. 2017). Different methodologies have been accounted for increasing the specificity of CRISPR/Cas9 to limit off-targeting events (Tycko et al. 2016). These includes truncations (Fu et al. 2014a, b) and extensions (Kim et al. 2015) at the 5′ ends of gRNAs, paired nickase Cas9 mutants (Ran et al. 2013a), combination of dCas9 to dimerization-dependent Fok1 nuclease (Guilinger et al. 2014), and design and use of higher fidelity variants of Cas9 (Slaymaker et al. 2016; Chen et al. 2017). Different methodologies control

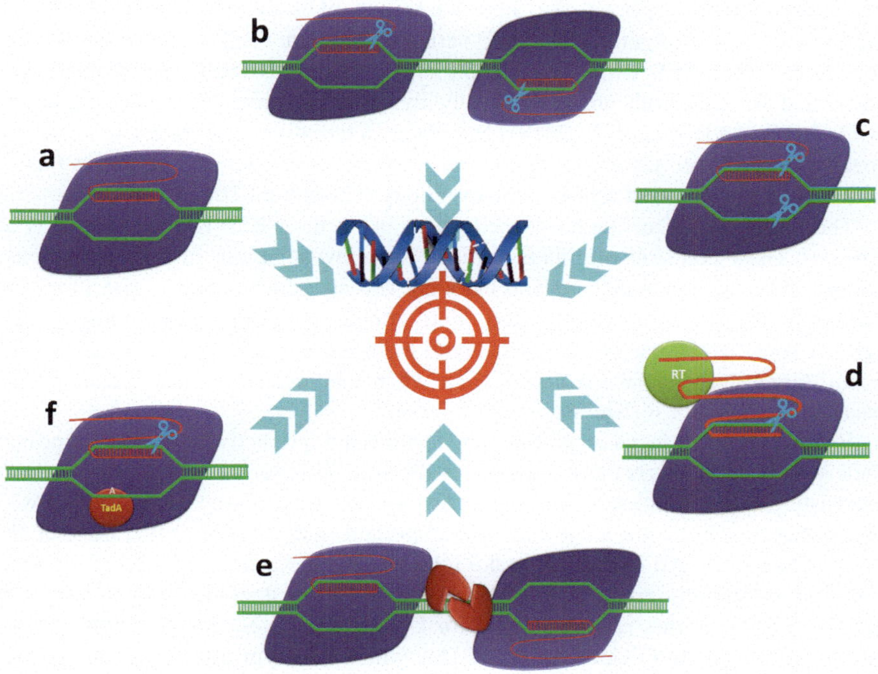

Fig. 10.1 Strategies for minimizing off-targeting using CRISPR technology. CRISPR toolkit has different tools which may be picked based on the user's objective. Many tools given in the figure may be chosen to minimize the chance of off-targeting. (**a**) dCas9 is not resulted in the creation of DSBs. (**b**) Cas9 with paired nickases is another tool which is used to target long regions having less chances of off-targeting. (**c**) Cas9, a basic and most frequently used tool for DSB-mediated targeted mutations. (**d**) Prime editing, a latest CRISPR tool which have pegRNA and reverse transcriptase to facilitate targeted insertions. (**e**) Cas9 with FokI nuclease, pair of Cas9 fused with FokI nuclease may be helpful in minimizing off-targeting due to long target site. (**f**) Base editing tool is helpful in producing SNP-based mutations at target sites

the CRISPR duration of activity in cells of eukaryotes, for instance, by transient Cas9 delivery and gRNA as a complex of ribonucleoprotein (RNP) by means of cationic lipids or electroporation (Zuris et al. 2015) or by coordinated expansion of a CRISPR/Cas9 inhibitor (Shin et al. 2017) and many other methods. Among these methods, transient RNP delivery has increased specificity of Cas9 and different techniques presently can't seem to be extensively adopted. There are vigorous techniques to deliver single guided RNAs (sgRNAs) with higher fidelity and specificity (Dellinger et al. 2011).

The endorsement of the crystal structure of Cas9 (Nishimasu et al. 2014) empowered researchers to normally structure mutant Cas9 proteins (upgraded Cas9 specificity) with higher reliability and specificities than wild-type Cas9 (WT-Cas9) (Slaymaker et al. 2016). Their structure depended on the hypothesis that devastating nonspecific interactions between a Cas9-RNA and target DNA would decrease off-target activity. Since the activity of on-target events is commonly higher than the off-target events, these variants of mutant Cas9 would show higher specificities than WT while holding on-target actions. Nonetheless, it has been accounted for that both are ineffectively dynamic at some target sites (Kulcsár et al. 2017; Anderson et al. 2018). Moreover, alternative approaches to increase specificity of Cas9 including two extra variants of Cas9, i.e., HypaCas9 and evoCas9, with improved activity and specificity (Chen et al. 2017).

The possibility of off-target consequences using the CRISPR/Cas9 system is a concern. These concerns originated initially from investigations of cell populations editing, although single-cell analysis has since revealed that the lack of specificity in this early research was exaggerated. Furthermore, designing Cas9 protein and sgRNA frameworks has boosted selectivity even more, resulting in trials in which no off-target cutting was identified, even when whole-genome sequence analysis was used to look for it (Kleinstiver et al. 2016).

Future study will focus on putting new features into predictive models to improve their accuracy. The target site sequence is the sole basis for current approaches for forecasting target efficiency and specificity. However, it is now widely acknowledged that the chromatin environment is important (Knight et al. 2015; Isaac et al. 2016; Chen et al. 2017) and can make influence on the activity of CRISPR/Cas9. Early investigations employing ChIP-seq to map the genome-wide binding of inactive Cas9 enzymes revealed a bias for DNAse sensitive areas (Wu et al. 2014; Kuscu et al. 2014; O'Geen et al. 2015), which are typically more accessible environments. Later studies demonstrated that high activity target locations were generally enriched for histone alterations associated with open-chromatin environments, supporting this theory (Chari et al. 2015).

10.2.1 dCas9 Specificity

Inactivated Cas9 (dCas9), which may be coupled with diverse domains of effectors for applications starting from programmable chromosome labelling to target epigenetic and transcription regulation, is utilized in wider range of applications in addition to DNA cleavage (Hsu et al. 2014). The target site for mechanisms of the activated Cas9 protein demonstrates that Cas9 at first scans the genome for sites of PAM (Jiang et al. 2015), bringing about a transient binding state balanced out by a five-base-pair seed sequence of the gRNA (Wu et al. 2014).

10.2.2 Broader Implications of Specificity

Cas9 genome specificity considers not just the total number of possible off-targeting sites for a targeted gRNA but also individual off-targeting physiological consequences. Moreover, the Cas9 nuclease specificity is not really limited to specificity of genome, particularly for in vivo applications. For instance, the specificity of cell type and tissue are additionally significant contemplations while

Table 10.2 Improvements in specificity of CRISPR/Cas9

Improvement	Description	Advantage	Disadvantage	References
BLESS	Biochemical ligation of adapters of next generation sequencing to exposed gDNA ends	No introduction of exogenous bait to cells	Sensitive to time of cell fixation	Slaymaker et al. (2016)
Targeted deep sequencing	Targeted amplicon NGS of known off-target sites	Sensitive and quantitative than alternative assays	Bias toward an off-target sites subset	Ran et al. (2013b)
Digenome-seq	Cell-free samples of gDNA are digested by Cas9 RNP in vitro with multiplexed guide RNAs	Applicable to any cell type as extracted gDNA digestion is performed	Costly, and have to be coupled with other methods for sites validation	Kim et al. (2016)
Single/paired nickases	An active site point mutation of one of the nuclease domains of Cas9 yields a targeted nickase	Less off-target edits. Nickases are also capable of mediating efficient HDR	Low efficiency in on target editing with various gRNAs	Cong et al. (2013)
SpCas9 PAM variant D1135E	Single-point mutation increases specificity	Genome-wide specificity is improved for various guides	Efficiency of on-target may get affected	Kleinstiver et al. (2015)
eSpCas9	gRNA-gDNA increased stringency complementation for activation of nuclease	Avoidance of off-target editing	Efficiency of on-target may get affected	Slaymaker et al. (2016)
SaCas9	3.2 kb Cas9 ortholog uses a guide RNA of 20–23 nt	Improvements in specificity with guides by BLESS	Low efficiency in on-target in various situations	Ran et al. (2015)
WGS of reference genome	WGS of the relevant animal model, cell line, and patient	Identification of creation of new 0ff-target sites by genetic variation	Remains costly	Yang et al. (2014)
Tru-guides	Short region of RNA-DNA complementarity results in less mismatches	Decrease in genomic off-target sites activity	Low efficiency in on-target editing with various gRNAs	Fu et al. (2014b)

targeting on hereditary diseases that principally influence certain organs or cells (Zincarelli et al. 2008). Future examinations could research whether combinations of synthetically changed single guided RNAs and high-fidelity Cas variations may yield much further specificity enhancements and may likewise empower progressively adaptable delivery control and movement in various kinds of tissues and cells (Table 10.2).

10.3 Comparison with Other Genome Editing Tools

Studies on the mechanisms of repair essential DNA damage and the subsequent changes in DNA structure have shaped the premise of targeted editing of genome (Guha et al. 2017). The idea of targeted editing of a genome was investigated by the improvement of a mega nuclease, which was at first made by the combination of chemically dynamic nuclease domain of FokI and I-SceI having a capacity of targeting a sequence of 18 base pairs. Zinc finger nuclease (ZFN) is a genome editing tool which utilizes zinc finger modules, each of which recognizes a 3 nt sequence of DNA. A combination of proteins is made from a variety of zinc finger modules and a domain FokI nuclease prompts a DSB at the targeted site.

Likewise, transcription activator-like effector nucleases utilize a comparative stage to that of zinc finger nuclease; aside from that, zinc finger proteins are supplanted by 14–24 TALENs and each of which recognizes an oligonucleotide of 1 bp by the base specificity repeat variable diresidue (RVD) (Li and Yang 2013). In spite of the fact that the innovation of artificially designed meganuclease which was followed by TALENs and ZFNs progressively expanded the efficacy of genome editing, targeted on various sites in the genome required re-designing or even reengineering of other proteins set. The complexity in re-engineering of protein and cloning in TALENs and ZFNs is difficult which is not extensively adopted by the researcher. In this regard, CRISPR is more flexible and easier to design with higher specificity and accuracy (Jansen et al. 2002). Additionally, it offers a multiplexed and upscaled editing of genome.

Unlike TALENs and ZFN, the CRISPR recognition is guided by tracrRNA and crRNA, which cleaves 3–4 bps upstream of the sequence of PAM instead of the particular location, and there is no DNA binding domain. As a result, guide RNAs may be simply made for targeting any sequence. CRISPR systems are simple to employ, allowing for the creation of genome-wide libraries. Because it relies on Watson-Crick RNA-DNA base pairing instead of protein-DNA interaction, it improves genomic targeting. It also employs a large number of parallel aligned RNAs to concurrently target many locations and is thus considered as a highly useful method. By direct injection of the RNA-encoding machinery of CRISPR into the embryo, whole creatures may be changed. The alternative approach to it is more difficult, needing HR to target embryo stem cell genes, which is then followed by the process of selection and culturing. CRISPR, on the other hand, has a larger chance of generating off-target effects. While relying on a single sgRNA that is uncomplicated, it results in lower specificity. TALENs, on the other hand, have a built fail-safe that necessitates dimerization of pairs of TALEN for cleavage. However, CRISPR has a substantial edge over ZFNs and TALENS and in terms of efficiency, simplicity, and cheap cost. It's worth noting that all three methods have limitations in terms of specificity, targeting, and gene targeting. As a result, its implementation is left to the scientist's choice focused on the application.

When compared to other genome engineering methods, CRISPR/Cas9 has significant advantages. The way it uses nucleic acid base pairing ensures its target specificity—a trait that also underpins the integrity of DNA transcription and

Table 10.3 Comparison among genome editing tools (Kumar and Jain 2015)

Features	Meganucleases	ZFN	TALEN	CRISPR/Cas
Efficiency in target recognition	Low	High	High	High
Level of experiment setup	Need expertise for engineering of protein Redesigning is required for new target site	Procedure is complicated and it needs high expertise	Relatively easy procedure	Fast and easy method for new target site designing
Efficiency of cleavage	High	High	High	High with multiplexing possibility
Components	Target recognition domain and nuclease domain	Nonspecific domain of Fokl nuclease, zinc finger domain	Nonspecific domain of Fokl nuclease, TALE domains	CrRNA and Cas9 protein
Off-target effect	Detected	Detected	Detected	Detected
Target sequence length (bp)	14–40	18–24	24–59	20–22

replication in the plant and animal worlds (and as a result, all molecules of DNA, regardless of their origin, are open to this type of editing). Multiplex genome editing in mammalian cells (editing of many separate locations of genome in one procedure) is performed for the first time, and it has far more efficiency (successes per effort). The components are simple to make: sgRNA is only about 100 nucleotides long and may be made using commercially available kits. The technique is based on a universal framework of Cas9 protein, which eliminates the requirement of creating a unique protein for each target DNA. Although TALENS and ZFNs have been widely utilized for gene editing in human, plant, and animal cells, there are still certain constraints that restrict their effectiveness. For instance, ZFN's specificity is poor, and it usually generates mutations that are not intended (Puchta 2017). Moreover, creating vectors for TALENS and ZFNs takes time and effort (Tang et al. 2017). As a result, since 2013, the focus has shifted to the utilization of CRISPR technology, and in recent times, various newly discovered CRISPR/Cas variations. CRISPR is an endonuclease guided by RNA that utilizes nucleotide base pairing to selectively targeted sequences of DNA. The comparison of salient features of CRISPR/Cas9 with other genome editing tools, such as TALENS and ZFNs, has been given in Table 10.3.

10.4 Choice of Delivery Methods

There are various factors that influence CRISPR/Cas9 effectiveness when it is applied for gene modification into different cell types. These important factors must be addressed for effective in vivo activity of CRISPR for gene therapy and targeted modifications in the cells. To take sophisticated and significant results, three Ds are critical to be considered: designing, delivery, and detection. The focus then shifts to delivery, which is the most difficult obstacle to possible in vivo CRISP application. Physical delivery methods such as electroporation and microinjection; nonviral delivery methods such as full-sized adenovirus, adeno-associated virus (AAV), and lentivirus; and viral delivery methods such as lentivirus, adeno-associated virus (AAV), and full-sized adenovirus have all been reported for efficient delivery of CRISPR/Cas9 (e.g., gold particles, polyplexes, and liposomes). Improvement in the delivery methods of the technology will enhance the potential of CRISPR in therapeutics and will make its spectrum broader. Up to this day, different methods of delivery have been used for the delivery of Cas9-gRNA system in the cells.

Genome editing via CRISPR is a hot topic right now, with a lot of research being done on it, including synthetic biology, agriculture, metabolic engineering, and molecular medicine. Several considerations, such as the high dosage of the chemical agent or enormous size of the plasmid, obstruct the transport of this complex. CRISPR/Cas9 and its constituents can be delivered to target cells via a variety of ways. It comprises physical, nonviral, and viral ways for delivering CRISPR components by ribonucleoprotein (RNP) or plasmid. However, immunogenicity, insertional mutagenesis, off-targets, and targeted delivery make in vivo CRISPR/Cas9 administration difficult for the researchers. Yet, research suggests that CRISPR/Cas9-RNP delivery may be capable of overcoming these obstacles. Various delivery methods of CRISPR reagents have been demonstrated for efficient delivery resulting in targeted genome modifications (Fig. 10.2).

Rather than treating disease symptoms, treatment of disease at the genetic level will necessitate careful consideration of progression of disease and developmental time. Researchers have a variety of options for the introduction of gRNA and Cas9 into cells in a research center setting. Cell lines, for example, can be transfected with lentiviruses that carry hereditarily encoded Cas9 and gRNA and then delivered via electroporation. For mammalian cell lines, previous approaches were not viable alternatives. While some diseases such as leukemia can be treated ex vivo, where changed bone marrow can be transferred into patient cell lines, many diseases will necessitate in vivo gRNA and Cas9 delivery.

There are some of extra challenges explicit to the Cas9 delivery because Cas9 protein is an extremely huge particle (around 160 kDa) (Jinek et al. 2014); additionally, the long backbone of phosphate of guided RNA is carrying negative charge. However, another significant issue is that, like other medication items, the correct dose of materials of CRISPR is a must to be delivered to the target location in the ideal duration to accomplish therapeutic adequacy. The complexity of CRISPR should likewise avoid cell degradation systems, including RNAses, proteases, and

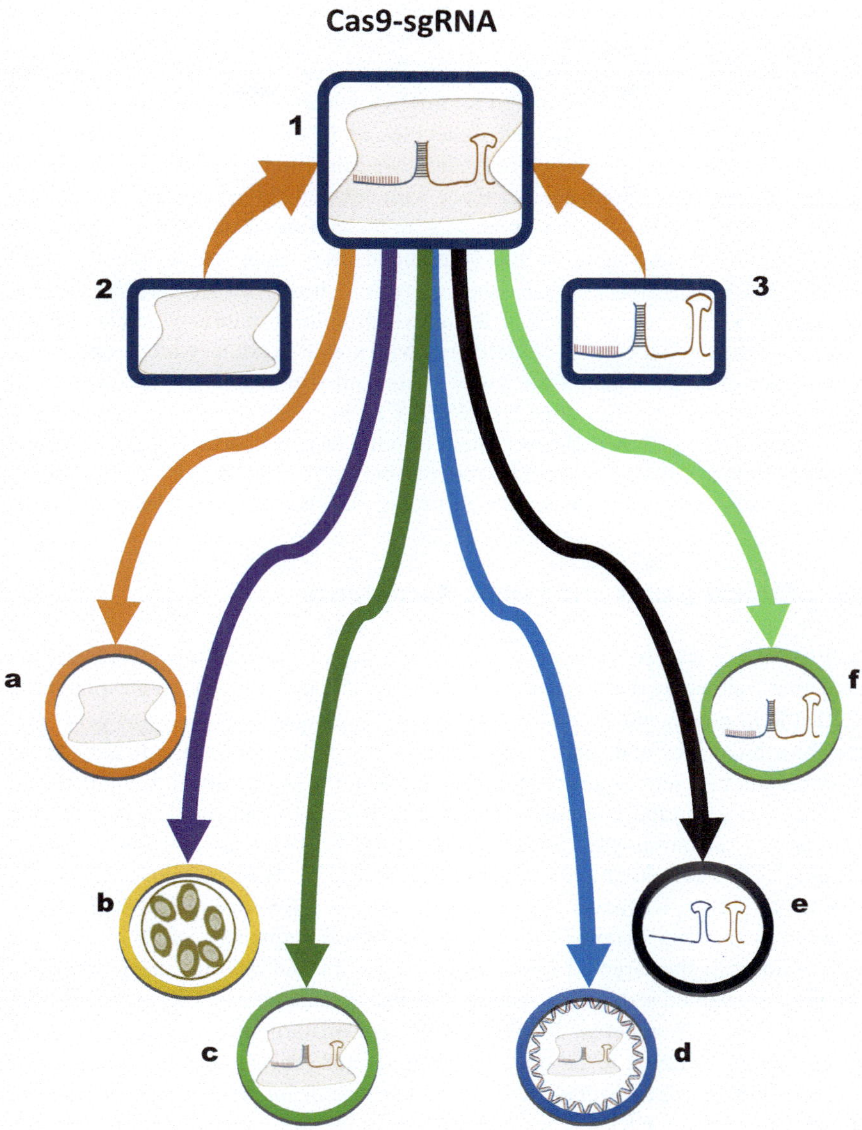

Fig. 10.2 Cas9-sgRNA delivery into cells. CRISPR system is the complex of Cas9 protein and gRNA. The detail of the figure is given below. (1) Showing complex of Cas9-sgRNA. (2) Showing Cas9 protein. (3) Showing trans-CRISPR RNA. Cas9, sgRNA, and Cas9-sgRNA complex may be delivered into cells using different methods. Some of the important methods have been enlisted in the figure. The detail is as below: (**a**) Cas9 protein delivery. (**b**) Cas9 expressing cell lines. (**c**) Delivery of ribonucleoprotein (RNPs). (**d**) Delivery in the viral vectors/plasmids. (**e**) Oligos. (**f**) Delivery of crRNA/tracrRNA complex as mRNA

lysosomes, and must be figured in an approach to limit an insusceptible reaction. A well-planned vehicle of delivery could serve to relieve these difficulties.

Cpf1 mRNA delivery has recently been shown (Li et al. 2017). The route of gRNA distribution is a unique concern for mRNA delivery. Because both gRNA and mRNA are naturally single-stranded molecules of RNA, they may frequently be delivered using the same delivery vector; however, the timing of delivery may be an issue (Jiang et al. 2017). Enhanced efficacy with delayed delivery of the gRNA may help in desirable, predictable, and precise genome editing at target sites. Chemical changes to the gRNA itself, on the other hand, may improve its stability after delivery (Yin et al. 2017). A crossbreed method has been utilized by several researchers for delivery of Cas9 in its native protein structure, wherein mRNA encoding for the Cas9 protein is co-delivered with a viral genome, which persistently generates the gRNA for keeping it away from both within and outside of the cell (Yin et al. 2016). The strength of CRISPR/Cas innovation is dependent on the translocation of the gRNA and Cas protein into the targeted cells. The development and use of novel delivery materials and vectors with ever-increasing specificity, efficiency, and safety will enhance the favorable performance of CRISPR delivery.

10.5 Ethical Issues and Public Acceptance

CRISPR/Cas9 is widely regarded as one of the most significant discoveries of the twenty-first century by the scientific world and related industry. The widespread use of CRISPR in healthcare, livestock, and the environment, however, has generated major bioethical, societal, and legal issues. One of the most significant ethical considerations in scientific research is that the benefits must outweigh the dangers. Risks must be given more attention because they have the potential to harm living things or the surroundings. The use of the CRISPR/Cas9 method has a risk of producing off-target alterations, which can be harmful. Human cells have a high rate of off-targeting, whereas mice and zebrafish have a low rate (Yang et al. 2013). One issue is that big genomes may include several sequences of DNA that are closely similar to the targeted DNA sequence. CRISPR/Cas9 has the potential to break these unwanted sequences, resulting in alterations that can lead to transformation or cell death (Zhang et al. 2015). Mutations of off-targeting have been reduced, but more research needs to be done, particularly for precise changes required for therapeutic treatments (Hsu et al. 2013). Another significant challenge is the safe and effective transport of CRISPR into difficult to infect or transfect tissues and cell types (Rodriguez 2016).

Commercial production of genetically modified crops including corn, cotton, and soybeans has been increased in several countries, including the United States, Canada, India, Argentina, and China, since the 1990s (Brookes and Barfoot 2014). Furthermore, no food GM crops have been grown commercially in a number of countries, including the majority of EU and non-EU countries (excluding Czech Republic, Spain, Slovakia, Portugal, and Romania) including Japan, and New Zealand (Lucht 2015). Poor understanding of genetically modified organisms, lack of faith in scientists and/or applicable laws, inadequate communication of risk-benefit, and ethical principles are all linked to a bad opinion toward GMOs (Tanaka 2013; Zilberman et al. 2013). These points must be carefully considered if one chooses to believe that genetic modification can tackle problem of food security by crops breeding with advantageous genes including higher production, enhanced nutrition value, diseases and pest resistance, and less susceptibility to changes in the environment such as greater or lesser temperatures and water shortage. Furthermore, plant breeding through genetic engineering may lead to a wider variety of foods.

Likewise, several scientists distinguish between clinical germline editing and fundamental research using CRISPR/Cas9 on embryos of humans. Although, there was no clinical use of germline editing that might provide any significant advantage over existing techniques at the time, other applications of CRISPR/Cas9, such as somatic cells gene therapy, should never be categorically dismissed or kept too limited. Basic research guaranteed valuable results pertinent to the treatment of human ailments, even if it was also connected to ethical concerns like egg cell donation, embryo wastage, or use of animal in laboratory.

Clearly, several rules regulating fundamental research and somatic cells gene therapy, including the present laws regarding embryo protection that prohibit any use of embryos for any scientific purposes in several nations such as Germany, might stifle research, development, and innovation. To avoid such a consequence, they employ comparator gene technique and the Asilomar meeting, appealing for a moratorium and demonstrating their readiness to confine research operations to acceptable areas and engage in meaningful public discussion. Rather than a general prohibition on research into any technology, they opt for a more precise restriction. However, it's possible that the boundaries drawn between somatic and germline editing, and between fundamental research and therapeutic application, are still too arbitrary. The advantages, dangers, and ethical concerns associated with using CRISPR/Cas9 for various treatments and basic research objectives, as well as their benefits, dangers, and ethical difficulties, must all be considered. Although it is reasonable for a researcher to place a high value on the possibilities afforded by research, this necessarily leads to a reduction in the weight given to the economic and ethical consequences for clinical practice that can only be disclosed by taking a more differentiated approach.

CRISPR system has offered us diverse range of applications in various fields due to its high accuracy, ease in use, and cost effectivity. Compared to ZFN and TALEN, precision and accuracy are much higher in CRISPR-based genome editing. Due to its high accuracy and preciseness, it has raised many social, moral, and ethical issues also. There are lots of concerns regarding social and ethical values of using these modern techniques such as CRISPR/Cas9 in human germline cells. Many of the scientists working on CRISPR-based genome editing has a general perspective that CRISPR system would at least be allowed for creating models of human disease, in understanding diseases molecular mechanisms and their development (Rossant 2018; Cathomen et al. 2019). In addition, scientists believe that the use of CRISPR-based editing should not be allowed for other purposes like eugenics or enhancement. It will be predicted soon that human embryos gene editing would not be possible, when their social, ethical, and application inconsistencies are considered (Duardo-Sánchez 2017; Macintosh 2019). Therefore, there exists a very high risk of intentional harm due to nontargeted heritable genetic mutations. Yet, scientists are curious and will use CRISPR/Cas9 system-based germline editing in the future after mitigation of the above-discussed challenges (Duardo-Sánchez 2017; Hirsch et al. 2019).

In a simpler way, CRISPR is a well-renowned technology, and it is not just worth to discuss its social, ethical, and public perceptions; however, its possible effects would not be given up. So, risks on nontargets or environment, i.e., intentional harm, ecological degradation, and use of this technology for genetic betterment of agriculture and animal-related products must also be considered (Hirsch et al. 2019). We already have discussed above that CRISPR edited organisms are considered GMOs, and once they are released (controlled or uncontrolled manner) into natural ecosystem, it is difficult to recall back those organisms, and it will raise high concerns. For example, the effect of Gm mosquitoes cannot be predicted when CRISPR/Cas9 system is used to protect mosquitoes from malarial parasites and thus resulting in infertility in female mosquitoes (Hammond et al. 2016).

One of the important concerns in agriculture is the public trust and acceptance of GM events produced with CRISPR/Cas9. Previously, GMOs produced with technologies other than CRISPR/Cas9 has faced many reactions by the public. The fact behind public mistrust might be the GMO production with CRISPR/Cas9, as their products are difficult to recall, may raise high concerns, ambiguities, and uncertainties (Carroll 2017; Shinwari et al. 2017). Therefore, clear explanations and declarations by the authorities should be made in transparent manner to avoid questions and misinterpretation by public before the release of such type of products.

Patenting is another issue worth consideration while using CRISPR/Cas9 technology as it can limit application of such technologies. Though patenting in a unilateral way could increase profits of companies dealing with biotechnological products, it also raises ethical issues. The cultural and environmental concerns incorporate property rights to hereditary assets, patents to hereditarily modified animals and plants, access to gene pool just as inquiries of seed and food sovereignty, effects on biodiversity, and the reversibility of selection (Dabrock 2009). GMOs are used only for therapeutic purposes; so, there is disagreement among scientific communities regarding patenting (Shinwari et al. 2017; Sherkow 2018). Moreover, one of the important aspects in patenting is commercialization and release of eco-friendly products into the environment (Rodriguez 2016; Shinwari et al. 2017).

To summarize, medicinal applications of the CRISPR/Cas9 system have aroused significant public, ethical, and social issues (Shinwari et al. 2017; Carroll 2019). Consumers have the right to know, what they are consuming, so, there should be an immediate response, and certain guidelines, legislations on regulation, and application of the content produced through these new technologies should be made by involving public to develop trust in them. Science community and other parties (ethical, legal, social, and government) should have a complete guide for future processing, containment, and use of modern techniques like CRISPR/Cas9 (Cathomen et al. 2019). This is the only way to develop a long-lasting policy that will help science community and their products and develop high public trust in new technologies. There is an extraordinary need to CRISPR innovation on different stages that incorporates support from specialists from moral, religious, social, administrative, and technological grounds to build up a durable arrangement regarding advantages and uncertainty about CRISPR innovation. CRISPR/Cas9 starts with numerous social and ethical issues from human points of view as well as for the environment. Hazard evaluations for natural and ecological concerns should be performed.

10.6 Future Prospect of CRISPR/Cas

CRISPR/Cas toolkit has various versatile tools with amazingly capable gene editing efficiency that has only recently uncovered a variety of applications. CRISPR is faster, more precise, simpler, and presumably less costly than traditional genetic editing technologies in farming and food systems. So far, the research has shown that successful CRISPR/Cas9-mediated gene modification necessitates the presence of single guided RNA with about 20 base sequence specific to the targeted DNA. In addition to agricultural development, genetic engineering has benefited fermentation techniques by producing strains with better functional qualities. *Lactobacillus* spp. strains are important microorganisms in the business related to food, primarily in psychobiotics, and probiotics (Mishra et al. 2017). Consequently, the usage of CRISPR technology in these microorganisms could significantly decrease their susceptibility to a variety of stress situations, increasing the profitability of the maturation process and thereby improving the quality of food. We still have some gaps to be filled in our understanding of the CRISPR/Cas9 framework in gene editing. Gene knock-in and knockout using CRISPR can boost homozygosity while lowering heterozygosity. There are several variants of Cas9 which have been established with enhanced on-target activity as well as with engineered PAM sites. CRISPR system with no limits of PAM sequence increases the target efficiency and DNA specificity and decreases off-target activity. Efforts should be made to introduce new suitable systems which can recognize diverse PAMs, such as SpCas9-NG and CasX. The large size makes editing difficult and prevents it from being packed into viral vectors for distribution to somatic tissues. Many efforts have been undertaken to increase efficiency using tiny CRISPR systems like SpCas9-NG, CasX, and others, which may also improve editing efficiency. To improve the efficacy of the virus-induced gene silencing (VIGS)-mediated sgRNA system, alternative systems such as DNA-free or temporary CRISPR systems have been created. These systems are easier to handle and cost-effective. For the enhancement of homozygosity and reduction of heterozygosity, high throughput techniques have been developed that can reduce cost, time, and resources for mutation detection.

Recently, base editing has been demonstrated for efficient genome editing in plants which can introduce highly expectable and precise changes in nucleotides at genome targets with no double-stranded break (DSBs) and donor templates of DNA. It is profoundly preferable to increase the efficiency and avoid undesirable editing. To overcome the effects of off-targeting in the genome, an inducible CRISPR mechanism has been introduced which can be induced by many kinds of external stimuli for editing the genome. Moreover, CRISPRi technique has been used for the regulation, repression, and silencing of target gene, which enhances the knockdown efficacy of the target gene. CRISPRa has also been used successfully to activate a gene of interest by histone acetylation and other epigenetic modification. Other approaches of CRIPSR have been used for the epigenetic modification which helps in the gene expression regulation via alteration in the chromatin structure and epigenetic marks, protein phosphorylation, and ubiquitination. Multiplex CRISPR/Cas9 system provides ease in the introduction of heterologous genes in the targeted loci as well as metabolic pathway engineering of organisms. Due to its simplicity, high fidelity, sensitivity, and effectiveness, CRISPR/Cas9 has become a versatile and highly applicable tool in contrast to previously used genome manipulation technologies. CRISPR technology researchers, Jennifer A. Doudna and Emmanuelle Charpentier, have been awarded with a Nobel Prize in 2020 which is the most significant achievement in the short history of the genome editing era. CRISPR and other developing gene editing tools have the potential to further change the entire landscape of life sciences, agriculture and molecular breeding.

References

Alkan F, Wenzel A, Anthon C, Havgaard JH, Gorodkin J (2018) CRISPR-Cas9 off-targeting assessment with nucleic acid duplex energy parameters. Genome Biol 19(1):1–13

Anderson KR, Haeussler M, Watanabe C, Janakiraman V, Lund J, Modrusan Z, Stinson J, Bei Q, Buechler A, Yu C (2018) CRISPR off-target analysis in genetically engineered rats and mice. Nat Methods 15(7):512–514

Anzalone AV, Randolph PB, Davis JR, Sousa AA, Koblan LW, Levy JM, Chen PJ, Wilson C, Newby GA, Raguram A, Liu DR (2019) Search-and-replace genome editing without double-strand breaks or donor DNA. Nature 576(7785):149–157

Bae S, Park J, Kim JS (2014) Cas-OFFinder: a fast and versatile algorithm that searches for potential off-target sites of Cas9 RNA-guided endonucleases. Bioinformatics 30(10):1473–1475

Brookes G, Barfoot P (2014) Economic impact of GM crops: the global income and production effects 1996–2012. GM Crops Food 5:65–75

Carroll D (2017) Focus: genome editing: genome editing: past, present, and future. Yale J Biol Med 90(4):653

Carroll D (2019) Collateral damage: benchmarking off-target effects in genome editing. Genome Biol 20:114

Casini A, Olivieri M, Petris G, Montagna C, Reginato G, Maule G, Lorenzin F, Prandi D, Romanel A, Demichelis F (2018) A highly specific SpCas9 variant is identified by in vivo screening in yeast. Nat Biotechnol 36(3):265–271

Cathomen T, Schüle S, Schüßler-Lenz M, Abou-El-Enein M (2019) The human genome editing race: loosening regulatory standards for commercial advantage? Trends Biotechnol 37(2):120–123
Chari R, Mali P, Moosburner M, Church GM (2015) Unraveling CRISPR-Cas9 genome engineering parameters via a library-on-library approach. Nat Methods 12(9):823–826
Chen JS, Dagdas YS, Kleinstiver BP, Welch MM, Sousa AA, Harrington LB, Sternberg SH, Joung JK, Yildiz A, Doudna JA (2017) Enhanced proofreading governs CRISPR–Cas9 targeting accuracy. Nature 550:407–410
Cho SW, Kim S, Kim Y, Kweon J, Kim HS, Bae S, Kim JS (2014) Analysis of off-target effects of CRISPR/Cas-derived RNA-guided endonucleases and nickases. Genome Res 24(1):132–141
Cong L, Ran FA, Cox D, Lin S, Barretto R, Habib N, Hsu PD, Wu X, Jiang W, Marraffini LA (2013) Multiplex genome engineering using CRISPR/Cas systems. Science 339(6121):819–823
Dabrock P (2009) Playing god? Synthetic biology as a theological and ethical challenge. Syst Synth Biol 3:47
Dellinger DJ, Timár Z, Myerson J, Sierzchala AB, Turner J, Ferreira F, Kupihár ZN, Dellinger G, Hill KW, Powell JA (2011) Streamlined process for the chemical synthesis of RNA using 2′-O-thionocarbamate-protected nucleoside phosphoramidites in the solid phase. J Am Chem Soc 133(30):11540–11556
Dow LE, Fisher J, O'rourke KP, Muley A, Kastenhuber ER, Livshits G, Tschaharganeh DF, Socci ND, Lowe SW (2015) Inducible in vivo genome editing with CRISPR-Cas9. Nat Biotechnol 33: 390–394
Duardo-Sanchez A (2017) CRISPR-Cas in medicinal chemistry: applications and regulatory concerns. Curr Top Med Chem 17(30):3308–3315
Frock RL, Hu J, Meyers RM, Ho YJ, Kii E, Alt FW (2015) Genome-wide detection of DNA double-stranded breaks induced by engineered nucleases. Nat Biotechnol 33(2):179–186
Fu Y, Foden JA, Khayter C, Maeder ML, Reyon D, Joung JK, Sander JD (2013) High-frequency off-target mutagenesis induced by CRISPR-Cas nucleases in human cells. Nat Biotechnol 31(9):822–826
Fu Y, Reyon D, Joung JK (2014a) Targeted genome editing in human cells using CRISPR/Cas nucleases and truncated guide RNAs. Methods Enzymol 546:21–45
Fu Y, Sander JD, Reyon D, Cascio VM, Joung JK (2014b) Improving CRISPR-Cas nuclease specificity using truncated guide RNAs. Nat Biotechnol 32(3):279–284
Gabriel R, Von Kalle C, Schmidt M (2015) Mapping the precision of genome editing. Nat Biotechnol 33(2):150–152
Guha TK, Wai A, Hausner G (2017) Programmable genome editing tools and their regulation for efficient genome engineering. Comput Struct Biotechnol J 15:146–160
Guilinger JP, Thompson DB, Liu DR (2014) Fusion of catalytically inactive Cas9 to FokI nuclease improves the specificity of genome modification. Nat Biotechnol 32(6):577
Hammond A, Galizi R, Kyrou K, Simoni A, Siniscalchi C, Katsanos D, Gribble M, Baker D, Marois E, Russell S, Burt A (2016) A CRISPR-Cas9 gene drive system targeting female reproduction in the malaria mosquito vector Anopheles gambiae. Nat Biotechnol 34(1):78–83
Hay EA, Khalaf AR, Marini P, Brown A, Heath K, Sheppard D, MacKenzie A (2017) An analysis of possible off target effects following CAS9/CRISPR targeted deletions of neuropeptide gene enhancers from the mouse genome. Neuropeptides 64:101–107
Heigwer F, Kerr G, Boutros M (2014) E-CRISP: fast CRISPR target site identification. Nat Methods 11(2):122–123
Hirsch F, Iphofen R, Koporc Z (2019) Ethics assessment in research proposals adopting CRISPR technology. Biochem Med 29(2):206–213
Hsu PD, Scott DA, Weinstein JA, Ran FA, Konermann S, Agarwala V, Li Y, Fine EJ, Wu X, Shalem O, Cradick TJ (2013) DNA targeting specificity of RNA-guided Cas9 nucleases. Nat Biotechnol 31(9):827–832
Hsu PD, Lander ES, Zhang F (2014) Development and applications of CRISPR-Cas9 for genome engineering. Cell 157(6):1262–1278

Isaac RS, Jiang F, Doudna JA, Lim WA, Narlikar GJ, Almeida R (2016) Nucleosome breathing and remodeling constrain CRISPR-Cas9 function. elife 5:13450
Iyer V, Shen B, Zhang W, Hodgkins A, Keane T, Huang X, Skarnes WC (2015) Off-target mutations are rare in Cas9-modified mice. Nat Methods 12(6):479
Iyer V, Boroviak K, Thomas M, Doe B, Riva L, Ryder E, Adams DJ (2018) No unexpected CRISPR-Cas9 off-target activity revealed by trio sequencing of gene-edited mice. PLoS Genet 14(7):1007503
Jansen R, Embden JD, Gaastra W, Schouls LM (2002) Identification of genes that are associated with DNA repeats in prokaryotes. Mol Microbiol 43(6):1565–1575
Jiang F, Zhou K, Ma L, Gressel S, Doudna JA (2015) A Cas9–guide RNA complex preorganized for target DNA recognition. Science 348(6242):1477–1481
Jiang C, Mei M, Li B, Zhu X, Zu W, Tian Y, Wang Q, Guo Y, Dong Y, Tan X (2017) A non-viral CRISPR/Cas9 delivery system for therapeutically targeting HBV DNA and pcsk9 in vivo. Cell Res 27(3):440–443
Jin S, Zong Y, Gao Q, Zhu Z, Wang Y, Qin P, Liang C, Wang D, Qiu JL, Zhang F (2019) Cytosine, but not adenine, base editors induce genome-wide off-target mutations in rice. Science 364:292–295
Jinek M, Jiang F, Taylor WD, Sternberg SH, Kaya E, Ma E, Anders C, Hauer M, Zhou K, Lin S (2014) Structures of Cas9 endonucleases reveal RNA-mediated conformational activation. Science 343(6176):1247997
Kim HJ, Lee HJ, Kim H, Cho SW, Kim JS (2009) Targeted genome editing in human cells with zinc finger nucleases constructed via modular assembly. Genome Res 19(7):1279–1288
Kim D, Bae S, Park J, Kim E, Kim S, Yu HR, Hwang J, Kim JI, Kim JS (2015) Digenome-seq: genome-wide profiling of CRISPR-Cas9 off-target effects in human cells. Nat Methods 12(3):237–243
Kim D, Kim S, Kim S, Park J, Kim JS (2016) Genome-wide target specificities of CRISPR-Cas9 nucleases revealed by multiplex Digenome-seq. Genome Res 26(3):406–415
Kim S, Bae T, Hwang J, Kim JS (2017) Rescue of high-specificity Cas9 variants using sgRNAs with matched 5'nucleotides. Genome Biol 18(1):1–6
Kim D, Kim DE, Lee G, Cho SI, Kim JS (2019) Genome-wide target specificity of CRISPR RNA-guided adenine base editors. Nat Biotechnol 37:430–435
Kleinstiver BP, Prew MS, Tsai SQ, Topkar VV, Nguyen NT, Zheng Z, Gonzales AP, Li Z, Peterson RT, Yeh J-RJ (2015) Engineered CRISPR-Cas9 nucleases with altered PAM specificities. Nature 523(7561):481–485
Kleinstiver BP, Pattanayak V, Prew MS, Tsai SQ, Nguyen NT, Zheng Z, Joung JK (2016) High-fidelity CRISPR–Cas9 nucleases with no detectable genome-wide off-target effects. Nature 529(7587):490–495
Knight SC, Xie L, Deng W, Guglielmi B, Witkowsky LB, Bosanac L, Zhang ET, El Beheiry M, Masson JB, Dahan M, Liu Z (2015) Dynamics of CRISPR-Cas9 genome interrogation in living cells. Science 350(6262):823–826
Komor AC, Kim YB, Packer MS, Zuris JA, Liu DR (2016) Programmable editing of a target base in genomic DNA without double-stranded DNA cleavage. Nature 533(7603):420–424
Konermann S, Brigham MD, Trevino AE, Joung J, Abudayyeh OO, Barcena C, Hsu PD, Habib N, Gootenberg JS, Nishimasu H (2015) Genome-scale transcriptional activation by an engineered CRISPR-Cas9 complex. Nature 517:583–588
Kulcsár PI, Tálas A, Huszár K, Ligeti Z, Tóth E, Weinhardt N, Fodor E, Welker E (2017) Crossing enhanced and high fidelity SpCas9 nucleases to optimize specificity and cleavage. Genome Biol 18(1):1–7
Kumar V, Jain M (2015) The CRISPR–Cas system for plant genome editing: advances and opportunities. J Exp Bot 66(1):47–57
Kuscu C, Arslan S, Singh R, Thorpe J, Adli M (2014) Genome-wide analysis reveals characteristics of off-target sites bound by the Cas9 endonuclease. Nat Biotechnol 32(7):677–683

Li T, Yang B (2013) TAL effector nuclease (TALEN) engineering. In: Enzyme engineering. Springer, pp 63–72

Li B, Zhao W, Luo X, Zhang X, Li C, Zeng C, Dong Y (2017) Engineering CRISPR–Cpf1 crRNAs and mRNAs to maximize genome editing efficiency. Nat Biomed Eng 1:1–10

Li D, Zhou H, Zeng X (2019) Battling CRISPR-Cas9 off-target genome editing. Cell Biol Toxicol 35:403–406. https://doi.org/10.1007/s10565-019-09485-5

Lin Y, Cradick TJ, Brown MT, Deshmukh H, Ranjan P, Sarode N, Wile BM, Vertino PM, Stewart FJ, Bao G (2014) CRISPR/Cas9 systems have off-target activity with insertions or deletions between target DNA and guide RNA sequences. Nucleic Acids Res 42(11):7473–7485

Lucht JM (2015) Public acceptance of plant biotechnology and GM crops. Viruses 7(8):4254–4281

Macintosh KL (2019) Heritable genome editing and the downsides of a global moratorium. CRISPR J 2(5):272–279

Mali P, Esvelt KM, Church GM (2013) Cas9 as a versatile tool for engineering biology. Nat Methods 10(10):957–963

Martin F, Sánchez-Hernández S, Gutiérrez-Guerrero A, Pinedo-Gomez J, Benabdellah K (2016) Biased and unbiased methods for the detection of off-target cleavage by CRISPR/Cas9: an overview. Int J Mol Sci 17:1507

Mishra B, Reiling S, Zarena D, Wang G (2017) Host defense antimicrobial peptides as antibiotics: design and application strategies. Curr Opin Chem Biol 38:87–96

Nishimasu H, Ran FA, Hsu PD, Konermann S, Shehata SI, Dohmae N, Ishitani R, Zhang F, Nureki O (2014) Crystal structure of Cas9 in complex with guide RNA and target DNA. Cell 156(5):935–949

Nishimasu H, Shi X, Ishiguro S, Gao L, Hirano S, Okazaki S, Noda T, Abudayyeh O-O, Gootenberg JS, Mori H (2018) Engineered CRISPR-Cas9 nuclease with expanded targeting space. Science 361:1259–1262

O'Geen H, Henry IM, Bhakta MS, Meckler JF, Segal DJ (2015) A genome-wide analysis of Cas9 binding specificity using ChIP-seq and targeted sequence capture. Nucleic Acids Res 43(6):3389–3404

Pattanayak V, Lin S, Guilinger JP, Ma E, Doudna JA, Liu DR (2013) High-throughput profiling of off-target DNA cleavage reveals RNA-programmed Cas9 nuclease specificity. Nat Biotechnol 31(9):839–843

Paulis M, Castelli A, Lizier M, Susani L, Lucchini F, Villa A, Vezzoni P (2015) A pre-screening FISH-based method to detect CRISPR/Cas9 off-targets in mouse embryonic stem cells. Sci Rep 5(1):1–7

Puchta H (2017) Applying CRISPR/Cas for genome engineering in plants: the best is yet to come. Curr Opin Plant Biol 36:1–8

Ran FA, Hsu PD, Lin C-Y, Gootenberg JS, Konermann S, Trevino AE, Scott DA, Inoue A, Matoba S, Zhang Y (2013a) Double nicking by RNA-guided CRISPR Cas9 for enhanced genome editing specificity. Cell 154(6):1380–1389

Ran FA, Hsu PD, Wright J, Agarwala V, Scott DA, Zhang F (2013b) Genome engineering using the CRISPR-Cas9 system. Nat Protoc 8(11):2281–2308

Ran FA, Cong L, Yan WX, Scott DA, Gootenberg JS, Kriz AJ, Zetsche B, Shalem O, Wu X, Makarova KS (2015) In vivo genome editing using Staphylococcus aureus Cas9. Nature 520(7546):186–191

Rauch BJ, Silvis MR, Hultquist JF, Waters CS, McGregor MJ, Krogan NJ, Bondy-Denomy J (2017) Inhibition of CRISPR-Cas9 with bacteriophage proteins. Cell 168:150–158.e110

Rees HA, Liu DR (2018) Base editing: precision chemistry on the genome and transcriptome of living cells. Nat Rev Genet 9(12):770–788

Rees HA, Komor AC, Yeh WH, Caetano-Lopes J, Warman M, Edge AS, Liu DR (2017) Improving the DNA specificity and applicability of base editing through protein engineering and protein delivery. Nat Commun 8:1–10

Rodriguez E (2016) Ethical issues in genome editing using Crispr/Cas9 system

Rossant J (2018) Gene editing in human development: ethical concerns and practical applications. Development 145(16):150888

Sherkow JS (2018) The CRISPR-Cas9 patent appeal: where do we go from here? CRISPR J 1(5):309–311

Shin J, Jiang F, Liu J-J, Bray NL, Rauch BJ, Baik SH, Nogales E, Bondy-Denomy J, Corn JE, Doudna JA (2017) Disabling Cas9 by an anti-CRISPR DNA mimic. Sci Adv 3(7):e1701620

Shinwari ZK, Tanveer F, Khalil AT (2017) Ethical issues regarding CRISPR mediated genome editing. Curr Issues Mol Biol 26:103–110

Slaymaker IM, Gao L, Zetsche B, Scott DA, Yan WX, Zhang F (2016) Rationally engineered Cas9 nucleases with improved specificity. Science 351(6268):84–88

Standage-Beier K, Zhang Q, Wang X (2015) Targeted large-scale deletion of bacterial genomes using CRISPR-nickases. ACS Synth Biol 4:1217–1225

Tang L, Zeng Y, Du H, Gong M, Peng J, Zhang B, Lei M, Zhao F, Wang W, Li X (2017) CRISPR/Cas9-mediated gene editing in human zygotes using Cas9 protein. Mol Gen Genomics 292(3):525–533

Tanaka Y (2013) Attitude gaps between conventional plant breeding crops and genetically modified crops, and psychological models determining the acceptance of the two crops. J Risk Res 16(1):69–80

Tsai SQ, Joung JK (2016) Defining and improving the genome-wide specificities of CRISPR–Cas9 nucleases. Nat Rev Genet 17:300–312

Tsai SQ, Wyvekens N, Khayter C, Foden JA, Thapar V, Reyon D, Goodwin MJ, Aryee MJ, Joung JK (2014) Dimeric CRISPR RNA-guided FokI nucleases for highly specific genome editing. Nat Biotechnol 32(6):569–576

Tsai SQ, Zheng Z, Nguyen NT, Liebers M, Topkar V-V, Thapar V, Wyvekens N, Khayter C, Iafrate AJ, Le LP (2015) GUIDE-seq enables genome-wide profiling of off-target cleavage by CRISPR-Cas nucleases. Nat Biotechnol 33:187–197

Tycko J, Myer VE, Hsu PD (2016) Methods for optimizing CRISPR-Cas9 genome editing specificity. Mol Cell 63(3):355–370

Wang X, Wang Y, Wu X, Wang J, Wang Y, Qiu Z, Chang T, Huang H, Lin R-J, Yee J-K (2015) Unbiased detection of off-target cleavage by CRISPR-Cas9 and TALENs using integrase-defective lentiviral vectors. Nat Biotechnol 33(2):175–178

Wu X, Scott DA, Kriz AJ, Chiu AC, Hsu PD, Dadon DB, Cheng AW, Trevino AE, Konermann S, Chen S (2014) Genome-wide binding of the CRISPR endonuclease Cas9 in mammalian cells. Nat Biotechnol 32(7):670–676

Yang H, Wang H, Shivalila CS, Cheng AW, Shi L, Jaenisch R (2013) One-step generation of mice carrying reporter and conditional alleles by CRISPR/Cas-mediated genome engineering. Cell 154(6):1370–1379

Yang L, Grishin D, Wang G, Aach J, Zhang CZ, Chari R, Homsy J, Cai X, Zhao Y, Fan JB (2014) Targeted and genome-wide sequencing reveal single nucleotide variations impacting specificity of Cas9 in human stem cells. Nat Commun 5(1):1–6

Yeh WH, Chiang H, Rees HA, Edge AS, Liu DR (2018) In vivo base editing of post-mitotic sensory cells. Nat Commun 9:1–10

Yin H, Song CQ, Dorkin JR, Zhu LJ, Li Y, Wu Q, Gupta A (2016) Therapeutic genome editing by combined viral and non-viral delivery of CRISPR system components in vivo. Nat Biotechnol 34(3):328–333

Yin H, Kauffman KJ, Anderson DG (2017) Delivery technologies for genome editing. Nat Rev Drug Discov 16(6):387

Zhang XH, Tee LY, Wang XG, Huang QS, Yang SH (2015) Off-target effects in CRISPR/Cas9-mediated genome engineering. Mol Ther Nucleic Acids 4:264

Zilberman D, Kaplan S, Kim E, Hochman G, Graff G (2013) Continents divided: understanding differences between Europe and North America in acceptance of GM crops. GM Crops Food 4: 202–208

Zincarelli C, Soltys S, Rengo G, Rabinowitz JE (2008) Analysis of AAV serotypes 1–9 mediated gene expression and tropism in mice after systemic injection. Mol Ther 16(6):1073–1080
Zischewski J, Fischer R, Bortesi L (2017) Detection of on-target and off-target mutations generated by CRISPR/Cas9 and other sequence-specific nucleases. Biotechnol Adv 35(1):95–104
Zuo E, Sun Y, Wei W, Yuan T, Ying W, Sun H, Yuan L, Steinmetz LM, Li Y, Yang H (2019) Cytosine base editor generates substantial off-target single-nucleotide variants in mouse embryos. Science 364:289–292
Zuris JA, Thompson DB, Shu Y, Guilinger JP, Bessen JL, Hu JH, Maeder ML, Joung JK, Chen Z-Y, Liu DR (2015) Cationic lipid-mediated delivery of proteins enables efficient protein-based genome editing in vitro and in vivo. Nat Biotechnol 33(1):73–80

The manufacturer's authorised representative in the EU is Springer Nature Customer Service Centre GmbH, Europaplatz 3, 69115 Heidelberg, Germany. If you have any concerns regarding our products, please contact ProductSafety@springernature.com

Printed and bound by CPI Group (UK) Ltd, Croydon, CR0 4YY
30/09/2026
02225248-0008